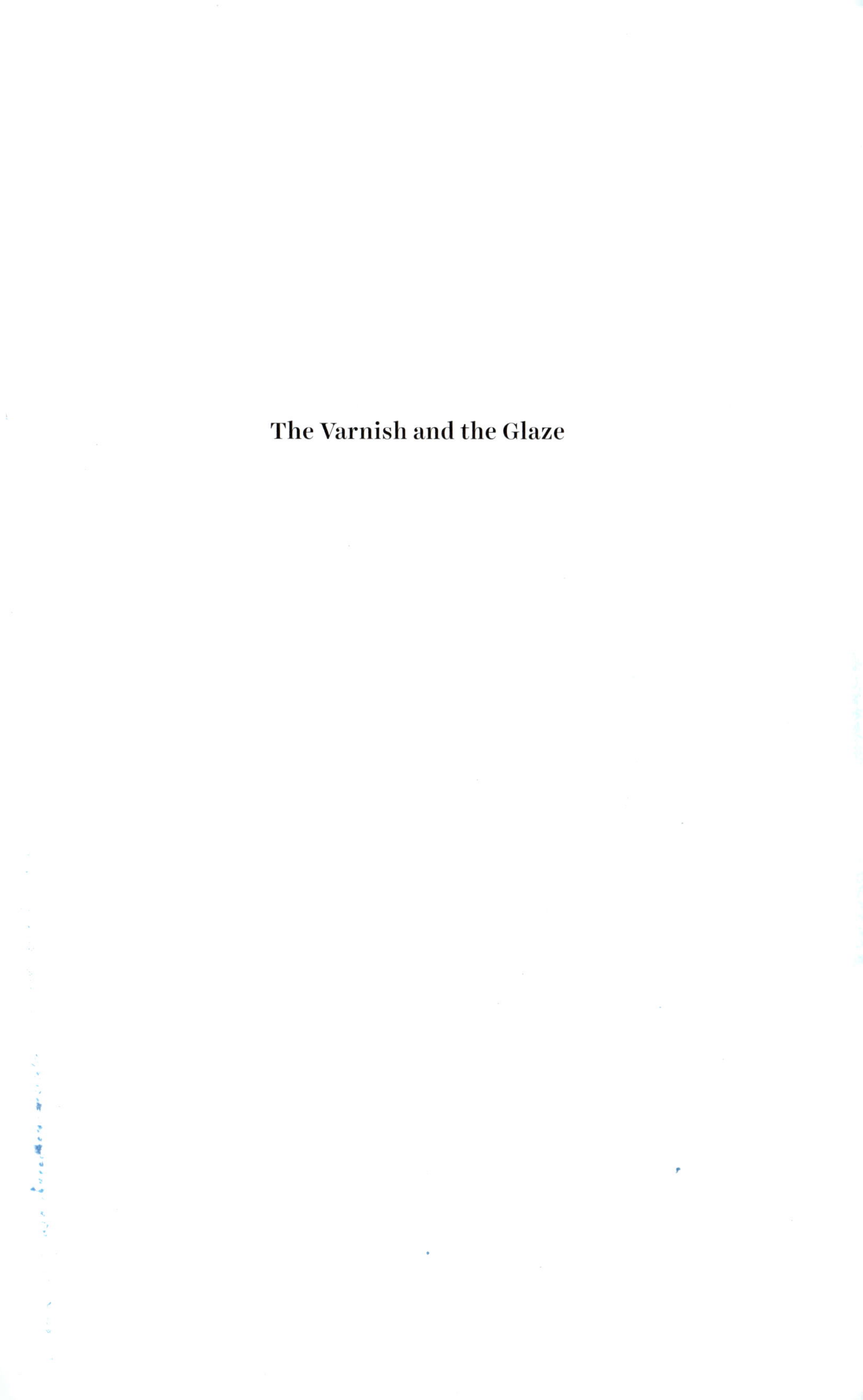

The Varnish and the Glaze

THE VARNISH & THE GLAZE

Painting Splendor with Oil, 1100–1500

Marjolijn Bol

The University of Chicago Press
Chicago & London

The University of Chicago Press, Chicago 60637
The University of Chicago Press, Ltd., London

Published 2023
Printed in China

32 31 30 29 28 27 26 25 24 23 1 2 3 4 5

ISBN-13: 978-0-226-82036-1 (cloth)
ISBN-13: 978-0-226-82263-1 (e-book)
DOI: https://doi.org/10.7208/chicago/9780226822631.001.0001

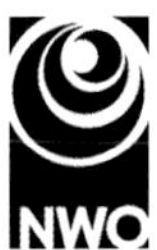

This publication is part of the project "Deceiving Stuff: Histories, Functions, Techniques, and Effects of Material Mimesis" (with project number 275-54-001) of the NWO Talent program Veni which is (partly) financed by the Dutch Research Council (NWO).

Library of Congress Cataloging-in-Publication Data

Names: Bol, Marjolijn, author.
Title: The varnish and the glaze : painting splendor with oil, 1100–1500 / Marjolijn Bol.
Description: Chicago : University of Chicago Press, 2023. | Includes bibliographical references and index.
Identifiers: LCCN 2022020186 | ISBN 9780226820361 (cloth) | ISBN 9780226822631 (ebook)
Subjects: LCSH: Varnish and varnishing—History—To 1500. | Glazing—History—To 1500. | Eyck, Jan van, 1390–1440. | Panel painting, Medieval—Technique.
Classification: LCC ND1530 .B65 2023 | DDC 751.409—dc23/eng/20220627
LC record available at https://lccn.loc.gov/2022020186

∞ This paper meets the requirements of ANSI/NISO Z39.48-1992 (Permanence of Paper).

Contents

Disclaimer

Some of the reconstructions presented in this book require specific health and safety precautions (e.g., heating flammable substances such as oils and varnishes). Readers who are interested in repeating some of the experiments or making their own experiments should always make sure to protect their health and the environment.

Introduction

The Paint That Glows

The medieval panel painter evokes the glow of precious stones, the radiant colors of stained glass windows, and the soft sheen of polished gold and silver. The same can be said of panel painters working in the fifteenth century. Yet their approaches to rendering these materials are markedly different. The medieval painter is best known for the two-dimensional nature of his pictures, including glass imitations of gems and the lavish application of gold leaf. The early Netherlandish painters created windows into worlds filled with gold and glistening gemstones depicted with paint alone. The spectacular verisimilitude in the paintings of Jan van Eyck (before ca. 1390–1441) and his contemporaries breaks in a revolutionary way with everything made before their time. This book explores this transformation in the mimetic ambition of painters around the "Eyckian turning point" by investigating the history of the two oil painting techniques used to depict everything that glistens and glows: the varnish and the glaze.

Both the varnish and the glaze have long histories of association with the art of Jan van Eyck. About a century after his death in 1441, Van Eyck was widely credited with two technical inventions because of his extraordinary visual realism. In 1550 Giorgio Vasari (1511–74), the first modern writer of artists' biographies, introduced the influential story that the renowned Flemish painter had invented a special varnish and oil paint. Vasari writes that Van Eyck became frustrated because he had to dry his varnished paintings in the sun, causing one of his panels to break at its joints. To prevent this from happening again, Van Eyck began exploring new varnish materials and discovered that some oils dried better than others. With linseed oil and walnut oil, specifically, he produced a varnish that did not need the sun to dry. After more experiments, Van Eyck also discovered that he could paint with these oils. According to Vasari, oil paint enabled Van Eyck and every painter after him to depict the visible world in all its shapes, textures, and colors.

Vasari's story about the invention of oil paint was repeated and retold until the end of the eighteenth century. Since then, more recently discovered written sources on paint technology, and the scientific examination of medieval artworks, have shown that at least three centuries before Van Eyck, painters used drying oils on a large scale to make both varnish and paint. But notwithstanding the refutation of Vasari's invention story, scholars still considered Van Eyck's remarkable ability to depict the visible world an issue of technique.

One of the most prominent theories attributes the appearance of Jan van Eyck's paintings to his discovery of a translucent paint, the so-called glaze. Ground with certain pigments, drying oils are the only binding medium that allows painters to make a light-transmitting paint layer of saturated color. Owing to its optical properties, a glaze permits light to penetrate the paint and, after hitting a reflective layer underneath, to travel back through the paint to the eye. In this way an oil painting illuminates itself from the inside out, almost like a translucent gemstone set on polished metal. In some of the earliest responses to the discovery that Van Eyck did not invent the oil medium, he instead became the inventor of this special translucent oil paint. In 1816 none other than Johann Wolfgang von Goethe (1749–1832) pointed out that whereas Van Eyck may not have invented a drying oil, he was the first to use it over the entire surface of his paintings, seeking those materials that allowed the reflection of light within his paint:

> Whatever may be said about the invention of oil painting, we do not doubt that Van Eyck was the first to mix with the colours themselves the oily substances which had hitherto been used to cover pictures after they were finished, and to have selected the most easily drying oils and the clearest and least opaque pigments, so as to allow the light of the white ground to shine through the colours over them as much as he liked. Since the whole force of colour, which by nature is dark, is not released by light reflected from it, but rather by light shining through it, this discovery and method answered the highest physical and artistic demands.[1]

Today most theories about the nature of Van Eyck's painting revolve around this idea that he was the first to discover or refine the glazing technique. But despite its presumed importance for our understanding of the revolution in fifteenth-century panel painting, the history of the use of the light-transmitting qualities of oil paint has not been written.

This book argues that, rather than being a fifteenth-century refinement, the techniques of varnishing and glazing were first developed to serve the mimetic ambition of the medieval painters, used to imitate the shine of gold and the translucency of gems, enamel, and other splendorous materials. It also shows that the history of varnishing and glazing is not just a history of panel painting. Both techniques were used and developed by a variety of artisans as part of the medieval material culture of mimesis and splendor: from artisans decorating metalwork and creating imitations of precious stones to scribes trying to spare their eyesight while poring over manuscripts. In turn, these artistic explorations of materials and their optical properties stimulated medieval authors of lapidaries, encyclopedias, and travel chronicles to come up with theories about the nature and genesis of transparent and translucent materials produced by nature, especially precious stones. The history of varnishing and glazing thus connects the medieval history of panel painting to other arts, to the history of knowledge, and to material culture more generally. In turn, it also shines new light on the long historiography, and especially Vasari's invention story, regarding the use of varnishes and glazes by Van Eyck and the early Netherlandish painters. The early Netherlandish painters were, in fact, deeply indebted to the medieval explorations of the oil medium, but they deviated from the purpose for which these techniques were once developed. Rather than imitating gold or gems by applying translucent paint and coatings, fifteenth-century panel painters aimed to reproduce the optical appearance of light and splendor in the natural world. As a result of these new ideas about the mimetic ambition of the picture maker, easel painters in the late fifteenth and sixteenth centuries would search for new materials and techniques to transform medieval varnishes and glazes into something more suited to the new ambitions of their craft.

Oil and Mimesis: Materials versus the Visible World

In the Middle Ages panel painting, sculpture, and many other crafts were largely defined by the desire to convey splendor like that of the goldsmith's precious work. The lavish use of gold and silver leaf, so characteristic of early medieval painting, is probably the strongest reminder of this and has received much study.[2] The special role the oil medium played in the painter's re-creation of the goldsmith's work, however, has largely remained unstudied. Yet besides gold and silver leaf, medieval painters used the translucency and transparency of oil paint and varnish to transform panel paintings to resemble precious metalwork with its glowing decorations. I call such optical replacement of translucent materials with oil paint *material mimesis.*

For at least two reasons, the crucial role of material mimesis in the history of panel painting has been poorly understood. Foremost, art historical studies of European painting usually focus on the *mimesis of the visible world,* or the artist's ability to represent objects, people, and events in an environment as they appear to the human eye.[3] Painters' ability to create a mimesis of the visible world eventually endowed painting with the status of a liberal art, and until the development of abstract art at the beginning of the twentieth century, that remained its principal aim. But this kind of mimesis had not always been painters' ambition. As this book will show through the case of varnishing and glazing, it was the mimesis of materials that lay at the heart of painterly practice before the fifteenth century.

A second reason we lack studies into the importance of material mimesis for the history of panel painting is that substitutes generally have a bad name and are considered "cheap" or fraudulent. In art history the technical term for material mimesis is *Ersatz,* a Germanism associated with inexpensive substances invented to replace scarce or unavailable materials during wartime (e.g., *Ersatzkaffee,* a substitute for coffee). As a result, embellishing a panel to make it resemble a goldsmith's work by replacing solid gold with thin leaves of metal and precious gems with glowing layers of paint can easily be considered cheating rather than a praiseworthy attempt at mastering an artistic task. Indeed, transforming one material into the appearance of another is often relegated to the realm of imitation, not art. In using the term *material mimesis,* therefore, I aim to avoid the negative connotations of terms such as *Ersatz* or "imitation," so that the history of this phenomenon in medieval painting can be studied in its own right. In addition, the term *material mimesis* clearly distinguishes it from other forms of mimesis, especially the mimesis of the visible world, which became important in the course of the fifteenth century.

Two ancient Fayum mummy portraits, made between AD 100 and AD 170, beautifully illustrate how the two approaches were once used side by side (figs. I.1, I.2). In the first portrait, the golden earrings, necklace, and hair ornament are evoked by material mimesis; thin leaves of gold capture the light the same way real gold jewelry would.[4] The second mummy portrait captures the sheen and reflectivity of gold in an entirely different manner; it replicates the optical qualities of gold with paint alone. In this example, the ancient portrait painter used a yellow paint to evoke the color of gold and added touches of white paint to summon the precious metal's specular reflections.

The specular reflection needs further clarification because, as this book will show, it is fundamental to understanding the changes in the art of panel painting in the fifteenth century. Ernst Gombrich first focused on this particular problem. In his seminal paper "Light, Form and Texture in Fifteenth-Century Painting" (1964), he almost foresaw the conclusions of technical research in the twenty-first century when he warned against the "overrated explanatory force of a phrase such as the meticulous observation of nature" to explain the changes in fifteenth-century painting.[5] Instead, Gombrich argues that one of the most important contributions of the early Netherlandish painters, and Van Eyck in particular, appears to have been the discovery of the specular reflection (Gombrich calls it the highlight). Specular reflections can be observed when light bounces off a surface at the same angle as the incident light ray (fig. I.3).

For this to happen, materials have to be perfectly smooth at a microscopic level, as are polished metals, a glass mirror, polished precious stones, and the wet surfaces of our eyes. These small white glimmers of specular reflections help the human eye interpret a surface as "glossy" (fig. I.4). Somehow this kind of reflection had been lost to medieval painting since its first depiction in antiquity. In fact the specular reflection cannot be found on a single painting before Van Eyck. But since shortly after the 1430s virtually no painted precious stone or other reflective material has been depicted without it.

The rest of this introduction sets out three things readers need to know before delving into this book. The first part gives a brief history of how Van Eyck became the master of glazing. I discuss the earliest attempts to identify binding media in paintings until more recent discussions of the role of the glazing technique in the work of Van Eyck. Since even today most art historical scholarship still associates the glazing technique with Van Eyck, it is important that we understand the origins and historiography of this attribution before beginning a fresh account of the history of the technique. The next section of this introduction discusses how this book uses art technological sources, especially recipes, and historical reconstructions to investigate the history of varnishing and glazing. And last, it defines the optics of varnishes and glazes. This rather technical discussion is crucial to understanding how premodern sources convey knowledge about the light-transmitting qualities of the oil medium.

How the Inventor of Oil Paint Became the Master of Glazing

In the last decades of the eighteenth century, scholars uncovered various manuscripts containing recipes for making varnish and oil paint. Because

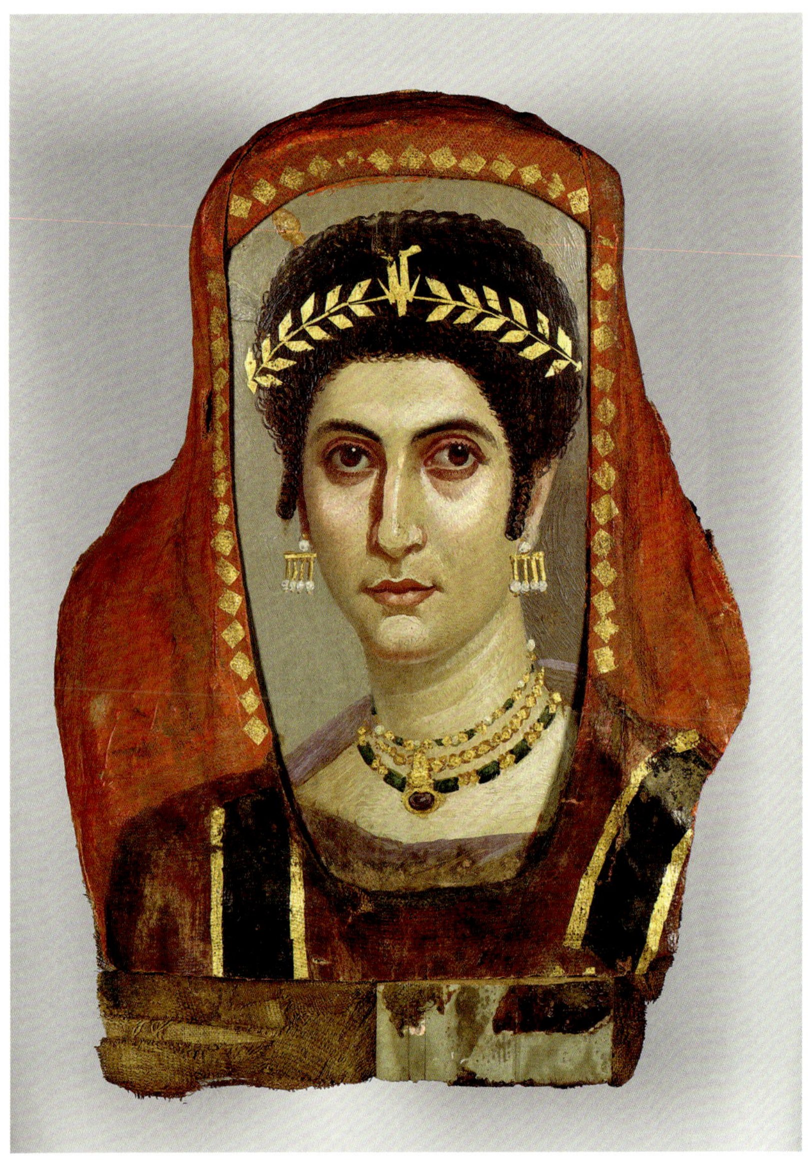

FIGURE I.1. Attributed to the Isidora Master (Romano-Egyptian, active AD 100–125), *Mummy Portrait of a Woman*. Encaustic on linden wood, gilt, linen. The J. Paul Getty Museum, Los Angeles. Image made available under Getty's Open Content Program.

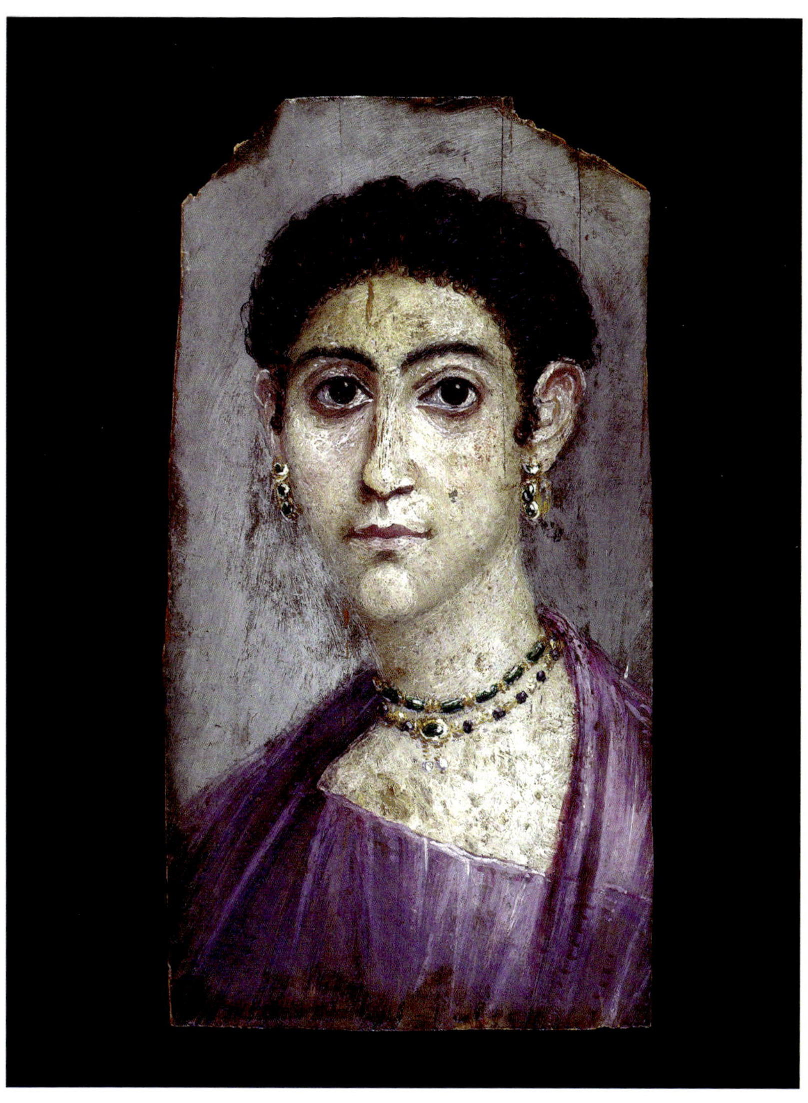

FIGURE I.2. *Mummy Portrait of a Woman*, ca. AD 100–120, Hawara (found). Encaustic on limewood. The British Museum, London. © The Trustees of the British Museum.

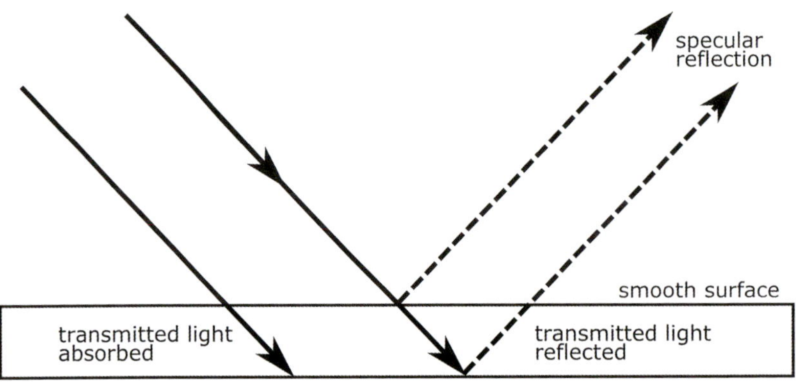

FIGURE I.3. Schematic representation of specular reflection. By Marjolijn Bol.

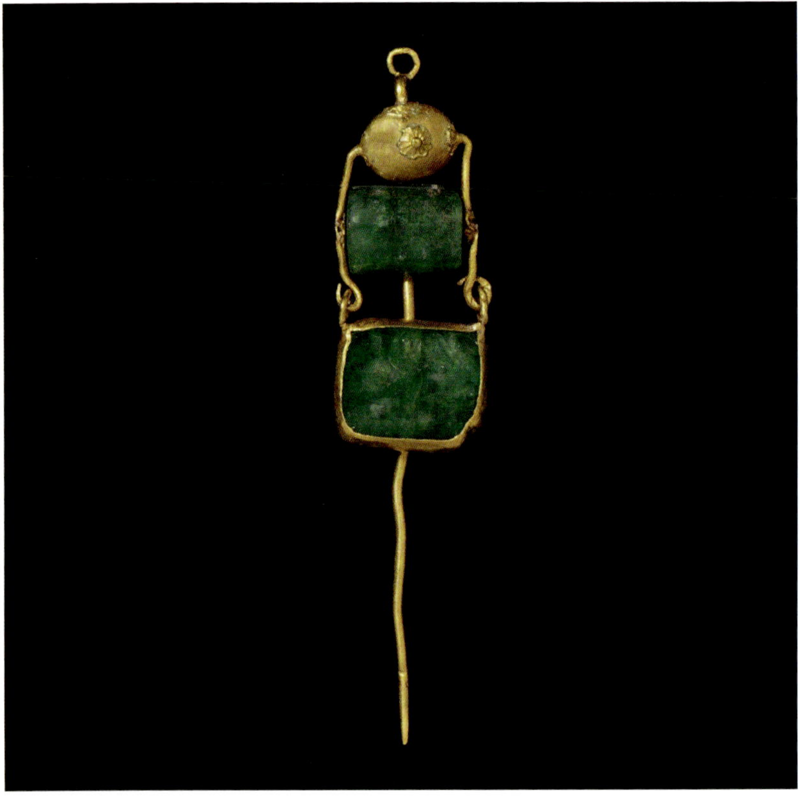

FIGURE I.4. "Gloss" or specular reflections on two emeralds set in gold as a hair ornament or pin, made first to fourth century AD. Victoria and Albert Museum, London. © Victoria and Albert Museum, London.

some of these treatises were written more than three centuries before Van Eyck allegedly invented the oil medium, their discovery prompted inquiries into the accuracy of Vasari's story. In addition, in the eighteenth and nineteenth centuries a renewed interest in the techniques of the old masters gave momentum to investigations into Van Eyck's special paint.[6] Apart from the general tendency in this period to admire and revive medieval history, the interest in the techniques of painters working before 1800 resulted from concerns about the durability of contemporary painting materials, products of the Industrial Revolution. Nineteenth-century chemists and contemporary painters admired the paintings of Jan van Eyck for their brilliance and their lasting beauty. They thought that unraveling the nature of his materials and technique would give contemporary painting a similar appearance and durability.[7] There opened up three new areas of research in which chemists and artists played a prominent role: research into sources on art technology; scientific examination of works of art; and historical reconstructions. These three approaches later proved fundamental for the development of the field now known as technical art history and make up an important part of the methodology of this book. Let me therefore first discuss what they have thus far contributed to the questions surrounding Van Eyck's technique, in particular his use of varnishes and glazes.

The English writer, connoisseur, and antiquarian Horace Walpole (1717–97) was one of the first to cast doubt on Vasari's story of the invention of oil paint. Walpole studied thirteenth-century archival documents recording the materials King Henry III (1207–72) ordered for decorating his court at Westminster. His discovery of numerous documents mentioning that oils were used to make the Westminster paintings led Walpole to speculate that "Where the discovery [of oil paint] was made I do not pretend to guess: the fact seems to be that we had such a practice [in England, before Van Eyck]."[8] About the same time, Gotthold Ephraim Lessing (1729–81), one of the most exceptional scholars of the European Enlightenment, found evidence in another type of source that drying oils were used long before Van Eyck: the "recipe collection." In his essay *Vom Alter der Ölmalerei aus dem Theophilus Presbyter* (1774), Lessing presents his discovery of a twelfth-century manuscript written by an anonymous monk under the pseudonym Theophilus.[9] In three books, the treatise of Theophilus provides readers with practical instructions for the arts of painting, glass painting, and metalwork. In the book on painting, Lessing discovered various recipes that explain how to use drying oils both for varnishing and for painting. Three centuries before Van Eyck, Theophilus's advice about the use of oil paint seemed definitive proof that the binding medium had not been invented in the fifteenth century. The Theophilus manuscript is just one example in a vast tradition of recipe treatises, and the study of recipe collections would come to play a prominent role in discussions about the Eyckian and pre-Eyckian painting medium.

In the same period, questions about the nature of Van Eyck's invention also prompted the scientific examination of works of art, instigated by advances in the new field of chemistry.[10] At first antiquarians realized that chemical analysis of paint samples could support what they had found in written sources. Walpole had already suggested that scientific examination might be fruitful in addressing questions concerning the binding medium of paintings, but it was the antiquarian, painter, and engraver John Thomas Smith (1766–1833) who first commissioned chemical analyses to explore the materials and techniques of medieval art. In 1800, work to extend Saint Stephen's Chapel, originally part of the Old Palace of Westminster, led to the discovery of an untouched fourteenth-century wall complete with paintings, stained glass, and sculptures. Since these medieval works were slated for destruction to enlarge the House of Commons (various fragments kept in the British Museum, London, are all that is left of the medieval decorations), Smith asked permission to investigate and to draw them for his *Antiquities of Westminster*, published in 1807. At the beginning of his project, Smith asked the apothecary and physician John Haslam (1764–1844) to perform binding-media analyses on the fourteenth-century wall paintings.[11] His method involved first removing the varnish with "impure aether (spiritus aetheris vitriolici of the London Pharmacopeia)" and then separating the binder the pigments had been prepared with. The remaining substance was decanted into water, allowing the binder to separate out. According to Haslam, this matter had the "peculiar smell of varnish," but he could not determine its exact composition.[12]

Based on his analyses, Haslam concluded that the chapel's walls had been painted entirely with oil-based media. Smith cites Haslam's full report in his *Antiquities of Westminster* and makes an important point of the fact that Haslam did not know of the archival documents that Smith found in subsequent years, which supported Haslam's claims about the materials used to paint the chapel:

> Although the preceding records have clearly pointed out the names of the several pigments and materials employed in the *Pictures* in St. Stephen's Chapel, yet in justice to a friend of Mr. Smith's, an *Analysis* of the colours is here introduced in a Letter, to shew the ingenuity and chemical acumen which could so correctly, and as it were prophetically, state every ingredi-

ent, full five years before those records were inspected: -by which the fact of their being painted *in oil*, is as decidedly established as by the records themselves.[13]

Having integrated his own findings in the Westminster archives with Haslam's analyses and the work done by Lessing and others, Smith concludes that oil painting must have certainly been known before Van Eyck.

Vasari's myth nevertheless proved persistent, and much of the research done in the nineteenth and twentieth centuries was geared toward demonstrating that, even though oil paint had been used before Van Eyck, it was not used for figurative painting.[14] Vasari had included some strange details about Van Eyck's using distillation and mysterious additives to invent his special varnish. New theories therefore revolved around the idea that Van Eyck's "novel technique" resulted from his invention of a specially prepared oil, his adding resins (secretions from plants) to his paint, his discovery of essential oils as a paint medium, or his use of secret egg-based emulsions. Nineteenth-century scholars in paint technology searched recipe collections for evidence that a sophisticated technology of oil painting developed in the fifteenth century and that medieval painters applied drying oils only as crude surface coatings.[15]

Because much of the previous research was undertaken to improve contemporary painting practice, by researchers who were often painters themselves, the new hypotheses were sometimes tested by practical experiment. The results were integrated into painting manuals for contemporary artists and, in a few cases, even resulted in new artists' paint media.[16] Most of the experiments focused on finding out how to reproduce the two qualities most admired in Van Eyck's work: its durability, that is, the ability to resist the test of time, and its translucent brilliance and glow. In his nine-volume *Traité complet de la peinture* (1829), the French painter Jacques-Nicolas Paillot de Montabert (1771–1849) argues that the transparent and enamel-like quality of Van Eyck's paintings proves he must have been the first to use a hard resin in his paint medium (*L'emaillé, la transparence de ses peintures semblant le prouver*).[17] And in his handbook *De la peinture à l'huile* (1830), the French painter Jean François Léonor Mérimée (1757–1836) praises the preservation of the works painted in the fifteenth century: "Van Eyck's pictures are in a higher state than those which have been painted two centuries later."[18] Mérimée concludes that the durability of Van Eyck's work resulted from his perfecting an oil-varnish medium he sought to "preserve the transparency and brilliancy" of his colors when dry.[19] The English painter Charles Lock Eastlake (1793–1865), best known perhaps as the first keeper of the National

Gallery of London, continues along similar lines in his *Materials for a History of Oil Painting* (1847). Like the previous authors, Eastlake believed Van Eyck's improvement must be sought in the admixture of resin to his paint medium. He points out that Van Eyck chose his materials with the "utmost efficacy" by a "process which, in every stage, was calculated for brilliancy of effect and durability." And whereas Paillot de Montabert likens Van Eyck's technique to the qualities of enamel, Eastlake compares its brilliance to stained glass windows: "The important attribute of *depth* was thus proved to be greatly within the power of the new art; and it is the more probable that Van Eyck founded much of his style on the principle of glass-painting."[20]

It was not until well into the twentieth century that chemical analysis became sophisticated enough to finally refute the idea that the special appearance of Van Eyck's works could be attributed to his invention or modification of the oil medium. A turning point in the discussion was Paul Coremans's publication of the results of the technical examination of the Ghent Altarpiece by the Central Laboratory of the Belgian Museums—undertaken to assist the 1950–51 restoration of the famous polyptych.[21] The instruments available to Coremans and his team allowed them to establish whether a paint medium was "oil" or "protein-based" (an egg binder). The paint samples from the Ghent Altarpiece showed that Jan van Eyck used a drying oil ground with pigments common long before the fifteenth century. But because Coremans could not yet establish what kind of drying oil Van Eyck had used to make his paint, or identify possible additions to the binding medium, his research did not completely resolve the nineteenth-century idea that Van Eyck might have used a mysterious ingredient to give his oil paint its special properties.[22]

In the last decade of the twentieth century, various new instrumental methods allowed for more precise identification of the individual constituents of the binding media of paints. This helped researchers to investigate the type of oil Van Eyck used and identify possible additions to his paint medium. In *Investigating Jan Van Eyck* (2000), scientists at the National Gallery of London substantiated and refined most of Coremans's results. With new methods for analyzing binding media, they discovered that Van Eyck used linseed oil to paint all his works in the National Gallery's collection.[23] Their analyses did not find any of the additives that earlier research believed held the key to Van Eyck's secret. Another seminal publication by the same museum, published in its 1997 *Technical Bulletin*, is devoted to the scientific examination of the National Gallery's large and representative collection of fifteenth-century German and early Netherlandish panel paintings.[24] Analyses showed that in most of the paintings examined linseed oil again appears to have been the medium of choice. In fact, one of the most important conclusions of the research was that, some small individual differences aside, all the paintings examined had been made with similar materials and by similar methods. In contrast to the nineteenth-century presumptions, therefore, scientific examination of paint samples showed that Van Eyck and his contemporaries used linseed oil and barely, if at all, manipulated the properties of their paint medium by adding other materials such as resins or egg. What is more, scientific examination of northern panel paintings made during the two centuries before Van Eyck showed that the oil medium was common during this time as well; linseed oil was found in most of the works examined.[25]

One might suppose that with these answers from science the matter of Van Eyck's technique would finally be laid to rest. But instead, the mystery of his sudden innovation resulted in a new technical myth; the great Flemish painter became the master of glazing. We have already seen that in the eighteenth and nineteenth centuries, the translucency of Van Eyck's paints was considered a crucial element of his technique. Since the 1950s, this has shifted to the idea that Van Eyck invented or perfected the glazing technique. In his seminal *Early Netherlandish Painting* (1953), Erwin Panofsky points out how Jan van Eyck's paintings enchant viewers because of his ability to capture in a different medium the splendor of the works kept in the treasuries of his patrons: "In a sense, he duplicated with the brush the work of the goldsmiths in metal and gems. His painting is jewellike in a quite literal sense, meant to recapture the glow of pearls and precious stones."[26] Four decades later we find a similar idea in Otto Pächt's *Van Eyck: Die Begründer der Altniederländischen Malerei* (1989). Pächt attributes the jewellike character of Van Eyck's paintings to his use of oils and the glazing technique:

> And this, it would seem, was achieved by an admixture of ingredients with a fat content, emulsions and no doubt also transparent glazes. The jewellike, sparkling quality of Van Eyck paintings has often been remarked upon, and it has often been said that oil painting, so called, was an attempt to rival the transparency and the light-filled quality of the stained glass in Gothic cathedrals. The paint through which the depicted radiance—the light that belongs to the pictorial world—is made concrete does not merely simulate light: it gives off a light of its own. It glows—and that is why we are involuntarily reminded of the phenomenon of refraction in precious stones.[27]

In this passage Pächt still echoes some of the nineteenth-century theories about Jan van Eyck's inventing a paint medium—once again underlining the endurance of Vasari's myth.[28] But when more recent literature credits Van Eyck with inventing or perfecting the glazing technique, the findings of the scientific examination of his paintings are now brought up as evidence. Glazes are no longer considered the result of mysterious mixtures but are seen as translucent paints made with certain pigments ground with a drying oil. In *From Giotto to Dürer* (1991), Dunkerton, Foister, Gordon, and Penny argue that while partial glazing was known in medieval times, "the full exploitation of the translucency of certain pigments in an oil medium" is the basis of the painting techniques of Van Eyck and other early Netherlandish painters.[29] And in his seminal *Colour and Culture* (1993), John Gage writes that "what Van Eyck brought to the technique [of oil painting] was essentially a complicated method of glazing transparent colors over a light ground" and that about "the precise origins of this refinement there is still little agreement."[30] Jeffrey Chipps Smith similarly suggests in his 2004 book *The Northern Renaissance* that even though it is known that oil paint was in use in fourteenth-century painting, "Van Eyck did perfect the application of glazes of pigments mixed in linseed oil."[31] And in an essay in the 2008 exhibition catalog *The Master of Flémalle and Rogier van der Weyden*, Jochen Sander argues that the crucial prerequisite for the innovations of the *ars nova* enabled the early Netherlandish painters "for the first time to paint with many layers using glazes generously," and that by means of these "multiple layers of transparent paint, their painting achieved a previously unknown, enamel like brilliance and intensity of color."[32] Even in the most recent publication on Van Eyck, the 2020 catalog accompanying the exhibition *Van Eyck: An Optical Revolution*, his special ability to use the glazing technique is highlighted in several of its articles.[33] If these examples seem numerous, many more could be cited. In fact, the glazing technique is nearly always mentioned when Van Eyck is introduced. This makes it all the more compelling to investigate the history of this special, translucent paint.

Sources on Art Technology and Historical Reconstructions

An important part of the argument I set out in this book is based on information gleaned from sources on art technology and historical reconstructions. It is therefore necessary to discuss in more detail how they inform this look into the history of the varnish and the glaze. In a broad sense, sources on art technology can be understood as any material surviving from the past that provides information about the history of the materials, tools, and techniques used to make works of art—ranging from realia to the work of art itself, images, texts, and audio-visual sources. Examples of art technological sources in written form include recipe collections, contracts, artists' correspondence and diaries, guild regulations, and technical manuals, to name but a few. This book integrates research into sources on art technology with information from those sources typically studied as part of the history of science, such as encyclopedias, books of minerals, and alchemical treatises.

The mutual dependence between art practices and learned traditions in the premodern and early modern periods has recently gained much momentum within the history of science, which is more broadly redefining itself as the history of knowledge.[34] This has made room for integrating studies on hands-on practical knowledge of artists, artisans, and other types of skilled workers into a broad framework of knowledge practices far beyond what we have today come to understand as "science," which of course for most periods of history is an anachronistic concept.[35] Pamela Smith, for instance, argues in favor of an "artisanal epistemology" to understand the development of knowledge traditions during the early modern period, and William R. Newman has shown how the history of alchemy must be understood in relation to the creative exploration of materials in the visual arts.[36]

Most of the information about the history of varnishing and glazing can be found in "recipe collections." The tradition of writing recipes can be traced back to ancient Mesopotamia, which saw the birth of the first writing systems. A recipe, simply put, is a set of instructions explaining how to make or prepare something. We probably know them best from cooking, but here we are interested in recipes that prescribe how to make and manipulate the materials used in various crafts. Such recipes range from the making and working of parchment, stone, glass, textiles, paper, pigments, and dyestuffs to the production of miniatures in books, metalwork, enamels, ceramics, woodworking, panel painting, glass painting, and much more. The earliest recipes that record craft practices date back to the seventeenth century BC and deal with the art of glassmaking. These recipes, written on cuneiform tablets, explain how to make a green glaze for pots with copper and lead and how to produce a green pot by mixing the clay with the green pigment verdigris.[37] Best known perhaps are a group of cuneiform tablets dating to the seventh century BC that also record the practices of glassmaking and of making artificial precious stones out of vitreous materials.[38] Craft recipes relevant to the history of varnishing and glazing are of much later date. They survive in a broad range of manuscripts that have come down to us since the eighth century AD. The treatise of Theophilus, mentioned above, serves well to introduce the number of issues that arise when studying such recipe collections.[39] First let's discuss how the practice of copying in the manuscript tradition may influence our understanding of this type of source.

Theophilus's *Schedula diversarum artium* has come down to us through a large number of handwritten copies, now kept in libraries all over the world. If there ever was an original version of this treatise, it has not been preserved or discovered. Something similar is the case for most medieval recipe collections; the "originals" are long lost, and we are left with numerous copies connected by a difficult genealogy that can span many centuries. Thus it is often not immediately clear to what time a certain treatise or recipe ought to be dated. A recipe found in a twelfth-century manuscript may reflect practices that were common decades or even centuries earlier. Thus a twelfth-century manuscript does not always record twelfth-century artisanal practice. In some cases these long histories of copying also introduced scribal errors that further complicate our understanding of the text as it has come down to us. Errors were especially common when the copyist was unfamiliar with the specialized topics and terminology at hand.

Second, there is the problem of language. Spelling was not standardized in the period studied here, and recipes often describe materials and processes with inconsistent terminology. A certain term could be used to refer to a variety of materials—as we will see in chapter 2, the Latin term *electrum* could refer both to the resin amber and to an alloy of gold and silver—or a variety of terms could describe one material: amber was not just known as *electrum*, but also variously called *glaesum* or *succinus*. Another problem that may complicate our understanding of historical recipes is that technical terms may have changed meaning over time or became obsolete.

The third issue that comes to light when studying recipes is authorship. As with the *Schedula*, most medieval recipe collections are anonymous, and because until the fourteenth century they were commonly written in Latin, we typically rely on internal evidence to find out *where* they were written. That most artisans would not have been able to read and write may even lead us to wonder why we have such an extensive written tradition of their practice in the first place. This problem is connected to our fourth issue with the study of recipes: readership. Were recipe collections intended for use in artisans' workshops? Or was their audience a more general one with an interest in art technology? It is now widely acknowledged that medieval recipes were likely not actively used by artisans as reference materials. This is further supported in that most manuscripts do not bear the traces of being used among the dirt of a workshop. It is more likely that, to preserve older knowledge traditions, recipe compilations were copied in monasteries and, from the twelfth century to the fourteenth, in secular *scriptoria* as well. The scribes copying recipes were therefore almost certainly familiar with the methods and materials of the art of book illumination, either because they practiced it themselves or because they worked closely with illuminators.

The bigger monasteries sometimes also had on their grounds workshops for other crafts, especially metalworking. In some cases scribes might therefore have had access to the materials and methods of other crafts as well. In light of these circumstances, Stefanos Kroustallis recently posed an interesting hypothesis about a possible context for the treatise of Theophilus. Kroustallis argues that the *Schedula* was "to serve as a text-guide for abbots and bishops to control workshop activities and carry over artistic enterprises, as well as to offer an essential theoretical and practical training in order to prepare future successors."[40] If this was indeed the case, Theophilus's recipes represent the state of the art of good practice in medieval religious art: from making stained glass windows and richly embellished liturgical objects to painting wooden doors and casting bells, the recipes in the *Schedula* are certainly representative of outfitting a church building, as Kroustallis points out.

In the course of the fourteenth century, recipes were increasingly written down in the vernacular, and with the invention of the printing press in the fifteenth century, they were among the early genres of the printed book. During this period recipes are more often organized to form so-called artists' manuals, providing readers with a coherent, ordered overview of the methods and procedures used within a particular craft. Another form in which recipes flooded the publishing markets in the early modern period is books of secrets. William Eamon has shown that the printed books of secrets were important agents in transmitting early modern knowledge traditions by "bringing together scholars, craftsmen, merchants, and humanists engaged in common pursuits."[41] Indeed, the dedications and introductions of treatises from this period show that collections of recipes were now widely read and produced for a variety of audiences, ranging from artisanal to domestic and from aristocrats to the learned.[42]

It is not my ambition in this book to resolve these issues with the interpretation of recipes; this would entail a different kind of project. Rather, I hope to show that the sheer number of treatises including recipes for making varnishes and glazes allows for a comparative approach that goes beyond the reliability of any one given recipe or any one given compilation of recipes. Studying a large body of recipes on varnishing and glazing thus allows us to establish traditions of artisanal practice and discern changes in these traditions over time. Combined with results from the scientific and visual examination of art objects and historical reconstructions, the comparative study of recipes for making varnishes and glazes reveals that even though we cannot always be sure whether certain recipes were used or produced in an artisanal context, many of them do bear witness to artisanal practice.

The second methodological pillar of this book is historical reconstruction. Historical reconstructions attempt to re-create the materials and or re-enact the methods of the artisan as they are described by recipes or deduced through the scientific examination of art objects. We can attempt to reconstruct the artisan's materials, from the manufacture of pigments and drying oils to the production and working of glass, leather, and yarn. Or we can try to re-create the artisan's techniques and processes, from dyeing leather and painting on glass to depicting a jewel in a painting, weaving tapestries, or applying enamel to metalwork.

However, with historical reconstructions too we are faced with various issues of interpretation. They can be divided into three categories: (1) the nature of the ingredients: even if identified and available, our modern-day equivalent may not match its historical counterpart because of changes in the natural variety, manufacture, or other historical properties of the material; (2) quantities and measurement: neither of these were standardized in the period studied in this book; and (3) the actual procedures: more often than not these are described in little detail and therefore are difficult to re-enact. Despite these difficulties, reconstructions prove useful in filling the gaps left by the art objects and the textual sources that have come down to us. In addition, they help us better understand recipes, aid in rematerializing objects we know only through descriptions in texts, and provide insight into the "original" appearance of objects that survive in poor or changed condition. By visualizing some procedures difficult to put into words, the historical reconstructions presented here also offer readers a better understanding of the often complex processes by which artworks were made.

Describing the Optics of Varnishes and Glazes

This book offers a history of materials and techniques and examines the terms used to describe them through various periods and in various languages. It is therefore important that its own terminology be clear and consistent, which merits the rather technical discussion that follows.[43] Figure I.5 shows a macro-level photograph of a green glaze. What we see, however, is a black square. How could this be? To unravel this optical riddle, we need to know something about the way light interacts with paint. Here I will introduce how this book defines and describes the optics of paint layers, and of varnishes and glazes in particular, by analyzing its definition of a "glaze": the terms in italics will be explained in more detail:

A glaze is a *smooth, translucent* coat of *light-transmitting paint* of *saturated color* that is made with *pigments* having a *refractive index* similar to that of their *binding medium*. It can be used to modify the *hue, value, saturation,* and *texture* of the surface it covers.

Color

The spectrum of visible *light* comprises a range of frequencies corresponding to the *colors* that can be perceived by the human eye. The interaction of white light with a material determines the color we perceive. Principally, light can interact with a material in three ways: light rays can be reflected, absorbed, or transmitted. In the case of white light illumination, the color we perceive depends on the frequency of the light wave that is reflected from the surface of the material or transmitted to our eye through the material. A green object, for instance, absorbs or transmits all the colors of the visible spectrum except the light of the frequency that we perceive as green.

Pigments, therefore, have color because they absorb certain wavelengths and reflect or *transmit* others. When all wavelengths are absorbed a pigment appears black, and when all wavelengths are reflected a pigment appears

FIGURE 1.5. Macro-level photo of a copper green (verdigris) glaze over underlayer of black paint. Photo by Marjolijn Bol.

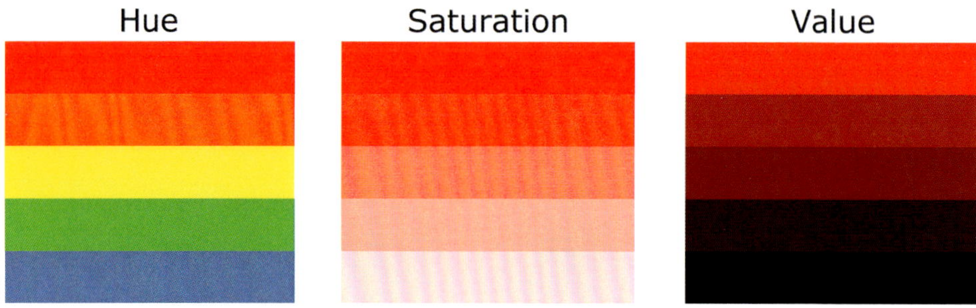

FIGURE I.6. Diagram depicting hue, saturation, and value. By Marjolijn Bol.

white. Since there is no transmission of light in either case, we can conclude that black pigments hide underlying layers by absorbing light, and that white pigments hide them by reflecting light.

To describe the colors of *pigments* and other materials, I use three terms that are often used to define color: *hue*, *value*, and *saturation* (fig. I.6). *Hue* is the basic position in the color spectrum, such as red, orange, yellow, green, blue, and possible intermediates such as yellowish orange and bluish green. *Value* (also called *tone*) is used to describe how light or dark a color is, independent of its hue. Value ranges from so light as to appear colorless to so dark as to look black. *Saturation*, one of the main characteristics of a glaze paint, is most difficult to define. It refers to the measure of the purity of a color; more precisely, it is defined as the colorfulness of an area (whether the area appears to exhibit more or less its hue) judged in proportion to its brightness.[44] Greatest saturation can be achieved when few other hues modify a spectral color. This means that a material appears most saturated when one specific wavelength is reflected or transmitted and all other wavelengths are absorbed.[45]

Scattering

For a colored material to reflect or transmit only its pure spectral color, *diffuse reflection* must be reduced to a minimum. The diffuse reflection of white light is also known as *scattering*. Because scattered white light contains all wavelengths, it causes desaturation when it mixes with the colored light reflected from a material. The scattering of light can be observed "at the surface" of a material and "within" it. When light rays scatter in random directions at the surface of a material, it is called *surface scattering* (fig. I.7). Surface scattering can be observed when a material is rough at the microscopic level, such as a woolen sweater, paper, or untreated wood. In contrast to diffuse reflection, the specular reflection I mentioned earlier is observable only at interfaces that are smooth at the microscopic level. The scattering of light within a material is called *internal scattering* or *subsurface scattering* (fig. I.7). Internal scattering describes the natural phenomenon of light scattering inside a material before being absorbed or leaving the material at a different location than the place it entered.[46] Except for metals, which are impervious to visible light, or some types of glass and liquids, which can be entirely light transmitting, most materials exhibit some degree of internal scattering. Materials that exhibit relatively high internal scattering, such as milk, marble, and skin, appear to be slightly illuminated from within. As with surface scattering, the internal scattering of white light in a material desaturates its color. To make a glaze, which according to the definition is of saturated color, the painter therefore needs to minimize the scattering of light. To avoid the scattering of light at the surface of his painting, he has to apply the paint as smoothly as possible. But how can the painter reduce the scattering of light within his paint as well? For this he needs a binding medium that transmits light the same way as the pigment used.

Light travels more rapidly in a vacuum such as space (or air as an approximation) than in solid or liquid materials. It thus is slowed down when it enters a new medium. This change in the direction of light is called *refraction* and is described by the *refractive index*. The refractive index (RI) describes how far a material bends light with reference to the velocity of light in a vacuum. The absolute refractive index of a given material is established by

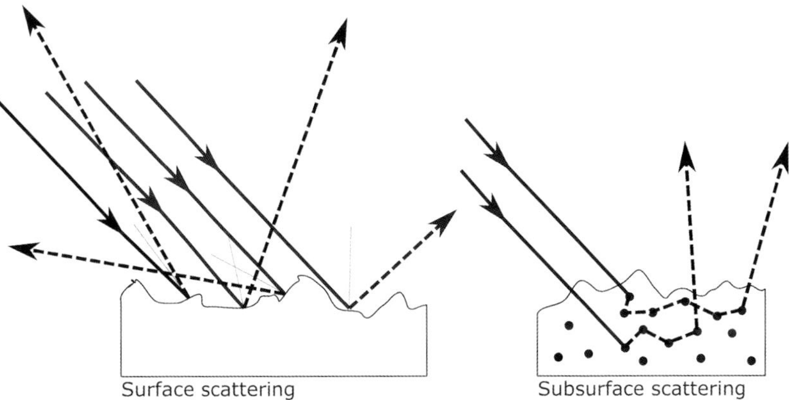

FIGURE I.7. *Left:* representation of diffuse reflection at the surface of a paint layer; *right:* diffuse reflection within a paint layer. By Marjolijn Bol.

dividing the speed of light in a vacuum by the speed of light in the material.[47] This means that when two materials transmit light at the same speed, they share the same refractive index, and this, as we will see, is important for making a glaze paint.

To create a paint that transmits the visible light without internal scattering, the painter needs a binding medium that matches the refractive index of his pigment. During the period studied in this book, artisans knew of only one type of binding medium that could achieve this: a drying oil.

Drying oils are liquid vegetable oils that react with oxygen and heat to form a transparent, solid film through a long process called polymerization. In the North, painters predominantly used linseed oil, while those south of the Alps typically preferred walnut oil. The drying property of these oils makes them suitable for painting, unlike nondrying oils such as olive oil. Cold-pressed linseed oil has a refractive index of about 1.47. This means that in this type of oil light travels 1.47 times slower than in air. Other binding media used by painters in the past, such as egg, glue, and gum, have a refractive index of about 1.34, close to the RI of water (1.33). Because the pigments available to premodern painters all have refractive indexes of 1.5 to 3, these binding media will produce a relatively opaque paint. Indeed, drying oils are the only paint medium approaching the refractive index of the pigments at the lower end of the scale and therefore the only one suitable for glazing.

The following pigments have a refractive index of about 1.5, meaning that, according to the definition, they are true glazing pigments: copper green (usually called verdigris in historical sources); the various red and yellow lake pigments (a naturally occurring dye precipitated onto a generally colorless substrate, often a form of hydrated alumina, see also chapter 5); and ultramarine blue. When these pigments are ground with a drying oil, light rays can travel through the pigment particles in the same way they travel through the oil. Scattering in the paint layer thus is reduced to a minimum, and light rays can be transmitted through the paint.

Let me explain this further with the example of a window. When a glass windowpane breaks into thousands of minute fragments, the colorless pieces, with all their rough edges, cause a large amount of scattering of white light. Light is no longer transmitted, and we cannot see through the once transparent window. Yet under the microscope the tiny glass particles will still be seen to transmit light. When we immerse the shattered glass in a liquid that has the same refractive index, the glass particles will become completely invisible (fig. I.8).

Something similar happens to pigments when they are ground with drying oils. When we look at bags and jars filled with pigments, none of

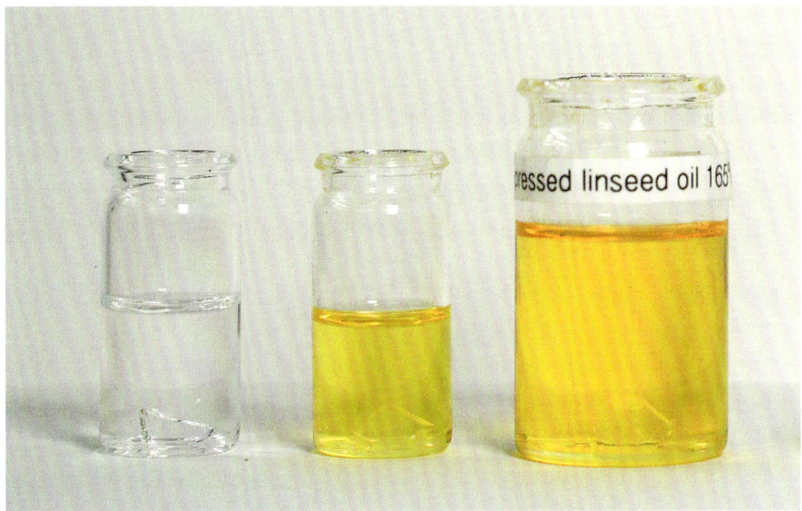

FIGURE 1.8. *Left to right:* Vial with glass and water; vial with glass and linseed oil; vial with alum and linseed oil. Since linseed oil, glass, and alum have similar refractive indexes, the glass and alum particles at the bottom of the vials with oil are close to invisible. Because of the greater difference in refractive index, the glass particle can be more clearly discerned in the vial with water. Photo by Marjolijn Bol.

the particles appear to be of saturated color, nor can their translucency be perceived. The pigments look this way because the tiny irregular fragments diffusely scatter the light in all directions, causing desaturation of color *and* decreasing their ability to transmit the visible light. But as with the small pieces of glass from the shattered window, pigments too can be seen to transmit light when we look at them under a microscope. This means that if we immerse a pigment particle in a binding medium that transmits light similarly to the way the pigment does—for instance, linseed oil—we can create a smooth, colored, translucent layer that, like a window, allows us to see through it.

Transparency and Translucency

The terms *transparency* and *translucency* require further clarification, since they are often applied loosely and interchangeably. Even asking the physical sciences for a definition does not yield a straightforward answer, for various disciplines explain the terms in different ways. The root of the English terms transparency and translucency is Latin. First used in antiquity, translucency (*translucidus*) is the older of the two and is derived from *lucere*, "to shine," and *trans*, "through." The term *transparere* was introduced to Latin in the later Middle Ages and comes from *parere*, "to appear," and *trans*, "through."[48] Nowadays transparency is generally used to define a quality of materials through which a clear image can be seen, whereas through something translucent only a diffuse image is visible. But when talking about the medieval understanding of the optics of varnishes and glazes, these definitions are too vague to maintain. For instance, it would be difficult to measure the level of detail that could theoretically be seen through a paint layer. Another problem is that colorless see-through materials, such as a picture varnish or a glass of water, and colored see-through materials, such as a glaze or a stained glass window, can both be called transparent.[49] Discussing the knowledge of such paint layers and other materials as revealed by medieval treatises soon becomes confusing.

For that reason I define materials that transmit *all* the (visible) light, and are therefore colorless, as "transparent" (in physics this is also called optical transparency). Completely transparent materials exist only in theory, but colorless glass, small volumes of water, and drying oils most closely approximate this category. Colored materials that transmit either a lot of light or some light are defined as translucent. This applies to oil when it is ground with certain pigments, some types of colored glass and enamel colors, various types of minerals, and so forth. When something can be seen through a transparent or a translucent material, such as a stained glass window, I will use "see-through." Note that by my definition a transparent material is always see-through, but a translucent material is see-through only in some cases. So, to reiterate, to make a transparent, colorless varnish the painter needs a binding medium that can transmit all the visible light without scattering. And to make a colored, translucent glaze paint the painter needs to grind a pigment with a binding medium that has a similar refractive index so that specific wavelengths can travel through the paint without scattering.

A Green Square?

We now have all the information needed to understand why the green glaze layer shown earlier appears invisible above the black paint (fig. I.5). When a glaze is applied to a black surface, all the visible light is transmitted *through* pigment *and* binding medium and absorbed by the black paint underneath. No light rays are reflected back to the eye, and so we cannot perceive the green color of the glaze that was applied. Certainly the total loss of color would have prevented painters from glazing over black paint, but perhaps

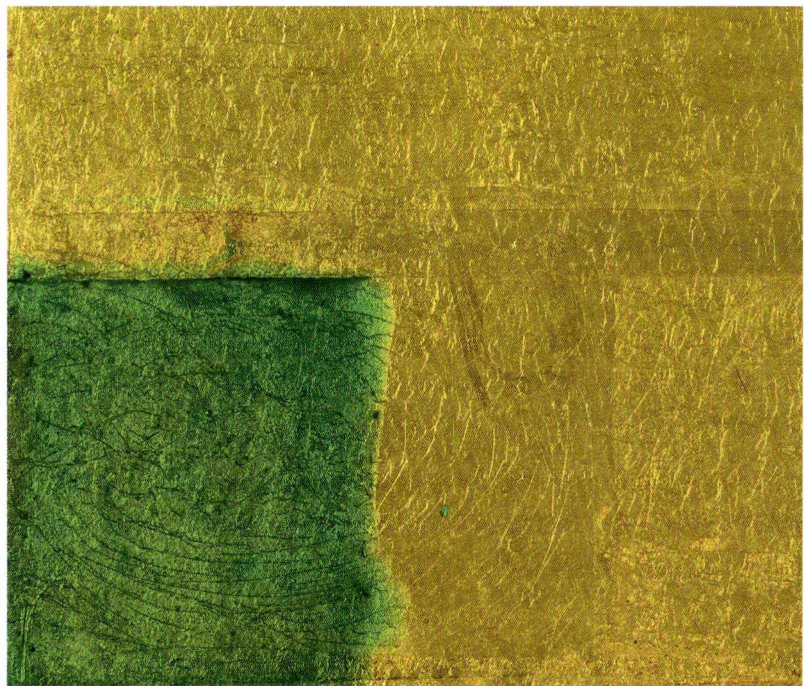

FIGURE I.9. Macro-level photo of a copper green oil glaze applied over gold leaf. Photo by Marjolijn Bol.

it is now easier to imagine the appearance of glazes on other surfaces. For instance, when a glaze is applied over a white ground layer or over polished metal such as gold or silver, the pure color of the pigment's hue is reflected from the white or metallic surface back through the glaze. This results in a paint layer that, as we saw earlier in this introduction, is often perceived to glow like precious stones, enamel, or stained glass. Indeed, the optics of such a "perfect" glaze applied to polished gold are very similar to the optics of precious stones set on metalwork (fig. I.9; compare fig. I.4). Here too, light travels through the stone, reflects on the metal underneath and travels back to the eye, as through a glaze, allowing us to perceive its translucency and saturated color.

Aim and Structure of the Book

By studying the discovery and use of the light-transmitting qualities of the oil medium, in this book I investigate how premodern sources convey an understanding of the rules of optics summarized above and explain why the exploration, manipulation, and imitation of transparent and translucent materials were vital to the arts at the time. As we have seen, Jan van Eyck's shadow looms large in discussions of the history of varnishes and glazes, not just in the sixteenth century but also in today's scholarship. Yet this book does not aim to solve the riddle presented by his art. Instead, I probe the history of varnishing and glazing to better understand the nature of medieval artistic production. By reconsidering the role of the varnish and the glaze in medieval art, however, the book also delivers a new perspective on the historiography of the sixteenth-century debate about Van Eyck and his alleged invention of the oil medium, the subject of chapter 1. It also allows us to pinpoint more precisely how the purpose of varnishing and glazing changed over the course of the fifteenth and sixteenth centuries, spurred by the changing ambition of panel painters from material mimesis to imitating the visible world.

Chapter 1 first turns to Vasari's famous myth that Jan van Eyck invented varnish and oil paint. Because of the ready acceptance of this account of Van Eyck's discovery, first as fact and later as revealing some part of the mystery of Van Eyck's painting technique, art historical scholarship has overlooked the close literary parallel between Vasari's myth and another famous discovery recounted by Pliny the Elder (AD 23–79) in his *Historia naturalis* (AD 79): the invention of varnish by the great Greek painter Apelles (fourth century BC). This chapter shows that to understand why Vasari introduced the myth of the invention of oil paint to the *Vite*, we need to interpret it in light of this ancient invention story.

Chapters 2 and 3 continue with the history of varnishing from 1100 to 1500. They examine how various sources including recipe books, encyclopedias, and other forms of premodern lore consider the optics, drying time, handling properties, and functions of varnish. Chapter 2 shows that, even though varnishing is best known for protecting paintings and other objects, knowing its history is crucial to understanding the early development of painting as an art of material mimesis, used to transform panel paintings and other polychromy on wood into the appearance of solid gold and lustrous enamel. Dealing with the history of varnishes in the fifteenth and sixteenth centuries, chapter 3 argues that the changes in panel painting brought by Van Eyck and his contemporaries instigated changes in varnishing practice as well. These material and technical changes were not introduced by Van Eyck, but rather were driven by the new ambition for painting that his art inspired; recipes show that painters explored new materials and techniques to influence the optical qualities, drying time, handling properties, and function of varnish so as to make the substance more suitable for imitating the visible world. It was on these changes in varnishing practice, as chapter 3 will show, that Vasari based the more technical parts of his invention myth.

Chapter 4 turns to the medieval desire for *splendor* to show how the early history of using drying oils in artisans' workshops is intertwined with the history of metalwork, gems, glass painting, and enameling. It shows that the materials used to make varnishes and glazes were also explored outside the workshops of panel painters, especially to make imitation gems. This chapter further considers how the artisanal pursuit of *splendor* inspired natural philosophers to theorize about the nature and genesis of examples of earth's splendor: precious stones. Natural historians, especially lapidarists, were influenced by the medieval art of material mimesis precisely because it so thoroughly understood refraction and reflection on the level of physical substances—of oils and pigments and of glass, gems, and metals. As a result, ideas about the genesis of gems and the nature of their optical qualities emerged together with the artisanal practices developed to reflect and refract light.

Chapter 5 focuses on how glazes were made. The study in chapters 2 and 3 of what craftsmen considered a "varnish" allows us to distinguish this substance historically from what artisans considered a "glaze." This distinction is especially important because varnishes were sometimes colored as well, and so they are often confused with glazes. Using recipes and historical reconstructions, this chapter shows that even though the materials and application methods of glazes were different, their general function was similar to that of varnish; medieval painters developed the glazing technique for various kinds of material mimesis in painting on wood. Studying the earliest history of glazing in relation to other crafts, especially those discussed in chapter 4, demonstrates that through a masterly knowledge of the optics of their materials, medieval painters were able to faithfully imitate the material characteristics of metalwork.

The final chapter then considers how painters dealt with the behavior of light on various surfaces in the premodern period. By comparing the medieval materials and techniques used to depict translucent motifs with those used by early Netherlandish painters from the 1420s onward, this chapter shows how knowledge about glazing and varnishing was adapted to new ideas about the ambition of painting to imitate the visible world.

It has often been suggested, both in historical sources and in contemporary scholarship, that the singular *splendor* of stained glass, saturated glazes, and glowing enamel was the particular pursuit of the arts of the North. Beyond the art of panel painting, *The Varnish and the Glaze* shows how the northern quest for deeply saturated colors resulted from the interplay of various practices of material culture, from the desire to color crystals to imitate emeralds and rubies to the use of oil-coated animal skins as translucent window coverings.[50] For the particular moment in the history of panel painting that is the subject of this book—one could call it the "Eyckian turning point"—it will show that in the materials and processes used to make paint, glazes, and varnishes, there lies hidden a cultural history of ideas about light, art, and mimesis that deeply influenced the development of Western painting.

1

Oil and Apelles

Vasari's Invention of a New Paint Medium

Giorgio Vasari is the father of many influential anecdotes, but one of his most enduring stories is about the invention of a fifteenth-century Netherlandish painter. Published for the first time in 1550, Vasari's *Vite*, containing the biographies of painters, sculptors, and architects from Cimabue (Cenni di Pepo) (1240–1392) up to his own time, provides us with an account of Jan van Eyck's invention of a new medium for painting with, a drying oil.[1] Even before the second edition of the *Lives* was published in 1568, Vasari's story was picked up by his contemporaries both north and south of the Alps. The idea that Van Eyck had invented a revolutionary new coloring method became firmly grounded in art history when Karel van Mander (1548–1606), the authority on the lives of the Dutch and German painters, included it in his 1604 *Schilder-boeck* (*Book on Painting*).[2]

Vasari's invention myth is typically studied as a potential key to understanding the particulars of Van Eyck's painting technique. But, as this chapter will show, Vasari introduced the story of the invention of oil paint for quite a different reason. Indeed, rather than a mere technical anecdote, he considered the discovery of the new binding medium a pivotal moment in the development of Italian painting. Vasari hoped that, with the help of oil paint, the painters of his own day and age had finally surpassed those of ancient times.

The sixteenth century is characterized by the deep admiration for the achievements of the ancients and, most famously perhaps, by the idea that this period was antiquity reborn—that it was, in fact, a "renaissance."

Adopting a biological model for history, Renaissance art theory almost inevitably became intertwined with the fear that, mirroring what had happened to the Greeks and the Romans, the sixteenth-century's achievements were also in danger of falling into ruin and oblivion. In the preface to the *Vite,* Vasari considers two things that may cause such decay and eventually the disappearance, or "death," of art: the absence of examples and the absence of writers. Examples and writers are necessary to record the steps of artistic progression for posterity:

> For since, of the works that are the life and the glory of the craftsmen, the first and step by step the second and the third were lost by reason of time, that consumes all things, and since, for lack of writers at that time, they could not, at least in that way, become known to posterity, their craftsmen as well came to be forgotten.[3]

One of the principal aims of Vasari's biographical project was to preserve the achievements of the artists from Cimabue up until his own time to prevent the *seconda morte* of art since antiquity.[4] As such, the *Vite* was designed to help future artists recognize the variations in quality between the works of different masters, so that "even when once more art has fallen into the same chaos or ruin; that these my labours, may be able to keep her alive." Vasari even pointedly adds: "if indeed my writings may be worthy of a happier fortune."[5]

To achieve this goal, Vasari planned to educate artists by means of a treatise about technique, which he included after the preface to the *Vite,* and through examples from good to better to best, as discussed in the various biographies. To further clarify the progress of the arts, Vasari divided the *Vite* into three ages, defined as *prima, seconda,* and *terza maniera.* Each age is introduced with its own *proemio* describing the artistic advancements of that period.[6] In the preface to the *seconda maniera,* Vasari elaborates on the motivation behind his work. He insists that it was not his "intention to make a list of the craftsmen and an inventory, so to speak, of their works." Instead, he believed that "the true soul of history" is "to observe the judgments, the counsels, the resolutions, and the intrigues of men," because this is what "teaches men to live and makes them wise."[7] Vasari therefore undertook the writing of the *Vite* so that future artists would be able to distinguish between the good, the better, and the best:

> And I have striven not only to say what these craftsmen have done, but also, in treating of them, to distinguish the better from the good and

> the best from the better, and to note with no small diligence the methods [*modi*] the moods [*arie*], the manners [*maniere*], the characteristics [*tratti*], and the fantasies [*fantasie*] of the painters and sculptors; . . . the causes and origins of the various manners and of the amelioration and that deterioration of the arts which have come to pass at diverse times and through diverse persons.[8]

In the preface to the *terza maniera*, Vasari explains how the efforts of the artists from the second age had led to five improvements: rule, order, proportion, draftsmanship, and manner. These improvements enabled the masters of the third age "to aspire still higher and attain to that supreme perfection." Starting with Leonardo da Vinci (1452–1519), the third age includes the biographies of the artists who worked in or close to Vasari's own time. These masters had achieved such perfection that they not only rivaled those famous painters, sculptors, and architects working in antiquity but were superior to them.

Through the ideas set out in the prefaces, Vasari thus transformed the *Vite* from a mere sequence of artists' lives into an educational tool recording the causes of regression and progression in the arts. With such a strong program, it is remarkable that, throughout the biographies, Vasari only twice refers to the ideas he so eloquently sets out in the prefaces.[9] In what follows, we will see why it is significant that one of these instances is found in the biography that tells the story of Jan van Eyck's invention.

Van Eyck's Invention of Varnish and Oil Paint

The *Vite* contains two references to Van Eyck's discovery of oil paint. The first is in the technical *proemio*, and a second, more detailed, account appears in the biography of the Venetian painter Antonello da Messina (ca. 1430–79), which Vasari included as part of the *seconda maniera*. Antonello's biography is introduced with the universal struggle of the painters of the second manner to improve the art of painting:

> When I consider within my own mind the various qualities of the benefits and advantages that have been conferred on the art of painting by many masters who have followed the second manner [*seconda maniera*], I cannot do otherwise than call them, by reason of their efforts, truly industrious and excellent, because they sought above all to bring painting to a better condition, without thinking of discomfort, expense, or any particular interest of their own.[10]

Starting with Cimabue and Giotto di Bondone (ca. 1267–1337), Vasari explains why the earlier painters struggled with the limitations of their binding medium. These first painters working with *tempera* (Vasari's term for an aqueous paint made with egg) on both panels and canvases knew that their pictures lacked a certain *morbidezza* (softness); a certain *vivacità* (liveliness) that would give more *forza* (grace) to their designs; a certain *vaghezza* (loveliness) in their coloring. And, finally, they lacked something that would *facilità nella unire* (facilitate blending) their colors, because with tempera they could only *tratteggiare* (hatch) with the point of the brush. Vasari insists that many painters, including Alesso Baldovinetti (1425–99) and Giuliano Pesello (1367–1446), searched for a method that would take away the previous limitations. They tried "liquid varnish" (*vernice liquida*) and "the mixture of other kinds of oils with tempera" (i.e., they tried emulsions of oils and egg): "But even if they had found what they were seeking, they still lacked the method of making their pictures on panel adhere as well as those painted on walls, and also [a way] of making them so that they could be washed without destroying the colours, and would endure any shock in handling."[11]

Not just the Italian masters were having these problems; the same desire to find a new technique was shared by "many lofty minds devoted to painting, being all the painters from Germany, France, Spain and other countries." This universal struggle of painters with their binding medium sets the stage for the technical advancement that is about to follow: Van Eyck's discovery of oil paint. And, indeed, the previous discussion of the limitations of using egg as a paint medium sums up everything the painter *can* do with oil colors.

Based on the 1550 edition of the *Vite*, the story of the invention of oil paint can be paraphrased this way: When Jan van Eyck puts one of his newly varnished paintings in the sun to dry, "as is the custom for panel paintings" (*come si costuma alle tavole*), it cracks. This event makes Van Eyck, who was "disgusted no less with this varnish than with working in tempera" (*e recatosi non meno a noia la vernice che il lavorare a tempera*), determined to invent a varnish that would dry in the shade so that the sun would never again harm his paintings (*cominciò a pensare di trovare un modo di fare una sorte di vernice che seccasse a l'ombra, senza mettere al sole le sue pitture*). After many experiments, Van Eyck "eventually discovers that linseed oil and nut oil dry better than other oils" (*alla fine trovò che l'olio di seme di lino e quello delle noci, fra tanti che ne provò, erano più seccativi di tutti gli altri*). Accordingly, he invents a varnish by boiling these oils with other mixtures that all the painters in the world desire (*Questi dunque bolliti con altre sue misture, gli fecero la vernice che egli stesso desiderava*). Afterward, having experimented with other substances, Van Eyck discovers that when these oils are mixed with colors they have a solid consistency, protect the painting from water damage when dry, and make the colors so brilliant as to give them luster without a varnish. Moreover, Vasari concludes, they can be blended infinitely better than with tempera (*E cosí fatto sperimento oltre a quella di molte cose, vide che il mescolare i colori con queste sorti d'oli gli dava una tempera molto forte, che secca non temeva l'acqua altrimenti; et inoltre accendeva il colore tanto forte, che gli recava lustro da per sé senza vernice; e quello che più gli parve mirabile era che si univa meglio che la tempera infinitamente*).[12]

With his invention, Van Eyck had ended painters' centuries-long struggle to overcome the limits of their binding medium. In the technical preface to the *Vite*, Vasari explains how important this technical innovation was:

> Working in oil has come later, and this has made many put aside the method of tempera [*È poi venuto il lavorar a olio, che ha fatto per molti mettere in bando il modo della tempera*]: inasmuch as to-day we see that the oil medium has been, and still is, continually used for panel pictures and other works of importance. . . . [T]he manner of painting in oil went on gaining in importance till the time of Pietro Perugino, of Leonardo da Vinci, and of Raffaello da Urbino, so much that it has now attained to that beauty which thanks to these masters our artists have achieved.[13]

To Vasari, the invention of oil paint explained one of the most prolific changes in the history of Italian panel painting. It was by virtue of the new paint medium that the masters of his own time, the third age, had attained their excellence. Even though we now have long known that oil paint was not an Eyckian invention, this part of Vasari's story reflects some historical truth; in the course of the fifteenth century, panel painters in Italy had shifted from working chiefly with an egg binder to predominantly using oil paint.[14] That the invention story addressed an actual change in Italian workshop practice—still known to Vasari's contemporaries—is likely part of the reason it gained so much momentum. But the invention story was not entirely without critique. During Vasari's own time, various writers already questioned its veracity.[15] This may be why, in the second 1568 edition of the *Vite*, Vasari adds historical facts about the use of oil paint and varnish before Van Eyck. The most important change in this respect is Vasari's discussion of Cennino Cennini's (ca. 1360 to before 1427) *Il libro dell'arte* in the life of Cennini's master Agnolo Gaddi (1350–96).

Cennini's Oil

Vasari was made aware of Cennini's handbook for painters in a 1564 letter from his friend Vincenzo Borghini (1515–80), artistic adviser to Cosimo I de' Medici and an important intellectual figure in Renaissance Florence. In his letter, Borghini points out that Cennini's *Il libro dell'arte* contains numerous references to painting with pigments ground with oil, yet it had been written some time before this medium was allegedly invented by Jan van Eyck.[16] Quite likely in response to this remark in Borghini's letter, the 1568 edition of the *Vite* contains an elaborate discussion of Cennini's treatise. But this exposition was not to show that the oil painting myth had been refuted. Considering the contents of *Il libro dell'arte*, Vasari argues that he does not have to expound on them at large, "there being to-day a perfect knowledge of all those matters which he [Cennini] held as great and very rare *secreti* in those times."[17] He also points out that a variety of important materials and techniques, well known during Vasari's own time, were not yet known to Cennini and his contemporaries:

> But I will not forbear to say that he [Cennini] makes no mention (and perchance they may not have been in use) of some earth-colours, such as dark red earths, cinabrese, and certain vitreous greens. Since then there have been discovered umber, which is an earth-colour, giallo santo, the smalts both for fresco and for oils, and some vitreous greens and yellows, wherein the painters of that age were lacking.[18]

Vasari continues that even though Cennini mentions colors mixed with oil "in order to make grounds of red, blue, green and in other manners," he does not mention the use of oil paint for figures—*non già per figure*.[19] With the remark that oil paint was used only in a crude manner and with a limited palette of pigments, Vasari implicitly defends his account of its invention by Van Eyck. And in a way he is right; Cennini's treatise mostly discusses painting on panel with an egg binder and does not advocate using oil paint for the entire surface of the painting, except when he speaks of wall painting.[20] Vasari's interest in maintaining the invention story may also be why the 1568 edition of the *Vite* reveals more details about the nature of Alesso Baldovinetti's experiments with varnish that, as we have seen, Vasari had first brought up in the 1550 biography of Antonello da Messina. Significantly, this new anecdote is included not in Baldovinetti's biography, but in the technical preface to the *Vite*. We read that Baldovinetti believed he had invented *un raro e bellissimo segreto* (a rare and beautiful secret) when he finished one of his frescoes with a liquid varnish prepared over fire. Sadly, he soon came to realize that his varnish was too strong to adhere to the wall.[21]

It is quite possible that the previous references to Cennini's and Baldovinetti's insufficient knowledge of oil paint and varnishing techniques were Vasari's attempt to give Van Eyck's invention a sounder historical foundation. But that still leaves us with an important question: Why would Vasari have attributed such an apparently quintessential invention to a fifteenth-century painter from the North in the first place? To find the answer we have to study Van Eyck's reputation in Italy before Vasari wrote the *Vite*.

Van Eyck in Italy

As early as the fifteenth century, Van Eyck's paintings were fervently collected in Italy.[22] In the *Vite*, Vasari explains that Van Eyck's popularity south of the Alps is part of the reason his new coloring method made its way from Flanders to Florence. In the technical preface he writes that Italian painters could admire the works of Van Eyck and other early Netherlandish painters in the collections of various influential patrons of the arts, notably Alfonso V, King of Aragon (1396–1458) and of Naples (as Alfonso I, r. 1442–58); Federico da Montefeltro (1422–82, duke of Urbino 1444–82); Lorenzo de' Medici (1449–92); and the wealthy Portinari family.[23] In the Life of Antonello da Messina, Vasari adds that Florentine merchants sent a picture by Van Eyck to King Alfonso I of Naples. Not only did this painting please the king very much for its new method of coloring, but "all the painters in the kingdom came together to see it" (*a la quale opera concorse tutto il regno*).[24]

One of these painters was Antonello da Messina, who, on a trip to Naples, had heard about this amazing work painted in oil that "could be washed, would endure any shock, and was in every way perfect" (*che si poteva lavare e che reggeva ad ogni percossa*).[25] These special features of Van Eyck's paintings, readers may have noticed, precisely summarize the struggles of Baldovinetti, Pesello, and their contemporaries. After seeing the famous painting, Antonello was so strongly impressed by the "liveliness of the colours" (*la vivacità de' colori*) and by its "beauty and harmony" (*e la bellezza et unione*) that he "put on one side all other business and every thought and went off to Flanders."[26] Vasari continues that Van Eyck, who initially had been rather secretive about his new discovery, now in old age decided to reveal the specifics of his new coloring method to Antonello. This part of the invention story was not based on historical facts either. Antonello da Messina, born about 1430, was too young to have met Van Eyck and learned from him personally, though it is true that he was one of the first painters south of the Alps to fully adopt both the oil medium and the northern style.[27] This may have been why Vasari regarded Antonello da Messina as a suitable candidate for introducing the oil painting technique to Italy.

Besides the painters, Van Eyck's works in Italy also sparked the imaginations of various fifteenth-century writers. These early texts not only establish Van Eyck's popularity south of the Alps, they contain all the elements out of which Vasari's invention story appears to have been born. One of the first writers to allude to Van Eyck's fame in Italy is Cyriacus of Ancona (Ciriaco de' Pizzicolli, 1391–1452), the productive recorder of Greek and Roman antiquities. He calls Van Eyck the "pride of painting" and points out that the early Netherlandish painters have a remarkable ability to capture the visible world:

> Garments prodigiously enhanced by purple and gold, blooming meadows, flowers, trees, leafy and shady hills, ornate halls and porticoes, gold really resembling gold, pearls, precious stones, and everything else that you would think to have been produced, not by the artifice of human hands but by all-bearing nature alone.[28]

Even more telling is the first biography of Jan van Eyck, written by Bartholomaeus Facius (ca. 1400–1457), chronicler of King Alfonso I of Naples, who, as we have seen, was a keen collector of Van Eyck's paintings. In *De viris illustribus* (*On Famous Men*, 1455), Facius attributes Van Eyck's excellence as a painter to his learnedness. Van Eyck's study of the works of the ancients and the writings of the great Roman encyclopedist Pliny the Elder (AD 23–79), in particular, resulted in many discoveries about the properties of paints:

> Jan van Eyck has been judged the leading painter of our time. He was not unlettered, particularly in geometry and such arts as contribute to the enrichment of painting, and he is thought for this reason to have discovered many things about the properties of colors/paints[29] recorded by the ancients and learned by him from reading Pliny and other authors [*putaturque ob eam rem multa de colorum proprietatibus inuenisse. quae ab antiquis tradita ex plinii et aliorum auctorum lectione didicerat*].[30]

In his translation of *De viris illustribus*, Michael Baxandall points out that by transforming Van Eyck into a learned painter, Facius imbues him with "the Apelles factor" (Apelles, the greatest painter from antiquity, was similarly renowned for his learnedness).[31] A remark by the French cardinal Jean Jouffroy (ca. 1412–73), bishop of Arras, should be considered in the same light. In 1468, while residing in Rome, Jouffroy wrote that Van Eyck surpassed all the painters from antiquity, including Apelles:

> Consider now the age in which Aetion, Nicomachus, Protogenes, [and] Apelles flourished, or those who surpassed them [*aut qui illos superaverunt*]—John of Bruges [Iohannem Brugensem], whose paintings you have seen in Pope Eugenius' palace, and dearest to me, Rogier of Brussels [Brucellensem Rogerium], whose pictures add lustre to the palaces of every king.[32]

The previous authors admire Van Eyck for his extraordinary visual realism and compare the magnitude of his achievement to that of the most famous painters from antiquity. But it is important to note that they do not yet connect his painterly skill to his use of oil paint. For this we have to turn to Antonio di Pietro Averlino (ca. 1400 to ca. 1469), the Florentine architect also known as Filarete. In his *Trattato di architettura* (1457–64), Filarete writes that the northerners, in particular Jan van Eyck and his contemporary Rogier van der Weyden, know well how to use oil colors (*I quail hanno adopterato ottimanente questi colori a olio*).[33] He even specifies the particular oil used as linseed oil—*l'olio si è di seme di lino*.[34]

Almost a century before Vasari wrote the *Vite*, therefore, all the ingredients for the invention myth are present in these writings, which it is worth noting are all from Italian soil. They assert his fame as a painter and connect Van Eyck's novel visual vocabulary to the oil painting technique. Building on this textual tradition that was almost certainly based on more common presumptions about the famous painter from the North, it appears that Van Eyck was the perfect candidate for the inventor of the binding medium that helped explain the Italians' switch from egg to oil paint in the course of the fifteenth century. Yet there is something more to these early writings that regard Van Eyck as an educated and cunning researcher into paints who studied Pliny and emulated Apelles—this has everything to do with the second reason the invention of oil paint was so important to Vasari.

Surpassing the Ancient Varnish

Vasari drew from a variety of classical authors for the ideas set out in the various prefaces to his work. Particularly formative for the art history in the *Vite* was Pliny the Elder's account of the history of ancient art.[35] As Pliny did in book 35 of his *Historia naturalis*, which divides art history into periods of rise, peak, and decline, Vasari organized his collection of artists' biographies around a scheme of progressive development.[36] But Vasari differs from Pliny in that he famously models his ideas about the nature of artistic progress on human development, resulting in the idea of art's (re-)birth, growth, and maturity through successive periods of technical invention and stylistic refinement.[37] In this way Vasari's historical model reveals how, after the death of the art of classical times, it had been reborn with Giotto and Cimabue, to eventually grow into its excellent and perfect state in his own time.[38] But to Vasari, something more than a renaissance may have happened. In the preface to the second age he points out that he derives pleasure from "seeing past events as present" (*vedere le cose passate come presenti*) and that what happens in his own time strongly resembles, perhaps even mirrors, the events that took place more than two thousand years earlier:

> Now, this must have happened to painting and sculpture in former times [antiquity] in such a similar fashion that, if the names were changed round, their histories would be exactly the same [*Ma nella pittura e scultura in altri tempi debbe essere accaduto questo tanto simile che, se e' si scambiassino insieme u noi, saqrebbono appunto I medesimi casi*].[39]

The story of the invention of oil paint takes a special place in this idea that contemporary history runs parallel to that of the ancients. In fact, it appears that Vasari may have actually changed some names around. One of Pliny's stories is suspiciously reminiscent of Vasari's oil painting myth; Apelles's invention of a secret varnish, the *atramentum*:

> [Apelles's] inventions in the art of painting have been useful to all other painters as well, but there was one which nobody was able to imitate [*unum imitari nemo potuit*]: when his works were finished he used to cover them over with a varnish [*atramentum*] of such thinness that its very presence, while its reflexion threw up the brilliance of all the colours and preserved them from dust and dirt, was only visible to anyone who looked close up [*cum repercussum claritates colorum omnium excitaret custodiretque a pulvere et sordibus, ad manum intuenti demum appareret*].[40]

There are several clear parallels between Vasari's myth and Pliny's. We have seen that before he discovered oil paint Van Eyck, like Apelles, discovered a special varnish that was desired by all the painters in the world. Apelles and Van Eyck both kept their new inventions secret.[41] And neither Apelles's invention nor Van Eyck's could be imitated—the latter was even guarding his discovery by not allowing anyone to see him at work. According to Pliny, Apelles's *atramentum* was so thin that it could be seen only from up close; and in Vasari's story the Flemish paintings that arrived in Italy were closely scrutinized, even smelled, to try to figure out how they were made. But just as with Apelles's inimitable varnish, Vasari reports, simply looking was not enough to unravel Jan van Eyck's working method.[42]

There are nevertheless two crucial differences between the stories. Apelles only invented a varnish; but after some further experimentation Van Eyck discovers that he can also paint with linseed and nut oil. And unlike Apelles, Van Eyck eventually decides to share his special technique with an ardent pupil, Antonello da Messina, who brought the knowledge of coloring in oil to Italy. These two differences are no coincidence; they are key to understanding why Vasari based his invention story on an ancient discovery.

Oil and the Moderns

To Vasari and his contemporaries it was vital to believe that their time was antiquity reborn. To prove that this was true for the arts, they compared the achievements of the artists of their own day to those of artists working in classical times. According to Pliny, Apelles had outshone all painters before his time and was superior to all those working after him. Apelles's works therefore became the gold standard against which the paintings of Vasari's time had to be measured.[43] Unfortunately this posed a serious problem to Vasari. Unlike architecture and sculpture, no ancient paintings, let alone works by Apelles, had come down to the sixteenth century. In the third preface to the *Vite*, Vasari points this out when, having just explained why he believes Michelangelo surpassed the ancients in sculpture and architecture, he muses that despite lack of proof he may also have surpassed them in the art of painting:

> And the same may be believed of his [Michelangelo's] pictures, which, if we chanced to have some by the most famous Greeks and Romans, so that we might compare them face to face, would prove to be as much higher in value and more noble as his sculptures are clearly superior to all those of the ancients [*Il che medesimamente per consequenzia si può credere le sue pitture; le quali, se per adventura ci fussero di quelle famosissime greche o romane da poterle a fronte a fronte paragonare, tanto resterebbono in maggior pregio e più onorate, quanto più appariscono le sue sculture superiori a tutte le antiche*].[44]

Lacking direct comparison, Vasari therefore needed another argument to prove that the painters of his own time had surpassed those of the ancient past. He found such evidence in the realm of technique. Vasari had explained in the technical preface that the perfect excellence achieved by the artists of the third age was indebted to the invention and dissemination of oil paint. At the end of the biography of Antonello da Messina, Vasari not only repeats this point but adds that the invention of oil paint is proof that the moderns had surpassed antiquity. The reason for this, he continues, is that there is no evidence in ancient sources that oil paint was known at this time:

> Both of them [Messina and Van Eyck] benefited and enriched the art; for it is by means of this invention [of oil paint] that craftsmen have since become excellent, that they have been able to make their figures all but alive. Their services should be all the more valued, inasmuch as there is no writer to be found who attributes this manner of colouring to the ancients [*quanto manco si truova scrittore alcuno che questa maniera di colorire assegni a gli antichi*]; and if it could be known for certain that it did not exist among them, this age would surpass all the excellence of the ancients by virtue of its perfection [*se e' si potesse sapere che ella non fusse stata veramente appresso di loro, avanzerebbe pure questo secolo le eccellenzie dello antico in questa perfezzione*].[45]

This desire—to have evidence that the ancient painters had been surpassed—is the reason Vasari modeled Van Eyck's invention of oil paint on Pliny's story of Apelles's invention of varnish. Van Eyck first invented a new kind of varnish so that Vasari's time equaled the age of Apelles, only to vanquish the ancients on technical grounds as well with his discovery of oil paint. The invention of oil paint provided Vasari with the much-needed historical evidence that the moderns had not just triumphed over the ancient sculptors and architects but, by virtue of a technique unknown even to the great Apelles, also triumphed over the ancient painters. The fifteenth-century writings about Jan van Eyck discussed above showed that a parallel between Van Eyck and Apelles was already perceived a century before Vasari wrote the *Vite*. It is therefore significant that the question whether the ancients knew how to use oil paint and varnish had precedent as well.

Two years before Vasari published the *Vite*, the Portuguese court artist Francisco de Hollanda (1517–85) wrote a treatise called *Da pintura antiga* (*On Antique Painting*, 1548).[46] In this treatise De Hollanda contests the idea that varnish was a modern invention and attributes its discovery to Apelles instead: "Varnish some believe was invented in our time, but it was discovered by Apelles, the famous painter, according to what we see in Pliny, in Book XXXV."[47] While declaring his own inexperience with the medium, De Hollanda insists that the ancient painters also knew how to use oil paint:

> Next comes painting in oil, which is very noble and ancient, done on panel with all the mixtures of colors that the illustrious painters invented. And the celebrated paintings of the ancients were painted in oil, something I have never learned nor done.[48]

Before stating that the ancient painters knew how to use oil paint, however, De Hollanda had praised the art of illumination not only as being extremely durable if well conserved, but also as constituting "the sole advantage we have over the ancients." It appears therefore that De Hollanda—who had been trained as a miniaturist in the footsteps of his father, António de Hollanda (ca. 1480–1557)—had an ulterior motive for attributing the use of varnish and oil on panel to the ancients; he intended instead to credit the art of illumination, an art not compatible with oil colors, with having surpassed antiquity.

That De Hollanda and Vasari discuss whether the ancients knew and used the oil medium at about the same time suggests that there must have been some discussion about the modernity of varnish and oil paint before Vasari published his treatise. This likely also explains why Vasari's invention story almost instantly landed on fertile ground and was picked up by various of his contemporaries.

A Modern Apelles

Even though Vasari indirectly compares the achievements of Jan van Eyck to those of Apelles—a comparison, as we will see later in this chapter, immediately understood by his contemporaries north of the Alps—Vasari did not consider the great Flemish painter to have vanquished Apelles in all respects. In the preface to the third age, Vasari insists that although the work of the artists of the second age was important for the advancement of the arts, antiquity would be "gloriously" surpassed only in the sixteenth century:

> Truly great was the advancement conferred on the arts of architecture, painting, and sculpture by those excellent masters of whom we have written hitherto, in the second part of the *Lives*. . . , not indeed complete perfection, but with so near an approach to the truth that the masters of the third age . . . were enabled by means of their light [painters of the

> second age], to aspire still higher to that supreme perfection which we see in the most highly prized and most celebrated of our modern works . . . that true excellence which, having surpassed the age of the ancients, makes the modern so glorious [*e discorrer succintamente donde sia nato quel vero buono, che, superato il secolo antico, fa il moderno sì glorioso*].[49]

Vasari had reserved the special honor of surpassing the ancients for one modern painter of the third age in particular: Raphael (Raffaello Sanzio da Urbino, 1483–1520). Comparing painters to Apelles became a commonplace in sixteenth-century art theory, and Vasari's *Vite* is no exception. In the various sonnets and epitaphs cited throughout the *Lives*, painters are compared to Apelles many times. Yet beyond such poems and elegies Vasari brings up Apelles only three times throughout the main text of the first edition of the *Lives*, and always as a historical figure. In one of these instances, Apelles is said to have possibly been surpassed by Raffaello da Urbino. In this passage, which notably can be found in the preface to the third age and not in Raphael's biography, Vasari writes that Raphael brought the art of painting to that complete perfection that characterized ancient times. He then repeats the wish that he expressed in the *Life* of Antonello da Messina: if only their works could be compared, Raphael's paintings might prove superior to those of Apelles and Zeuxis:

> But more than all did the most gracious Raffaello da Urbino, who, studying the labours of the old masters and those of the modern, took the best from them, and having gathered it together, enriched the art of painting with that complete perfection which was shown in ancient times by the figures of Apelles and Zeuxis; nay, even more, if we may be as bold to say it, as might be proved if we could compare their works with his [*che ebbero anticamente le figure di Apelle e di Zeusi, e più, se si potessi dire o mostrare l'opere di quelli a questo paragone*].[50]

In the second (1568) edition of the *Vite*, Vasari underlines Raphael's status as the modern Apelles a second time in the newly included biography of Timoteo da Urbino (1469–1523). Timoteo da Urbino, who was first trained as a goldsmith, is said to have afterward rigorously applied himself to the art of drawing and painting, learning the most difficult things without the guidance of an appointed master. He became so infatuated with his new profession that he learned many *segreti della pittura* by watching other painters work. In this way Timoteo acquired such a pleasing manner that it approached that of his famous compatriot Raphael, who is referred to as the *nuovo Apelle*.[51] Indeed, when Jan van Eyck invented oil paint, and after Antonella da Messina brought the new method of coloring to Italy, the modern painters had merely been given a tool—albeit a fundamental one—to surpass the ancients. But to Vasari, Apelles had really been surpassed only during Vasari's own time, an achievement he attributes to Raphael.[52]

The Flemish Apelles

In today's scholarship, the implicit parallel between Van Eyck's invention of oil paint and Apelles's secret *atramentum* has gone unnoticed. It is therefore all the more striking that Vasari's contemporaries almost instantly picked up this reference to the most famous painter from antiquity. The Italian writer and painter Giovanni Battista Armenini (1533–1609), for example, explicitly connects the two stories when he writes about Van Eyck's invention in his art theory treatise *De' veri precetti della pittura* (*On the True Precepts of the Art of Painting*, 1586). Armenini cautiously points out that Van Eyck's use of oil paint is something considered unknown to the ancients. But contrary to what Vasari had in mind, Armenini regards Apelles's varnish as possible evidence negating Vasari's myth:

> Many assert that the inventor of oil painting was a certain Fleming, John of Bruges. It is believed that the ancients had no knowledge of this method, though some say that Apelles used a liquid like varnish at the end of his works, with which he enlivened all the colors, coating some of them more and some less, as he saw the need [*si stima che apresse de gli Antichi non ci fuße mai questo modo, ancorche alcuni dicono che Apelle usasse nel fine delle opere sue un liquor come vernice, col quale egli ravivava tutti colori riscoprendoli col più, e col meno secondo che di quelli egli vedeva esserli di bisogno*].[53]

The Milanese painter and art theorist Giovan Paolo Lomazzo (1538–92) makes a similar remark in his 1590 *Tempio della pittura* (*Idea of the Temple of Painting*). But unlike Armenini, Lomazzo is convinced that the oil medium was not known in ancient times. In chapter 13, "On the Seven Parts of or Types of Color," he writes that Leonardo da Vinci used oil paint for nearly all his work, "a manner of coloring first discovered by Jan van Eyck, since the ancients certainly did not know about it" (*la qual maniera di colorire sù ritrovata prima da Gio. da Bruggia, essendo certa cosa che gli antichi non la conobbero*).[54] Lomazzo continues that the ancients were nevertheless concerned with the durability of their work. The great Protogenes of Cauno, for instance, applied four paint layers to his pictures so that "if one layer fell off, there would still be another" (*accioche cadēdo una restasse l'altra*), and so did Apelles, whose Venus, we read, was preserved up until the time of Augustus. Since Vasari, durability had become ascribed to the oil medium as one of its special merits and, curiously, Lomazzo here appears to refer to this when he introduces the invention myth; even though the ancients sought to make durable art and their works were known to have been long-lasting, the oil medium must have still been a modern discovery.

North of the Alps as well, the implications of Vasari's account of the invention story were immediately understood. Here all traces of Armenini's hesitation disappear; inspired by Vasari's story, art theorists from the North transform Van Eyck into *the* "Flemish Apelles." The first example can be found in a poem on Jan van Eyck's Ghent Altarpiece, written by the Flemish painter and poet Lucas de Heere (1534–84) in honor of the 1559 meeting of the knights of the Golden Fleece in the Saint Bavo church. The poem, titled *Den Vlaemschen Apelles* (*The Flemish Apelles*), consists of twenty-four verses and was displayed opposite the renowned altarpiece. In 1565 it was published in De Heere's *Den hof en boomgaerd der poësien* (*The Court and Orchard of Poetry*).[55] Its first rhymes start out with the well-known and lively description of the panels of the polyptych, and the poem concludes with a short description of the life of Jan van Eyck. In a 1972 publication about De Heere and his writings on art, Jochen Becker argues that the verses of De Heere's poem draw a parallel between the life of Van Eyck and that of Apelles.[56] De Heere, for example, writes about Van Eyck's grace and reports that he was much loved by his respectable patron Philip the Good of Burgundy (1396–1467). In this way he implicitly likens Van Eyck to Pliny's descriptions of Apelles's grace and the way he was loved by his patron Alexander the Great (356–323 BC).[57] The poem also brings up two things that have no equivalents in Pliny's account of the life of Apelles. De Heere points out that a certain Italian author (referring to Georgius Vasarius in the margins) wrote that Jan van Eyck had invented oil paint and brought it to Italy, and in the next verse he marvels how remarkable it is that Van Eyck had perfected art without precedent: *Wat conste vantmen oynt (de waerheit t'orconden) / Soo perfect int eerste als dese const' excellent? / Van welcke belyden alle verstandighe monden, / Datmen heuren Meester noynt en heeft ghekent.*[58] With these verses De Heere aimed to show that Van Eyck's life did not just run parallel to that of the great Greek painter but surpassed it; Apelles had no knowledge of oil paint, and he had it easier because, unlike Van Eyck, Apelles had masters to learn from. In calling Van Eyck the "Flemish Apelles," De Heere therefore did not just indicate that he was a great painter, he wanted to show that Van Eyck's life imitated and emulated the greatest painter from antiquity. It appears that

Lucas de Heere perfectly understood the implications of Vasari's invention story and, like Vasari, implicitly modeled Van Eyck's achievements on those of Apelles to show that the latter had been surpassed. But unlike Vasari, De Heere transformed Jan van Eyck from a mere bearer of technique into the contemporary Apelles.

De Heere was not the only writer from the North who understood that Vasari's invention story put Van Eyck's achievements on a par with those of Apelles. Marcus van Vaernewyck (1518–69), the renowned chronicler of Flemish history and art, makes a similar argument in his *Den spieghel der Nederlandscher audtheyt* (*The Mirror of Netherlandish Antiquity*, 1568). Van Vaernewyck's work is of particular interest here because he is the first author (a decade before Armenini) to explicitly point out the connection between Vasari's invention myth and Pliny's story of Apelles's *atramentum*. Van Vaernewyck writes that according to Georgius Vasarius Aretinus the Italians were ignited with love (*onsteken waren tot liefde*) after having seen paintings by Van Eyck in Naples, Florence, and Urbino.[59] They recognized him as the first to paint with linseed oil because "the ancient painters Apelles / Parrhasius / Zeuxis and others / prepared their paints with just water / or, instead of oil, with eggs prepared for the purpose" (*want die oude Schilders als Apelles / Parrhasius / Seuxis ende andere / plochten haer verwen te temperen alleene met water / oft in stede van olye met eyeren daertoe gheprepareert*).[60] Van Vaernewyck adds the salient detail that Van Eyck discovered his drying oil because he was "incredibly industrious and learned / and knew how to clarify his oils so that the colors could be kept immortal" (*door zijn groote industrie ende gheleertheyt / ende dat hy die olye wist the purgieren om die coleuren onsterffelich te onderhouden*).[61] Like Lukas de Heere, Van Vaernewyck observes that it is remarkable that painters such as Van Eyck became famous in a time when their art was still unknown in Flanders, as "can be seen in old paintings and glass windows / that display a great coarseness" (*zoot wel blijckt aen die oude gheschilderde tafelen, ende glaesvensteren / Die een groote plompicheyt vertooghen*).[62] This makes Van Eyck's achievement all the more praiseworthy, because he perfected his art without examples. And unlike Vasari, who had pointed out that Van Eyck's invention was a mere technical one, Van Vaernewyck argues that Van Eyck did not just invent oil paint, but brought the art of painting to its greatest heights as well:

> They [the Van Eyck brothers] came from (as was said before) a raw and rugged land, where they could not see examples of their art / this is why they (for this adventure) need to be highly esteemed / they invented this art in this land from nothing / and not just invented: but also brought it

> to the greatest heights [*Oock quamen zy (so voorseyt is) uit een zeer ruyt landt, daer zy ooc gheen exempel oft voorbeelt van haerder conste en zaghen / waeromme zy (ter avontueren) niet min te achten en sijn / dan oft zy van nieuws dese conste in dese landen gheinventeert hadden / ende niet alleene ghebonden: maer oock ten hoochsten ghebracht*].[63]

In 1572 the Flemish humanist, poet, and painter Dominicus Lampsonius (1532–99) published an ode to Jan van Eyck as part of his series of engraved portraits, *Pictorum aliquot celebrium Germaniae inferioris effigies* (*Portraits of Some Celebrated Artists of the Low Countries*). Lampsonius's short poem (fig. 1.1) lets the artist speak about his *novum repertum*. Like De Heere and Van Vaernewyck, Lampsonius noticed that Vasari's invention story seems to imply that Apelles may not have known about oil paint:

> **Jan van Eyck, painter**
>
> I am he who first taught to mix joyful colours from the pressed oily seed of flax [*Ille ego, qui laetos oleo de semine lini*], with my brother Hubert. Bruges, flourishing with wealth, was astounded by this new discovery, perhaps unknown in the past to Apelles himself [*Atque ipsi ignotum quondam fortassis Apelli*]. Soon afterward our uprightness spread widely through the whole world.[64]

However, Lampsonius adopts a more careful attitude than his compatriots when he tentatively adds that *fortassis* (perhaps) Apelles did not know how to mix colors with linseed oil. When Karel van Mander, in his 1604 *Schilder-boeck*, includes a Dutch translation of Lampsonius's verse in his biography of Jan van Eyck, he adds a short but telling note. In the poem's margin, Van Mander insists that the "perhaps" is not necessary, for it is certain that in ancient times oil paint was not known (*Het misschien hoefde niet: want t'is gewis datmen by d'Antijcken van geen Oly-verwe en wist*) (fig. 1.2).[65]

Van Mander included similar remarks in some of the biographies of the ancient painters whom, unlike Vasari, he had decided to add to his *Schilder-boeck*. In the life of Aristides of Thebes, Van Mander writes about an unfortunate cleaning attempt that the artist's painting did not survive. For this reason (oil paintings *can* be washed with water, something Vasari had also pointed out), Van Mander believes this anecdote proves that oil paint was not known to the ancients—*d'Antijcke van geen Olyverwe wisten*.[66] Van Mander includes the same anecdote that Lomazzo brought up in the biography of the painter Protogenes in relation to the invention of oil paint. Van Mander writes that it is remarkable that Protogenes applied four layers to his

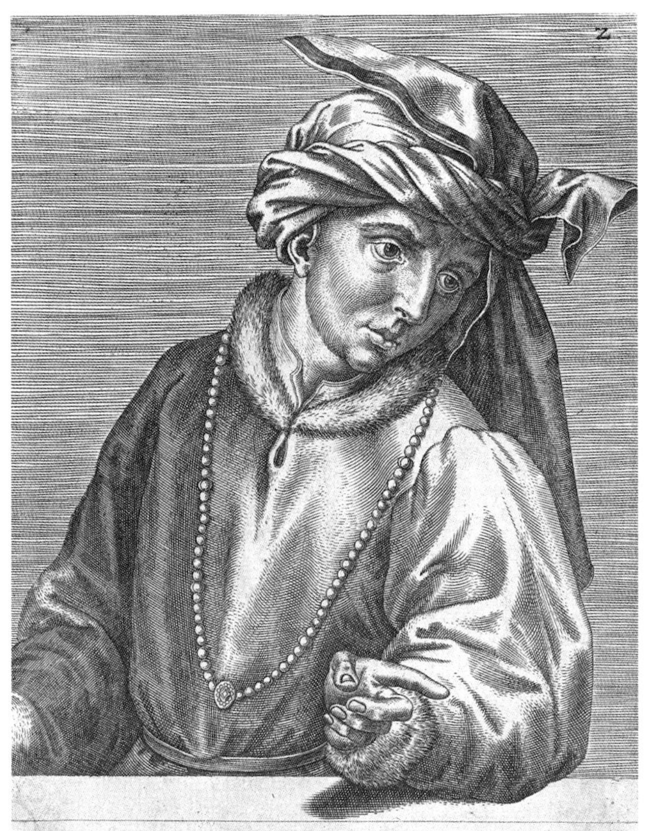

FIGURE 1.1. *Portrait of Jan van Eyck*, engraving, 1572. Plate 2 of Domenicus Lampsonius, *Pictorum aliquot . . . effigies*. Rijksmuseum, Amsterdam. Public domain.

Ian en Hubrecht van Eyck, Rogier van Brugge. Fol. 203.

Bidt Godt voor my, die Const minnen,
Dat ick zijn aensicht moet ghewinnen,
En vliedt zonde, keert u ten besten:
Want ghy my volghen moet ten lesten.

Het zijn eertijts t'Antwerpen in Coper-druck uytgegheven eenighe Conterfeytselen der vermaerde Nederduytsche Schilders / en voor eerst die van dees uytnemende ghebroeders/als d'oudtste constighe oeffenaers der edel Schilder-const in den Nederlanden: waer neffens/ oft onder/zijn seer constighe Carmina, oft ghedichten/in Latijn/door hoogh gheleerden Poeet Dominicus Lampsonius van Brugge / Secretaris van den Bisschop van Luyck/ en niet alleen een seer groot beminder onser Const/ maer oock daer in heel verstandich/en ervaren. Welcke ghedichten ten love deser constige Mannen/ ick in onser spraeck hebbe willen hier oock by voeghen.

Aen Hubertus van Eyck/ Schilder.

O Hubert met u Broer, de wel verdiende loven
Van onse Sang-Goddin, nu corts u toegheschoven, Thalia.
En zijnse niet ghenoech, voeght dese noch daer an,
Dat u leerknecht, u Broer, door u hulp u verwan.
Dit leert het werck te Ghent, welck in soo liefdich blaken
Coningh Philips bevingh, te doen een nabeeldt maken
Daer van, en door Coxy met wel gheleerde handt,
Om dit te schicken t'huys in t'Spaensche Vaderlandt.

Ioannes van Eyck/ als self sprekende/ seght.

Ick die de wijs' eerst wees, dat blijde verwe wert
In Lijn-oly ghemengt, met mijnen Broer Hubert,
Heeft Brugghe spoedigh rijck verwondert dees nieuw vonde,
Die voormael niet misschien Apelles vinden conde:
Doe weygherde haer niet eer langh ons deuchtsaemheyt,
De Weerelt wijt, en breedt te worden overspreyt.

Het misschië hoefde niet: want t'is gewis, dat men by d'Antijckë vaa geen Oly-verwe en wist.

Het leven van Rogier van Brugghe/ Schilder.

ALeer de wijt vermaerde stadt Brugghe in afgangh en verminderinge is ghecomen/ door dat A°. 1485. den Coophandel van daer ghweken is nae Sluys en Antwerpen / ghelijck t'gheluck en d'avontuer deser Weerelt wanckelbaer is: Soo hebben in dese bloeyende Stadt/ten tijde/en nae het leven van Ioannes, noch eenighe fraey edel gheesten gheweest. Onder ander eenen Rogier ghcheeten/die een Discipel is ghewoorden van den voornoemden Ioannes. Nochtans schijnt wel/dat het is gheweest ter tijdt/doe Ioannes al redelijck oudt was ghewoorden: want Ioannes zijn Oly-verwe Const en vindinghe tot in zijnen ouderdom heeft verborghen ghehouden/ niemant latende by hem comen daer hy wrocht/maer heeft eyndlijck dese zijn Const zijn Discipel Rogier deelachtich gemaeckt. Van desen Rogier zijn te Brugge in Kercken en huysen veel dinghen gheweest te sien: Hy was cloeck van teyckeninge/ en van schilderen seer gracelijck/ soo van Lijm-verwe/ Ey-verwe / als Oly-verwe. In desen tijt had men de maniere/te makë groote doecken/met groote beelden in/die men ghebruyckte om Camers mede te behangen/als met Tapijtserije/en waren van Ey-verwe oft Lijm-verwe ghedaen. Hier in was hy

Cc iij een

FIGURE 1.2. Page from Karel van Mander's *Schilder-boeck* (Haarlem: printed by Jacob de Meester for Passchier van Westbusch, 1604), fol. 203r, showing the poem by Dominicus Lampsonius to which Van Mander added a note with the remark that the *misschien* ("maybe") can be left out concerning the invention of oil paint by Van Eyck. Public domain.

painting with an *Ey oft Lym-verwe* (egg or glue paint) so that his work would be more durable—*dit stuck te langher mocht in wesen, en gheduerich blijven.* He insists that this technique would have been better suited for oil paint but, *alas*, they did not know about it—*doenlijcker geweest van Olyverwe, daer sy niet van en wisten.*[67]

Van Mander begins the lives of the illustrious Netherlandish and German painters (*Het Leven der Doorluchtighe Nederlandtsche, en Hooghduytsche Schilders*) with the biographies of Jan and his mysterious brother Hubert van Eyck. Van Mander's account of Jan's invention of oil paint is largely based on that of Vasari, but he makes a few telling changes and additions.[68] The first can be found in his introduction to Jan and Hubert's biography, which resembles Vasari's conclusion to the life of Antonello da Messina. But where Vasari only carefully hinted that oil paint might have been unknown to the ancients, Van Mander now presents it as an undisputed fact: "For what was never granted to either the ingenious Greeks, Romans or other peoples to discover—however hard they tried [*hoe seer soeckende*]—was brought to light by the famous Netherlander from Kempen, Joannes van Eyck."[69]

At the end of his account of the invention myth, Van Mander explains that the discovery of oil paint was necessary to ensure that the honorable art of painting could more closely approach nature and its appearance. He believes that if the ancient painters could be resurrected, they would marvel at the paintings made during Van Mander's time:

> Our art needed only this noble invention to approximate to or be more like nature in her forms. Had the ancient Greeks—Apelles, Zeuxis and others—been brought back to life again here and seen this new technique [*en dese nieuw maniere ghesien*] they would certainly have been no less astonished than the warlike Achilles [*sy en hadden wis niet min verwondert gheweest, dan den strijdtbaren Achilles*] or other warriors of antiquity would have been if they were to hear today's fierce thundering guns of war which the alchemist Bartholdus Schwarz, a monk from Denmark, invented in the year 1354. Or perhaps not less than the ancient writers were they to observe the useful art of book printing which Haarlem, with sufficient justification, prides itself on having first invented.[70]

Embellishing Vasari's biography of Van Eyck to suit his own agenda, Van Mander asserts that Van Eyck's discovery was of such magnitude that it should be considered as momentous as other modern inventions unknown to the ancients, notably gunpowder and the printing press. This last idea was most likely inspired by a recent print series depicting a number of "modern" inventions unknown to the “ancients.” The prints were designed from the 1580s to the 1590s for the Florentine nobleman Luigi Alamanni (1558–1604) by the Flemish artist Jan van der Straet (1523–1605), also known by his Latin name Johannes Stradanus. Stradanus had been active as an artist in Florence since about 1550 and worked closely with Vasari as his workshop assistant during the 1560s. He also worked with Vasari on the decoration of the twelve rooms, the Quartiere degli Elementi, of Cosimo I de’ Medici’s Palazzo Vecchio. Vasari’s close friend Cosimo Bartoli had provided part of the iconographic program for this project.[71] A few decades later, when he had long established himself as an independent artist in Florence, Stradanus designed the *nova reperta* series depicting twenty modern inventions and discoveries, including the compass, the discovery of America, the mechanical clock, and, indeed, the inventions mentioned by Van Mander: the printing press, gunpowder, and the oil colors used by Jan van Eyck (fig. 1.3).[72]

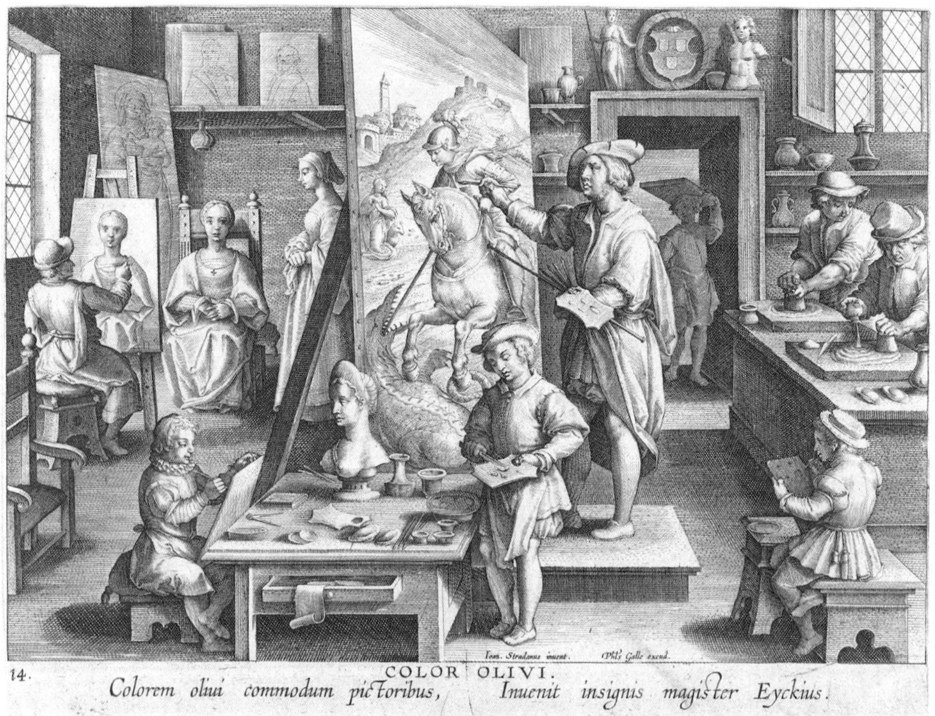

FIGURE 1.3. Jan Collaert I after Jan van der Straet, called Stradanus, *New Inventions of Modern Times [Nova Reperta], The Invention of Oil Painting,* plate 14, ca. 1600. Engraving. Metropolitan Museum of Art, New York. Public domain.

This last well-studied print is typically examined to find out if and how it reflects contemporary workshop practice and, more generally, to see how it can be understood within the larger *nova reperta* tradition.[73] We can now add to this that Stradanus's print can be fully understood only when we study it as part of the particular debate introduced by Vasari in the *Vite*.

Stradanus's print shows that the increasing interest in proving that the achievements of "modern times" improved on those of antiquity, together with the rising status of painting as a liberal art, gave new significance to the idea that the oil medium was unknown to the ancients. Indeed, as both Vasari and Van Mander pointed out, it was only by virtue of the oil medium that the "modern" painters could accomplish their new goal—to represent the three dimensions of the visible world as convincingly as possible on a flat surface. Even though the ancients had done this, the modern painters could do it better because of their novel method of coloring. Stradanus's print shows that by the end of the sixteenth century this new manner of painting had become important enough to be included next to some of the most momentous inventions of modern times, from the introduction of distillation and clockwork to the invention of eyeglasses and the compass. Van Mander appears to have gratefully picked up on this fact to emphasize the importance of the achievements of the greatest painter from the North.

Let us look at one last writer who ingeniously transformed Vasari's invention story to fit Netherlandish art history: the Dutch painter and art theorist Samuel van Hoogstraten (1627–78). In the second chapter of his *Inleyding tot de hooge schoole der schilderkonst* (*Academy of Painting*, 1678), which deals with "the various characteristics of paintings and painting techniques," Van Hoogstraten works the story of Apelles's varnish directly into the storyline of Van Eyck's invention of oil paint. Contrary to Vasari—who wrote that Van Eyck decided to invent a new varnish because his painting got damaged by drying in the sun—Van Hoogstraten insists that Van Eyck was so displeased that he could not wash his paintings made with egg or glue binders that he decided to varnish his work with distilled oils (*met eenige olyen daer toe gedistilleert*) as Apelles had done. But even after discovering Apelles's secret varnish, Van Eyck was not satisfied. He continued his experiments and eventually invented the best drying varnish ever made, which also happened to be useful for mixing with pigments:

> Just as it is said that Apelles used to varnish with a varnish that was so thin and smooth that it seemed that upon touching your hand felt as if though it was smudged by it [*met een vernis, die zoo dun en glad was, dat, wanneermenze aenroerde, men zich inbeelde dat de hand daer af als besmet wiert*], it

> kept the paintings from dust and gave them a beautiful luster, and no one was able to imitate it. No one, that is, but Jan van Eyck managed to varnish this way [*Ook niemant, als van Eik, is dit vernissen wel gelukt*], and everyone was astounded by the shine of this work, [*zoo dat een yder verwondert was over de glans van zijn werk*]; but even then he was not satisfied until, after long alchemical research, he discovered that linseed oil and nut oil, mixed and cooked with some other substances, resulted in the best and most drying varnish ever made [*nae, een lang, en alchymistisch onderzoek, de lijnzaet- en nootoly, met eenige andere stoffen vermengt en gezooden, voor de droogenste en beste vernissen uitvond*]: yes, finally pigments could even be better mixed and worked with it than with the liquids previously known. And this is how oil paint was invented in the Netherlands, first used in Bruges by Johan van Eyk, about the year 1410.[74]

For Van Hoogstraten, the invention of oil paint had made way for the invention of a varnish suitable to paint with. Using varnish as a paint medium had indeed become popular during Van Hoogstraten's time, but as will become clear in chapter 5, it seems to have been much less common in the fifteenth century. Indeed, the myth of the invention of oil paint appears to have lasted as long as it did because, north *and* south, it was reworked time and again to promote a particular local artisanal history. To Vasari, it explained the switch from egg to oil in the history of Italian panel painting, which not only allowed the art of painting to advance to the third age but gave modern panel painters a tool with which they could surpass the ancients. Chapters 2 and 3 show that the varnish Van Eyck struggled with according to Vasari perfectly describes the varnishing practices recorded in medieval recipes. That Vasari grounded the invention myth in long-established varnishing traditions lent further credibility to the story. When the invention myth was embraced by art theorists in the North, proving its veracity gained new significance. If Van Eyck had invented oil paint, he could be shaped into his new role as the "Flemish Apelles." This, of course, would put the art of painting in the Low Countries firmly on the map.

2

The Color of the Sun

Varnishing Practice before 1450

> Every painting coated with this varnish becomes bright and decorative and completely durable.
>
> THEOPHILUS PRESBYTER,
> *Schedula diversarum artium*, ca. 1100

A varnish is commonly understood as a final coating that, depending on its ingredients, dries into a colorless transparent film or a colored translucent film. Varnishes are perhaps best known for their use on paintings, but in the past they were applied to a multitude of objects, including metalwork, other types of woodwork, and glass. The purpose of a varnish is twofold; it may be used to protect an object from the atmosphere and mechanical damage, and it may also have an important optical function. These two functions were described in the earliest account of painter's varnish discussed in chapter 1. In Pliny's story, Apelles's secret *atramentum* protects his works from "dust and dirt" (*pulvere et sordibus*) and brings out "the brilliance of all the colors" (*repercussum claritates colorum omnium*) while at the same time giving them "somberness" (*austeritatem*).[1] Pliny does not give us an account of the materials and methods Apelles used to make and apply his varnish, an omission that, as we have seen, Vasari referred to in trying to prove that the oil medium might not have been known to the ancients.

In more recent times, the story of Apelles's *atramentum* led to much speculation about the color and optical properties of the ancient surface coating. In the 1950s Pliny's story even fueled the "cleaning controversy" about the National Gallery of London's policy of the complete removal of varnish in cleaning the old master paintings in their collection.[2] Because everywhere else in the *Natural History* Pliny uses the term *atramentum* to refer to black substances such as writing ink and shoemaker's black (*ater* means black), scholars have suggested he might have picked this name for Apelles's varnish because it too was dark.[3] The main reason for bringing up Apelles's allegedly dark varnish during the cleaning controversy was that painters of later times, influenced by Pliny's account, might have tried to re-create the varnish used by the great Greek painter. If painters had indeed attempted to emulate Apelles and his dark varnish or, for any other reason, had used dark or colored varnishes and glazes to influence the final appearance of their paintings, this would have had major implications for decisions on partial/selective or complete varnish removal in restoration campaigns.[4] If dark varnish was the painter's "original intent," removing such a coating (and replacing it with a clear and colorless varnish) would become a much less straightforward decision for the modern paintings conservator/restorer.[5]

The varnish certainly presents an interesting problem in the history of painting. Its technique appears to be at least as old as the art of painting itself, and its application would have had significant influence on the appearance and preservation of the artwork, and by extension on our historical understanding of it.

Yet the problem with varnishes is precisely that their historical value became limited to their ability to protect and enhance what lies below their surface. As varnishes age, they increasingly turn yellow and may exhibit other damage such as a network of cracks (*craquelure*),[6] cloudy areas that are usually caused by moisture penetrating ("blooming"), and various other kinds of degradation. Eventually a varnish may become so dark and damaged that it obscures the surface it was meant to enhance and protect.[7] When this happens varnishes are usually removed and replaced.[8] Most artworks have had their varnish removed on various occasions throughout their life, and if original varnishes do survive they no longer allow us to appreciate the optical effect they were once used for.[9] As a result, there is barely any physical evidence that helps us examine the appearance and ingredients of early surface coatings and, accordingly, their role in the aesthetic appearance of the works they were applied to. But even though most original varnishes have been lost, we do have numerous recipes and other written sources that describe their material characteristics, handling properties, and functions. These recipes, as this chapter aims to show, are key to understanding the role of varnish in the mimetic ambition of the medieval artisan. In trying to evoke the appearance of enamel, gold, and other splendorous surfaces on wood, oils and dyestuffs were explored for their ability to produce thick, lustrous transparent or translucent surface coatings. And since not many such layers have been preserved on medieval art objects or, where they have, no longer give us an impression of their former luster and shine, analyzing a large group of varnish recipes provides a surprisingly detailed window into the medieval ambition to create mimetic splendor.

Varnishes: A Short Introduction to Their Material Characteristics

Oil-Resin Varnishes

Before the advent of synthetic substitutes in the twentieth century, the varnishes used can be divided into three categories: oil-resin, water-based, and solvent-based.[10] Of these three types, written sources suggest that oil-resin varnish was the most common during the period studied in this chapter. To make an oil-resin varnish, natural resins are dissolved in a hot drying oil such as linseed oil. Natural resins are sticky, flammable, organic substances of vegetable origin (except for shellac, an insect secretion). They harden on exposure to oxygen and are insoluble in water but partly or wholly soluble in alcohol, drying oils, or volatile oils such as oil of turpentine. In the Northern Hemisphere, the main sources of resins are the conifers, especially cypresses and pine trees (families Cupressaceae and Pinaceae). Even if it is coincidental, the natural function of resins shows a remarkable parallel with one of the functions of varnish. Modern-day botany has shown that by coating parts of young leaves and stems, resins defend some plants against desiccation, high temperatures, and ultraviolet radiation. Resins' stickiness also helps keep certain insects or other organisms from attacking plants.[11] This last property was explored by premodern artisans as well; resins were often used to make glue.[12]

The plant resins most commonly used to make varnish in the period studied here include mastic (*Pistacia lentiscus* L.), sandarac (*Tetraclinis articulata* [Vahl] Masters), juniper resin (from the various species of *Juniperus*), turpentine oleoresins (from the various genera of Pinaceae), and the solid residue that is left after heating turpentine resin to vaporize its volatile components (oil of turpentine), known as "rosin" or "colophony." When resins are fossilized, meaning they were exuded by trees tens of millions of years ago (probably coniferous resins, but not from any family extant today), we call them amber (fig. 2.1).[13] Amber is extremely hard and is insoluble unless heated.

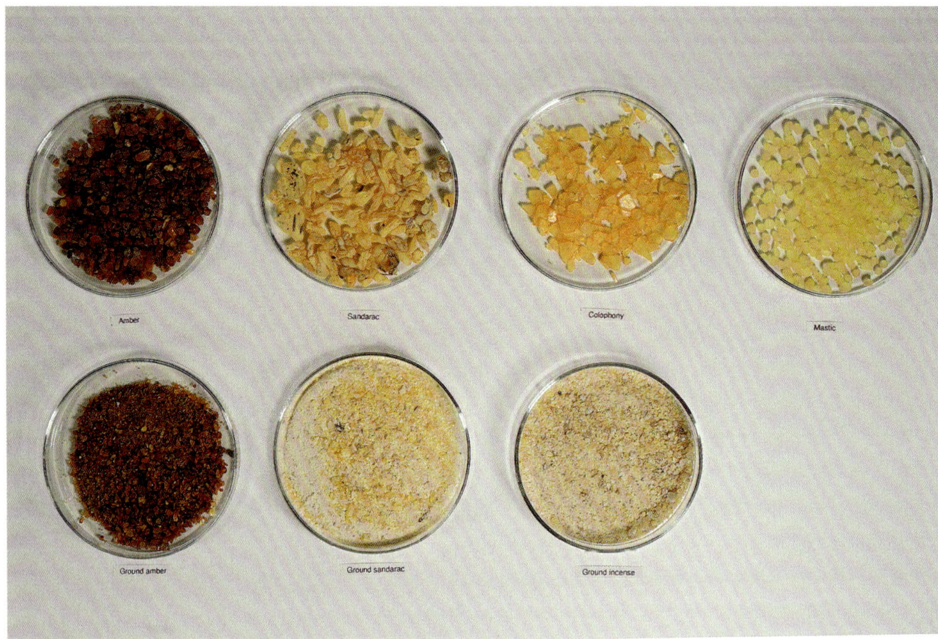

FIGURE 2.1. Various of the resins used to make varnish during the period studied in this chapter. Notice that in color sandarac and incense look alike and that compared with incense, sandarac (and amber) has a "brighter glitter" (*fulgorem clariorem*), as Theophilus also points out in his varnish recipe. Photo by Marjolijn Bol.

Resinous secretions are a complex mixture of various chemical compounds that, besides resin acids, consist of volatile oils. When the volatile oils evaporate through exposure to oxygen, the remaining material becomes tougher and harder. Natural resins also harden through polymerization. Unlike drying oils, they do not polymerize through cross-linking but form linear bonds instead. This "chain-growth polymerization" is reversible, and in the case of resins this means they soften on heating and solidify on cooling. Chemistry calls such a material "thermoplastic."[14] When resins are liquefied by applying heat, they can be mixed with other materials, such as drying oils in the case of varnish. Such an oil-resin varnish first hardens through evaporation of the volatile diluent, after which it hardens and dries further because of the cross-linking drying oil. This produces an exceptionally tough film.[15]

The optical effect of an oil-resin varnish depends on its ingredients, its preparation, and the surface it is applied to. To explain this further, let us briefly study what happens when a thick oil-resin varnish is applied to an oil painting.[16] We saw in the introduction that the scattering of light reduces the saturation of paint and paint's ability to transmit visible light. An oil-resin varnish of high viscosity can help reduce the scattering of light on a painting in two ways. First, owing to its thickness, an oil-resin varnish may cover irregularities in the paint layer. Covered with varnish, the painting's surface is now smooth at a microscopic level, so surface scattering is reduced. Second, a varnish made with natural resins shares a similar refractive index (RI ca. 1.53) with the oil medium (RI ca. 1.47). This helps reduce the scattering of light between varnish and paint layer as well. Because the refractive index of a drying oil increases as it ages, this effect becomes more pronounced over time. By reducing both surface scattering and internal scattering of light, a thick oil-resin varnish therefore increases saturation and gloss ("specular reflection"), darkens colors, and brings out the details of the painted surface. Not too different, in fact, from Pliny's at first sight rather cryptic description of the optical effect of Apelles's *atramentum*.

Water-Based and Solvent-Based Varnishes

Resins are often confused with "gums," one of the materials used to make the water-based varnishes in our second category. Like resins, gums are viscous secretions from trees or shrubs that harden on drying. But unlike resins, gums are made up of complex sugars and therefore are soluble in water.[17] When plant exudations contain both gum and resin, they are known as gum-resins. If these exudations contain a relatively high amount of volatile oil, making them somewhat soft, we call them oleo-gum resins.[18]

Even though gums may have sometimes been used to make surface coatings for panel paintings and other wooden objects, they are best known for their use as the binding medium for the watercolor paints used in manuscript illumination. Gum arabic, which is exuded by the acacia tree (*Senegalia senegal* [L]. Britton), was most widely used for this purpose. Another water-based medium that scribes used to bind their pigments, and that was also used to make varnish, is the liquid that settles out if beaten egg whites are left to stand, also known as glair. Glair varnishes are less resistant to mechanical damage than oil-resin varnishes. And owing to their lower refractive index (RI ca. 1.34) and lower viscosity, they do not increase the saturation and gloss of a painting the way an oil-resin varnish does—rather, they may reduce these qualities. Glair varnishes are therefore thought to have been applied as an intermediate coating (an oil varnish was later applied on top) or as a temporary coating that was removed as soon as the painting was dry enough to receive an oil-resin varnish.[19] Glair varnishes would have been particularly suited for this purpose because they can be removed with water when fresh or, later, can be brushed off.[20] We will see in this chapter that next to their use on panel paintings, glair varnishes also played an important role in making yellow coatings that were used in book illumination.

Our final varnish category comprises the solvent-based varnishes. These varnishes are made with a natural resin that is dissolved in a volatile oil such as turpentine oil. When the solvent evaporates, the resin hardens into a solid layer. Varnishes based on solvents are more brittle and less hard than oil-resin varnishes. Their main advantage is that they are more easily removed when they age because they do not contain the insoluble cross-linked network that oil-resin varnishes have. Since solvent-based varnishes were largely introduced in the course of the sixteenth century, I will discuss them in chapter 3. The previous chemical and physical properties of varnish and its ingredients were of course not known to medieval artisans in the detail described here. But as will become clear from what follows, artisans nevertheless knew perfectly well how to manipulate the physical properties of their materials to create a varnish tailored to the ambitions of their craft.

Brightening, Decorative, and Durable: Functions of Medieval Varnishes

Early Accounts of Varnish

Ancient texts suggest that melting resins in an oil medium might have had its root in medical practice. For instance, in book 24 of the *Natural History* Pliny mentions that "all resins can be dissolved in oil," and that this substance can be used "for the treatment of wounds and for poultices." The medical uses of such preparations are to "close wounds, to act as a detergent, and to disperse gatherings."[21] Likewise, the drying properties of oils such as linseed and walnut were recognized and used for a variety of medical purposes as well. Considering how well established the use of oils and resins was in ancient medicine, it may not be a coincidence that the earliest reference to drying oils' being used by painters appears in a medical text. Discussing the medicinal benefits of various oils, the Greek physician Aëtius of Amida (mid-fifth century to mid-sixth century) mentions that walnut oil is also used outside the medical field for its ability to dry. He points out that gilders and encaustic painters use it as a protective coating for their work:

> Walnut oil [*Nucis oleum*] is prepared like that of almonds, either by pounding or pressing the nuts, or by throwing them, after they have been bruised, into boiling water. The [medicinal] uses are the same; but it has a

use besides these, being employed by gilders and encaustic painters [*quod inaurantibus aut inurentibus conducit*], for it dries and preserves gildings and encaustic paintings for a long time [*siccat enim et ad multum tempus inaurationes et inustiones continet et adservat*].[22]

Unfortunately, Aëtius does not explain how this particular coating would have been prepared or applied. Nor can we be certain whether walnut oil was applied alone or mixed with other ingredients. Approximately two centuries after Aëtius wrote his medical treatise, collections of recipes on the technology of the arts provide us with a more constant flow of written accounts about varnish. The earliest varnish instructions survive in various related manuscripts known as the *Mappae clavicula* tradition of recipes.[23] The oldest of these manuscripts is Codex Lucensis 490, kept in the Biblioteca capitolare feliniàna in Lucca, Italy. This manuscript, known as the *Compositiones ad tingenda musiva* (after one of the recipes it contains) or *Compositiones variae*, can be dated to the late eighth or early ninth century and contains various recipes for making pigments, metallurgy, gilding, mosaics, for dyeing textiles, and indeed for making varnish.[24] Contemporary with the *Compositiones variae* is another, larger collection of craft recipes, known by its title as the *Mappae clavicula*.[25] The earliest surviving manuscript, today kept in the Humanist Library of Sélestat (Sélestat MS 17), is from the tenth century. Its recipes must nevertheless have been compiled at least a century earlier, since a now lost manuscript was listed in the 821–22 library catalog of the Benedictine monastery at Reichenau.[26] The Sélestat manuscript contains most of the recipes from the *Compositiones variae* and many more besides. Another twelfth-century copy of the *Mappae clavicula*, kept in the Corning Museum of Glass (MS5), contains about one-third new recipes while preserving most of the recipes from the Lucca and the Sélestat manuscripts.[27]

The recipes belonging to the *Mappae clavicula* tradition are thought to have survived from a much older core of recipes going back several centuries and, since the texts contain many Greek words, might have originally been translated from earlier Greek sources, quite possibly as part of an alchemical tradition.[28] Cyril Stanley Smith, the British metallurgist who translated and edited the recipes contained in the *Mappae clavicula* manuscripts, points out that many of the recipes do not make sense and seem to have been corrupted through centuries of copying. Smith goes as far as to say that "none of the surviving manuscripts of the *Mappae* are the work of an author who knew the technical realities behind the words."[29] Since these compilations might have had their origin in late antique alchemy, Robert Halleux and Paul

Meyvaert question whether they should be viewed in the context of medieval workshop production at all.[30] It is therefore difficult to establish if and how the varnish recipes contained in these manuscripts can be used to study craft traditions from the ninth century to the twelfth, or whether they reflect the state of the art of a much earlier period of technology. But however garbled these recipes may have come down to us, they are fundamental to the history of varnishing set out in this chapter; the manuscripts belonging to the *Mappae clavicula* tradition contain the earliest surviving practical instructions for making transparent and translucent coatings through the admixture of drying oils, resins, and, in a few examples dyes. The recipes in the *Mappae clavicula* contain all three types of varnish that we will encounter in later sources: uncolored varnishes for pictures and sculptures, colored varnishes for gold leaf, and colored varnishes to make tin and silver look like gold. We will see that despite their complicated genealogy, these early recipes are not far removed from varnish recipes that are more clearly embedded in a medieval artisanal context. And, intriguingly, they also reveal what was initially considered the most important function of varnish: to make the object it was applied to "bright."

"The Bright": Varnishing Pictures and Sculptures

One of the varnish recipes that can be found in the *Compositiones variae* and in all later related manuscripts explains how to make a coating that can be applied "over colors" (*super colores*).[31] Next to "linseed oil" (*lineleon*), the recipe lists a large variety of resins, gums, and other materials used to make the substance. These ingredients are to be pounded and sifted and put in a brass dish with linseed oil. They are then cooked "without flame" (*sine flamma*) in such a manner that none of the ingredients can "escape" (*nonexeat foras*). The caution to heat varnish without letting it contact flames was meant to make sure the extremely flammable oil-resin mixture would not catch fire. After you are done cooking it, the now liquid oil-resin mixture is strained through a linen cloth. We learn that when this substance has been applied to an object, it should be dried in the sun (*et pone ad sole*).[32] The recipe concludes that with this varnish you will be able to "brighten every painting or carving" (*et qualibet opera picta aut scappilata inluciddare super debas*).

Unlike Aëtius, this eighth-century recipe does not mention that the varnish thus made might be protective. Instead, it refers only to its optical function through the rather particular use of the Latin adjective *lucidus*, which is also used in the recipe's title: *De lucide ad lucidas*. The adjective *lucidus* (from *lux*, *luc*, "light") is common in medieval Latin and often occurs in the craft recipes studied in this book. In almost all these instances, *lucidus* should be translated by the English terms "bright" or "lucid." In our varnish recipe, however, the adjective *lucidus* is changed into the noun *lucide*, uncommon at this time, and the equally rare verb *lucido*. This means that the recipe with the title *De lucide ad lucidas* explains how to make "the brightening substance [*lucide*] [used] to elucidate [*lucidas*]." The title of an identical recipe from a later manuscript reads *lucida quomodo fiant super colores*. This is difficult to translate but, staying as close as possible to its original phrasing, it can be rendered as "the brightening substance that [you] should apply on colors."[33]

That these varnish recipes transform the adjective *lucidus* into a noun and verb uncommon in the Latin language at this time suggests that a specific Latin term to denote the practice of applying surface coatings to artworks did not yet exist. It also suggests that *lucide* was used this way because it best described the substance and its function: to make the object it was applied to "bright." The lucidity of the coating itself seems not to have been of main concern, since the wealth of materials that ought to be mixed and melted together would have resulted in a rather dark liquid.

Of the various varnish instructions in the *Mappae clavicula* tradition, only one recipe suggests that surface coatings might have also been used to protect objects. The recipe in question explains how to treat "a painting so that it cannot be destroyed by water" (*ut pictura aqua deleri non possit*).[34] To "waterproof" a painting, it must be covered with an oil called *cicinum*, implying castor oil.[35] According to the recipe, castor oil should be applied to the painting in the sun, and when it is dry "it is fixed so tightly that it can never be destroyed" (*et ita constringitur, ut nunquam deleri possit*).[36]

Castor oil is procured from the seeds of the castor oil plant (*Ricinus communis* L.).[37] Different from linseed oil, castor oil is barely reactive when exposed to oxygen, so it does not dry. This can be changed, however, when castor oil is heated to about 250–80°C; heating dehydrates the castor oil, giving it excellent drying properties.[38] Since castor oil is not frequently mentioned in premodern recipes as a varnish or paint material, it is difficult to establish if the effects of dehydration (heating) on this oil were known. Whatever the case, the method of applying and drying the castor oil varnish in the sun, as this recipe recommends, was an important characteristic of medieval varnish application that, as we will see later in this chapter, is mentioned in nearly every varnish recipe.

"The Bright": Varnishing Gold Leaf and Tin Foil

The *Mappae clavicula* contains two recipes that explain how to make a varnish for gold leaf. The first recipe, titled *De lineleon*, explains how to make an oil-resin coating for gold leaf that has been applied to a “raw skin” (*pellem crudam*).[39] To make this substance, “linseed oil” (*lineleon*), “gum” (*gumma*), “resin” (*resina*), and “saffron” (*grocum*) are cooked together in an earthenware pot.[40] Different from the picture varnish, this substance is given a yellow color by adding saffron, one of the more expensive materials available to the medieval artisan.[41] Saffron is a yellow dyestuff made from the dried stigmas of the crocus flower (*Crocus sativus* L., fig. 2.2).

Most natural dyestuffs extracted from plants and other living sources are different from pigments; they are soluble in the liquid medium they are immersed in, whereas a pigment is by definition insoluble.[42] This means that saffron would have dissolved in the oil-resin mixture that makes up the varnish described here. Saffron can thus be used to dye the varnish yellow so that the visible light can still pass through without scattering; light enters the yellow varnish and, after reflecting off the gold, travels back through the varnish to the eye (fig. 2.3). The second recipe in the *Mappae clavicula* for varnishing gold leaf shows that this was indeed the most important function of yellow varnish; to enhance the brightness of the precious metal underneath.

The recipe in question is quite similar to the recipe for picture varnish discussed above. Not only does it bear a similar title—“How to Make the Bright” (*confecti[o] lucide*)—but it is prepared in a similar fashion by cooking together a wealth of resins and gums in various steps.[43] After all varnish materials have been united by cooking, they need to be strained through a

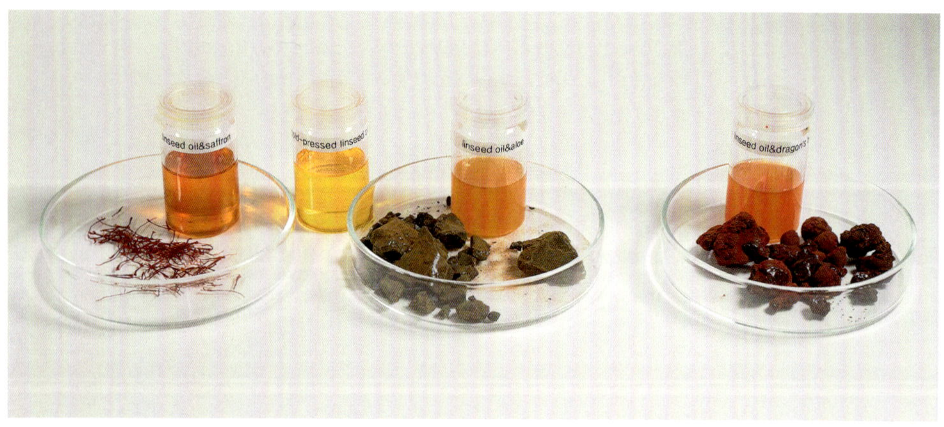

FIGURE 2.2. Three vials with yellow varnishes. *Left to right*: linseed oil varnish made with saffron; the resinous substance that comes from the aloe plant; and dragon’s blood. The second vial from the left shows regular cold-pressed linseed oil as a reference for the resins’ influence on the color of the oil. Photo by Marjolijn Bol.

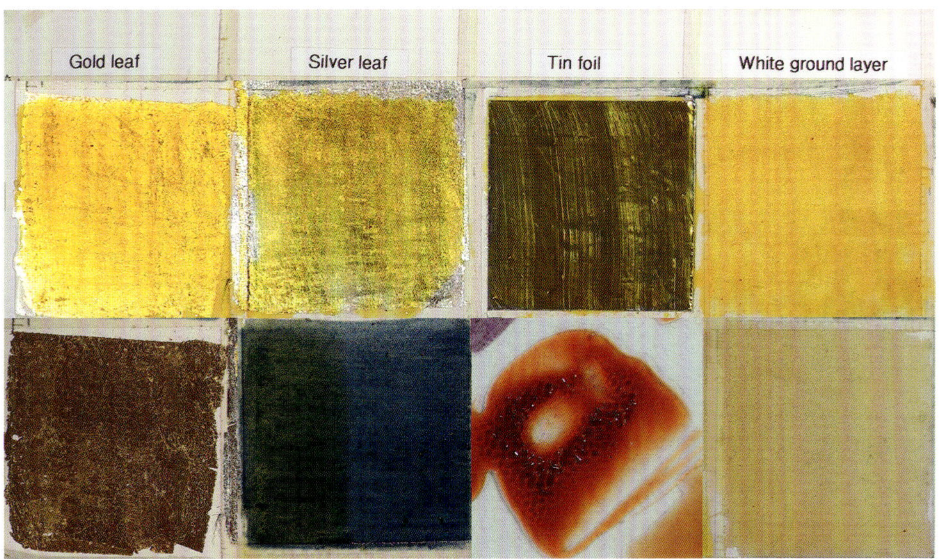

FIGURE 2.3. *Top row, left to right:* sandarac-saffron-oil varnish on gold leaf, silver leaf, tin foil, and a white ground. *Bottom row, left to right:* unvarnished gold leaf; linseed oil-sandarac varnish over azurite glaze; wrinkled madder lake glaze; and linseed oil-sandarac varnish over a white ground. Photo by Marjolijn Bol.

linen cloth to remove impurities, again similar to the way colorless varnish was prepared. The recipe shows its concern with the siccative properties of the yellow liquid when it recommends that, in case it does not dry, the resin "mastic" (*Pistacia lentiscus* L.) should be added: "Add as much *masticae* as you want, . . . and it will be corrected."[44] The substance thus made, so the recipe concludes, can be used "to make gold leaf bright" (*quomodo fieri debet at petalum aureum*).[45]

Besides enhancing the luster of gold, another important function of yellow-colored varnishes was to make other metals look like gold. The *Mappae clavicula* compilation contains one such recipe that describes how to make a yellow oil-resin varnish to give tin foil the appearance of gold (fig. 2.3).[46] The recipe, titled "How to Color Tin Leaf" (*Tinctio stagneae petalae*), explains how linseed oil, saffron, orpiment (a golden yellow arsenic sulfide mineral), rain or freshwater, and *gummi* have to be heated together to make a yellow liquid.[47] This liquid should be picked up with a sponge (*tollesque cum spongia*) and applied to the tin foil. When this has dried, a second layer of the yellow varnish must be applied by the same method. Finally, the recipe includes the rather remarkable direction that when the second coat of varnish is dry it must be "rubbed with an onyx stone until it is radiant" (*et desiccatam cum onichino defrica, ut splendeat*).[48] That yellow varnishes may have been rubbed to make them even shows how important the smoothness of this layer was to artisans at the time.

Quite a few medieval polychrome objects that would have been embellished by such gold-colored tin foils survive, but their decorations, unfortunately, are typically not in the best condition. The Museu nacional d'art de Catalunya has in its collections various altar frontals that were made in Spain during the thirteenth and fourteenth centuries and that would once have shone with elaborate tin foil decorations. A particularly striking example is an altar frontal from the church of Sant Pau de Esterri de Cardós, which is dated 1225 by an inscription. The central scenes of the frontal are three-dimensional stucco reliefs that evoke the appearance of low relief, that is, repoussé metalwork (fig. 2.4).

These reliefs would have originally been covered with tin foil—now largely lost—which was likely made to look like gold by a method similar

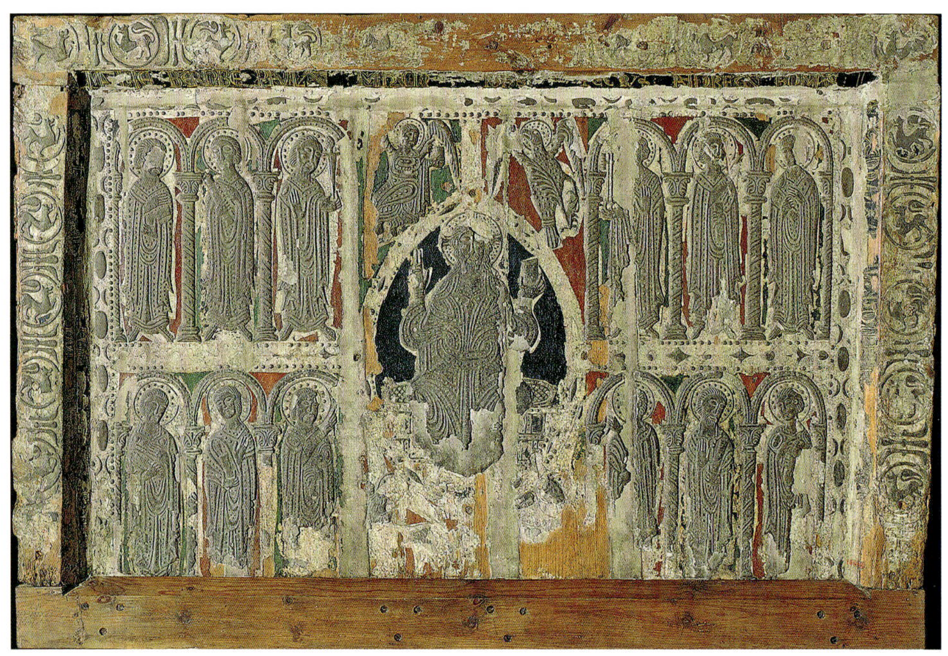

FIGURE 2.4. Altar frontal from the church of Sant Pau de Esterri de Cardós, 1225 (dated by inscription). Stucco reliefs and remains of metal plate on cloth-covered wood. Museu nacional d'art de Catalunya, Barcelona. Website of the Museu nacional d'art de Catalunya of Barcelona, www.museunacional.cat.

to the ones described in the previous recipes. This is supported by the fact that, despite their often deplorable condition, on quite a few of these early painted wooden objects traces of their original yellow varnish have been demonstrated by technical analysis. In her research reported in "White and Golden Tin Foil in Applied Relief Decoration," Josephine Darrah lists many examples—all made from 1240 to 1350—where yellow varnishes were used to make applied relief brocade.[49] Darrah's analysis shows that these varnishes had been made with a drying oil and pine resin. Lucretia Kargère and Adriana Rizzo found similar results when they did binding medium analysis of the varnish substance that had been applied to the tin foil decorations on a twelfth-century French polychrome sculpture in the Metropolitan Museum of Art (fig. 2.5).[50]

What makes the history of these yellow varnishes especially interesting, therefore, is that they appear to have been preserved more often than the varnishes applied to colors.[51] Considering that the recipes for picture varnish and yellow varnish are largely the same save for the use of dyestuffs, technical research into surviving yellow varnishes could be used to speculate about the properties and ingredients of the long-lost transparent, colorless varnishes.

Theophilus's Picture Varnish

For one of the first recipe collections that is more clearly embedded in medieval artisanal practice, we have to turn to the *Schedula diversarum artium*, the twelfth-century treatise by the pseudonymous monk Theophilus Presbyter, briefly discussed in the introduction. Many of the *Schedula*'s practical instructions are no longer copied outside their original context, but they are clearly connected to existing and developing workshop traditions.[52] The treatise of Theophilus is a landmark in the recipe tradition for another reason as well; it was the first compilation of recipes that made an effort to present its practical instructions by craft as an organized whole. There are many manuscripts containing copies and fragments of the *Schedula*, but the two oldest (ca. twelfth century) and most complete are kept in the Herzog August Bibliothek in Wolfenbüttel (MS Guelph. Gudianus lat. 2° 69) and the Austrian Nationalbibliothek in Vienna (MS 2527).[53] Cyril Stanley Smith and John G. Hawthorne, and the German goldsmith Erhard Brepohl, argued in their respective editions of the *Schedula* that the recipes in book 3 of the treatise present intimate knowledge of metalwork; nearly all the recipes are practicable, and many can be successfully reconstructed.[54] They therefore conclude that the *Schedula*'s book on metalwork must have been based on

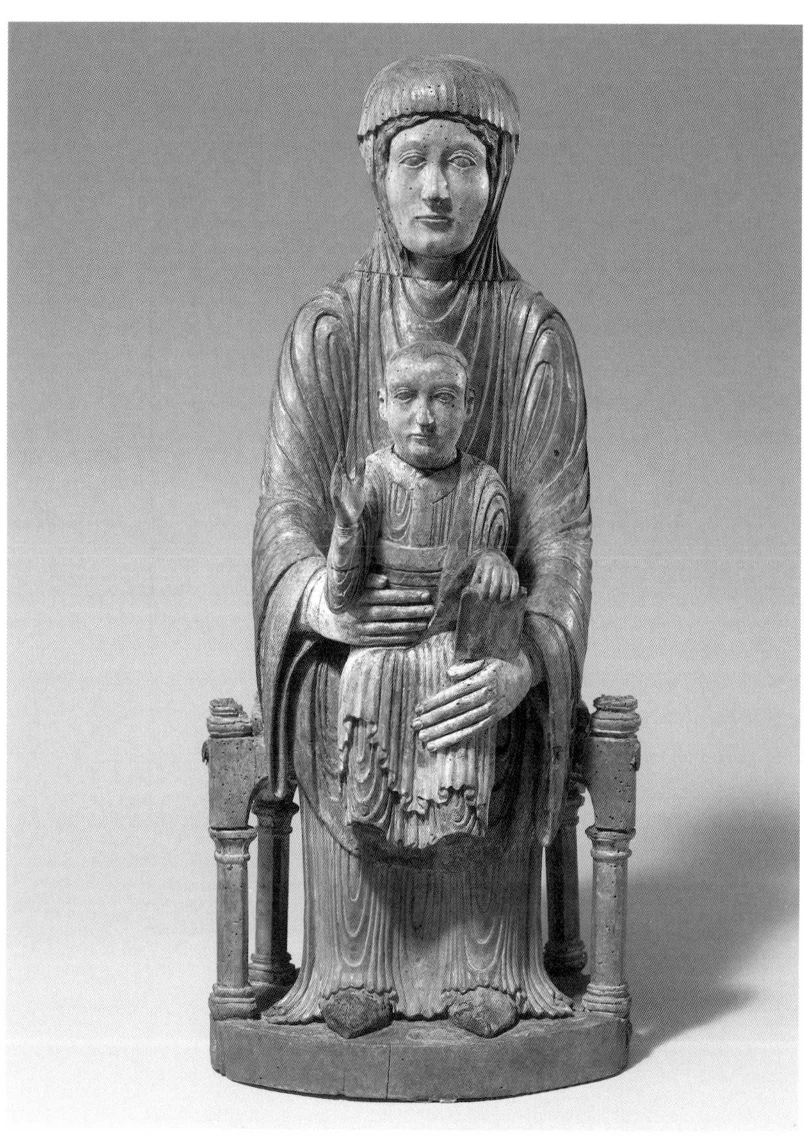

FIGURE 2.5. *Enthroned Virgin and Child*, made in Auvergne, France, ca. 1150–1200. Walnut with gesso, paint, tin leaf, and traces of linen. Metropolitan Museum of Art, New York. Public domain.

the monk's practical experience as a goldsmith. Because books 2 and 3 include more errors and some improbable recipes, Smith suggests that here the author was probably writing from "intimate observation" and "critical inquiry of his fellow artisans."[55]

The two recipes for making an oil-resin varnish (*De glutine vernition*) can be found in book 1 of the *Schedula*, which deals with the art of painting. Unlike the earlier varnish recipes, Theophilus's varnish is made by heating together only two ingredients: one part resin and two parts linseed oil by weight. The first recipe explains how a "finely powdered *gummi* that is called *fornis*" should be carefully heated over a fire with "linseed oil" (*oleum lini*) until it is reduced to two-thirds of the original amount. You have to be careful, Theophilus warns, because if the flame touches the oil-resin mixture it is "extremely dangerous and hard to extinguish."[56] In the second part of the varnish recipe, "some of the above-mentioned *gummi fornis* that the Romans call *glassa*" is melted in a closed pot with a small hole. This last advice is reminiscent of the varnish recipe we encountered in the *Mappae clavicula*, where the ingredients were also not allowed to "escape" during heating.

To test whether the resin has become liquid, Theophilus recommends using a "thin iron rod fitted with a handle."[57] As soon as the resin draws a kind of thread when the iron rod is withdrawn, it can be added to the pot of linseed oil that has been heated separately. This mixture of linseed oil and resin is then heated, and the rod is used once more to test the consistency of the liquid by applying some to "wood or stone." As in the recipes from the *Mappae clavicula* tradition, Theophilus uses the term *lucidus* to describe the brightness and luster that the substance thus made gives to the paint layer it is applied to.[58] In addition, Theophilus brings up two other functions of his varnish; it makes the painting decorative, and it makes it durable: "Every painting [*pictura*] coated with this varnish [*glutine*] becomes bright [*lucida*] and decorative [*decora*] and completely durable [*durabilis*]."[59]

That Theophilus recommends using two-thirds oil to one-third resin, which is accordingly reduced until a third is left (in the first recipe), suggests that the resulting varnish would have been extremely viscous. The special method of application such a thick oil-resin varnish would have required is consistently described throughout the recipes that have survived from this period.

In the Sun and with Your Hands: The Application of Varnish

Almost every varnish recipe recorded from the eighth century to the fifteenth explains that oil-resin mixtures ought to be applied and dried in the sun. As we have seen, the first of such varnish instructions appear in the *Mappae clavicula*. The recipe for making *lucide*, for instance, explains that after the painting has been varnished it needs to be placed in the sun to dry (*Et pone ad sole. Desicca illa*).[60] And the recipe for coating a painting with "castor oil" explains that the varnishing needs to take place in the sun as well. But save for the sponge application in the *Mappae clavicula* recipe discussed above, we do not yet learn how these viscous varnishes were applied. For the earliest account of varnish application, we have to return to Theophilus's *Schedula*. In a recipe titled *Quotiens idem colores ponendi sunt* ("How many times the same color may be applied"), Theophilus explains how to apply varnish in the sun and with your hands:

> When the painting is completed and dried and the work has been carried out into the sun, carefully spread over it the varnish [*glutine vernition*], and when this begins to run with the heat, rub it gently with the hand. Do this three times and then leave it until it is thoroughly dry.[61]

In Theophilus's varnish instructions the sun not only is important to drying the varnish but is critical to making the substance flow better during application by warming both it and the object it is applied to. Because the natural resin in Theophilus's varnish is a thermoplastic that becomes softer when exposed to heat, the warmth of the sun would have lowered its viscosity, making it easier to rub it onto the painting smoothly. Theophilus explains that by this process three layers of varnish are to be applied, resulting in a thick coat covering all irregularities of the painted surface.

Comparable instructions can be found in numerous other recipe compilations. These recipes demonstrate a long continuous tradition of varnishing with similar materials and methods. Studying the small variations in these recipes gives us more information about why applying three layers in the sun with your hands was critical to varnishing practice at this time.

The first recipe I will single out gives more details about why you must use your hands to apply the varnish. The recipe is in a manuscript compiled in 1431 by Jehan LeBègue (1368–1457), a clerk in the Chamber of Accounts, a notary, and secretary to Charles VI of France (1368–1422). In his manuscript, LeBègue copied various recipe compilations, including the *Schedula* of Theophilus, the *De coloribus et artibus Romanorum* by a certain Eraclius, two treatises by the miniature painter Jehan Alcherius, and *De coloribus faciendis* by Petrus de Sancto Audemaro (Peter of Saint-Omer). LeBègue also included a number of recipes that he himself had collected from contemporary artisans. By his own words, LeBègue was not an expert in the things he copied, and according to Merrifield, who first translated and published LeBègue's manuscript, this inexperience becomes clear from the numerous mistakes he made throughout.[62] But though LeBègue lacked an artisanal background, he was well connected to Parisian artists and met with a number of them who visited Paris from abroad. We may therefore suppose that, despite his not specializing in the things he wrote about, many of the recipes he collected do in fact reflect artisanal practice at the turn of the fifteenth century.[63]

The varnish recipe of interest here is in the collection of recipes LeBègue gathered from contemporary artisans. The recipe, written in French, explains how to make a "good liquid varnish for painters" (*A faire bonne vernix liquide pour peintres*).[64] First, a resin that LeBègue calls *glasse aromatique* is melted by a procedure that closely resembles the method described in Theophilus's second varnish recipe. As soon as the *glasse aromatique* has become liquid, you pour onto it two parts linseed, hemp, or nut oil (*oile de lin, ou de chanvre, ou de noix*) that has previously been heated in a separate pot. This substance is then cooked for an hour on a hotter fire and afterward stored in a clean vessel. The varnish must be applied with your fingers, because "if you were to do it with a brush it would be too thick and would not dry" (*car se vous fasiez du pincel il seroit trop espez et ne pourroit sécher*). The relatively thin application of varnish was necessary because when a drying oil, or oil-resin coating, is applied too thickly, it can start to "wrinkle." Wrinkling occurs when the surface of the film becomes higher in viscosity (because it is drying), while the bottom of the film is still relatively fluid. The surface layer is pulled into a pattern of wrinkles, greatly affecting the varnish's ability to transmit the visible light and making it less transparent (fig. 2.3).

For one of the most detailed descriptions of applying varnish, we turn to Cennino Cennini's *Il libro dell'arte*. *Il libro dell'arte* was composed during the last decade of the fourteenth century;[65] all surviving manuscripts are later. Two copies of the treatise survive in manuscripts dating to the fifteenth century (Florence, Biblioteca Medicea Laurenziana, MS 23, plut. 78, and Florence, Biblioteca Riccardiana, MS 2190), and a third copy survives in a manuscript dating to the eighteenth century (Rome, Vatican Library, MS 2974).[66] *Il libro dell'arte* is best known for its elaborate discussion of the art of painting on panel with an egg binder, but it also includes a variety of recipes for painting on walls, glass, and parchment and, as we saw in chapter 1, also deals with using drying oils for both panel and wall painting.

Cennini's recipe on varnishing details only its application; it does not explain how to make the varnish that is here called *vernice liquida*, like the varnish in LeBègue's recipe. It may be that because *vernice liquida* could be bought ready-made at apothecaries at this time Cennini did not consider a description of its preparation necessary.[67] We learn that before you start varnishing you have to place your painting in the sun (*la tua ancona al sole*) and be sure to wipe "dust and other nuisance" from its surface. To avoid the accumulation of dirt on the wet varnish, Cennini recommends working on a day that is not too windy,

> because dust is fine, and you would not be able, even with deft technique, to get it clean every time the wind carried it into your work. It would be good if you could be in particular grassy meadows or on the sea, because then the dust could not bother you.[68]

As in Theophilus's varnish recipe, Cennini explains that once both varnish and painting have been warmed in the sun, the panel should be laid flat. You then use your hands to spread the varnish all over, "thin and thoroughly" (*sottilmente ebene*). If you do not want to use your hands, "a little piece of really soft sponge" (*un pezzoletta di spungnia ben gientile*) dipped in the varnish can be rolled over the painting. One should "varnish methodically and remove and add as required." Using a sponge to apply varnish might have been common practice as well, since it was also included in the *Mappae clavicula* recipe for giving tin foil a yellow color.

The emphasis on applying varnish thinly should not lead us to conclude that the viscosity of oil-resin varnishes was an unwanted side effect, however. In the recipe for making *lucide* in the *Mappae clavicula* compilation, for instance, we read that if your varnish comes out too thin you should cook it down until it becomes thick (*et, si rara venerint, decoque ea usque dum spissa fiant*).[69] Indeed, the ability of a viscous varnish to cover surface irregularities, increasing saturation and gloss, was what medieval artisans were after. The search for a varnish that could give "brightness" to the colors underneath also explains why it was applied in no fewer than three coats. With such a thick layer of varnish, the painted surface would have been brought out in all its color and detail.

Varnishing Gold, Silver, and Tin

Theophilus's *glutine vernition* was also used to make an imitation of gold on tin foil, as in the *lucide* recipe in the *Mappae clavicula*. Theophilus's method is different, however; he does not advise making a yellow varnish but recommends "dipping" a varnished piece of tin into yellow dye. Theophilus's instructions for making "golden tin" are in a recipe about "tin foil" (*De petula stagni*).[70] He first explains how to prepare the tin by beating it so it is thin enough to apply to a wooden panel. After beating the foil and rubbing it with a boar's tooth to make it shine, it must be coated with a transparent, colorless varnish (*glutine vernition*) that, as in the way it is applied over a painted surface, is spread over the tin with your hands and afterward dried in the sun. When the varnished pieces of tin foil are dry, they are immersed in a mixture of saffron, the yellow scrapings from young wood cut in April, and "plenty of old wine or beer." According to Theophilus, this liquid makes the pieces of varnished tin "take on the color of gold" (*aureolum colorem sufficienter trahant*). Finally, the golden color is sealed in by giving the tin another coat of *glutine vernition* in the same way as before. When this has dried, the pieces of tin are stuck to the panel with hide glue (*glutine corrii*).[71] At this point the golden tin is ready to be painted. For this Theophilus recommends you take the colors you want to apply and "carefully grind them with linseed oil without water" (*oleo lini sine aqua*). With these oil paints you can "mix the colors for the faces and draperies as you made them before with water, and vary the colors of the animals, birds, or foliage as you please."[72] Theophilus likely added the remark about not using water when grinding pigments with oil because the first part of his book on painting is devoted to detailed descriptions of how to grind various pigments with an aqueous binding medium used in manuscript illumination.

A fifteenth-century recipe included in a Portuguese text written in Hebrew characters (Parma, Biblioteca Palatina, MS De Rossi 945) demonstrates a similar close relation between the picture varnish and the yellow varnish.[73] Even though this manuscript, known by its title as *O livro de como se fazem as cores* (*The Book on How to Make Colors*), is mainly concerned with the preparation of inks and dyes and their application in manuscript illumination, it also contains a recipe for making a yellow oil-resin varnish. The varnish (*vernis*, written in Hebrew characters) is made with linseed oil and a resin called *garash/graça de nobra*.[74] Before these two ingredients are put together, they are boiled vigorously in separate pots.[75] To test whether the *garash* is cooked, you stir it with a stick; if it flows off evenly it is ready, but if it adheres to the stick it needs to be cooked longer. In addition to this information on how to determine whether the resin is liquid enough, the recipe in *O livro de como se fazem as cores* also explains how to test if the oil is warm enough to be mixed with the liquid resin. To test the temperature of linseed oil, use a hen's feather; if the feather swells after being put in the oil, it is ready. The oil should now be taken from the fire and poured over the *garash* while constantly stirring both. We learn that "if you want to make this varnish gold color, you should separate half or as much as you want to make." In this case, aloes are ground in a mortar, then poured into the varnish and cooked. This last step would indeed have given the varnish a golden color, since the evaporated juice of the leaves (not to be confused with the gel) of certain species of aloe, such as *Aloe vera* (L.) Burm. f., produce a brownish yellow extract that can be used to give the varnish a yellow hue (fig. 2.2). The recipe continues by saying you should test the golden varnish by applying some on tin or silver. If it is "not good enough"—here likely meaning not yellow enough—you should add more aloes. The recipe concludes that when it is ready the substance can be removed from the fire and stored, "and thus it will remain good."

A similar concern for achieving a particular color and consistency of the golden varnish for tin and silver is shown in a recipe that survives in a fourteenth-century manuscript written in Middle High German (Erfurt, Wissenschaftliche Allgemein-Bibliothek, Amplonius fol. 49, 1368–70). The recipe describes how to make a golden color (*goltfarwe*) by cooking equal parts linseed oil (*linoel*), the resin *fernis*, and "aloe that can found in the pharmacies" (*aluwe . . . den findet man eyn der apteken veyle*).[76] In case the mixture "comes out too thick" (*wert es zcu dicke*), you should add more oil, but "if it is too thin" (*wenn es zcu dunne si*)—probably meaning not colored enough, as in *O livro de como se fazem as cores*—you should add more aloe. The recipe concludes that whatever is coated with this varnish "looks as if it is golden" (*daz stet, also es guldin sey*).

Thus far the recipes discussed here suggest that the ingredients and methods used to make golden varnish are largely congruent with the ingredients and methods used to make picture varnish. Two additional varnish recipes contained in a fourteenth- and a fifteenth-century manuscript show that the method of application was largely the same as well. The golden varnish was also applied in several coats, using your hands, like the picture varnish. The first recipe that details how to apply golden varnish is in a fourteenth-century Icelandic text compiled by an anonymous priest, titled *Líkneskjusmíð* ("Image Maker," Reykjavík, Arnamagnean Institute, MS AM 194.8°). We learn that the golden varnish—here called *gullfargi*—needs to be spread over silver with your fingers "as evenly as you can, so that the silver takes on a red color, and this the first application."[77] When the first coat of *gullfargi* is dry, the varnish is applied once more in the same manner. The recipe explains that you can make it "as red as you think fitting," but you should "take into account the fact that the color gets somewhat lighter when it dries."[78] Significantly, the dyestuffs used to make yellow varnishes indeed give the substance a bit of a reddish hue (fig. 2.3). It is also noteworthy that, as with Cennini's *vernice liquida*, *Líkneskjusmíð* does not explain how to make *gullfargi*. It might be that, like picture varnish, golden varnish could be bought ready-made. The *Liber diversarum arcium* (Montpellier, Bibliothèque interuniversitaire, fonds ancien, MS H 270), a fifteenth-century manuscript with an earlier core dating to about 1300, also includes two recipes that explain how to make and apply golden varnish, here called *doratura* (probably a compound from *de auratura* "to make golden").[79] The first *doratura* is a mixture made from "hepatic aloe" (*aloes epatici*), "linseed oil" (*oley linose*), and "saffron" (*croci*) that serves as a gold-colored varnish for improving gold, "and tin and silver the most."[80] The second recipe for making *doratura* introduces the resin *rasa* to the mix.[81] We also learn that if you cannot find hepatic aloe, you can substitute "horse aloe" (*aloe cabalino*) and "dragon's blood" (*sanguinem draconis*, a bright red resin from various plant species of the genus *Dracaena*; fig. 2.2).[82] This *doratura* needs to be applied "with the hand, thinly two or three times." But, so we learn, "on real gold [*in auro vero*] a somewhat thinner color is drawn once" (*tantum semel subtiliter color ille trahatur*).[83] That the varnish for gold leaf indeed should be somewhat thinner than the golden varnish for tin and silver—which had a similar consistency to picture varnish—is also shown by an elaborate recipe for varnishing gold leaf found in Eraclius's *De coloribus et artibus Romanorum*.

Auripetrum

Like the treatise of Theophilus, *De coloribus et artibus Romanorum* was written in a monastic context, and parts of this work as well were therefore more directly embedded in medieval artistic practice. In the first book of his treatise, Eraclius even emphasizes his own experience with the recipes he collected: "Indeed, I write nothing to you that I have not first tried myself (*nil tibi scribo quidem, quod non prius ipse probassem*)."[84] Although this certainly does not mean Eraclius had tried out all the recipes—it is likely that this expression was meant to give authority to the work—*De coloribus* does contain various interesting recipes that we need to take into account in this history of varnishing. *De coloribus* consists of three books. The first two books contain twenty-one stanzas written in metric verse dating from the tenth century. The third book is written in prose and dates from the twelfth century. The surviving manuscripts that contain *De coloribus* are all later; one of the oldest, now kept in the British Library (MS Egerton 840A), is from the thirteenth century, and a second manuscript concerns the copy LeBègue transcribed in the fifteenth century. This last manuscript includes the third book of Eraclius—not included in the earlier British Library manuscript—which contains the recipe for varnishing gold leaf.[85]

Eraclius first warns not to use "pure varnish" (*non de puro vernicio*) if you want to apply a coating to gold leaf.[86] Instead, you should use a substance called *auripetrum* that is "mixed with oil and a little *vernicio*" (*mixto tamen cum oleo modico vernicio*). In another recipe in book 3 of his treatise, titled *De auro petro*, Eraclius explains how to make this liquid.[87] You grind the dry bark of a certain tree called *vesprum* (collected in March or April), and if you do not have this you can use *incaustum* (?) or the dried and ground bark of *nigra spina*. It is impossible to establish with certainty the species of the plants Eraclius refers to here, but several trees and shrubs common on the European continent would have provided a variety of yellow dyes that, like saffron, are soluble in oil. The ground-up tree bark is to be soaked overnight in oil made from linseed (*oleum de lini semine factum*), and the next day this substance is "boiled over fire (*ad ignum bullies*) as long as you may think proper, but not much." You then strain the liquid through a cloth into another jar and boil it again with myrrh (*mirra*, soft oleo-gum resin extracted from *Commiphora* spp.), and aloe (*aloe*). After straining this as well, you add a resin called *vernix*—or, if you do not have *vernix*, a resin called *glassa*—and cook it on the fire one more time.[88] When this has cooled "you can put it away in a vessel to preserve it as long as you like" (*ad servandum in vase, quanto tempore volueris, repone*).

Now that we know how the gold-colored liquid called *auripetrum* had to be prepared, we can return to the recipe in which Eraclius explains how to varnish gold leaf with it. According to Eraclius, *auripetrum* mixed with linseed oil and the resin *vernicio* may be used to cover gold, and any part of the gypsum (*gipsei*) that appears through the gold. He recommends mixing *auripetrum* with more oil than *vernicio* because the varnish for gold must be thinner than picture varnish (*ne nimis sit spissum vernicietur super aurum*). It would indeed be desirable to use a slightly thinner coat of varnish for gold leaf, because too viscous and sticky a liquid could damage the delicate leaves of gold if you rub them too hard during varnish application. Finally, Eraclius concludes his recipe with the remark that if, instead of gold leaf, you want to varnish figures and other colors, you should use pure varnish or thick oil instead (*puro vernicio, vel de crasso oleo*).[89]

Some of the findings from the technical examination of the thirteenth-century Westminster Retable (fig. 2.6) are of interest in the context of this varnish recipe. Recently the Westminster Retable has been the subject of a comprehensive conservation project, so we know a lot about the materials and techniques that went into making it.[90] This large work—it is almost a

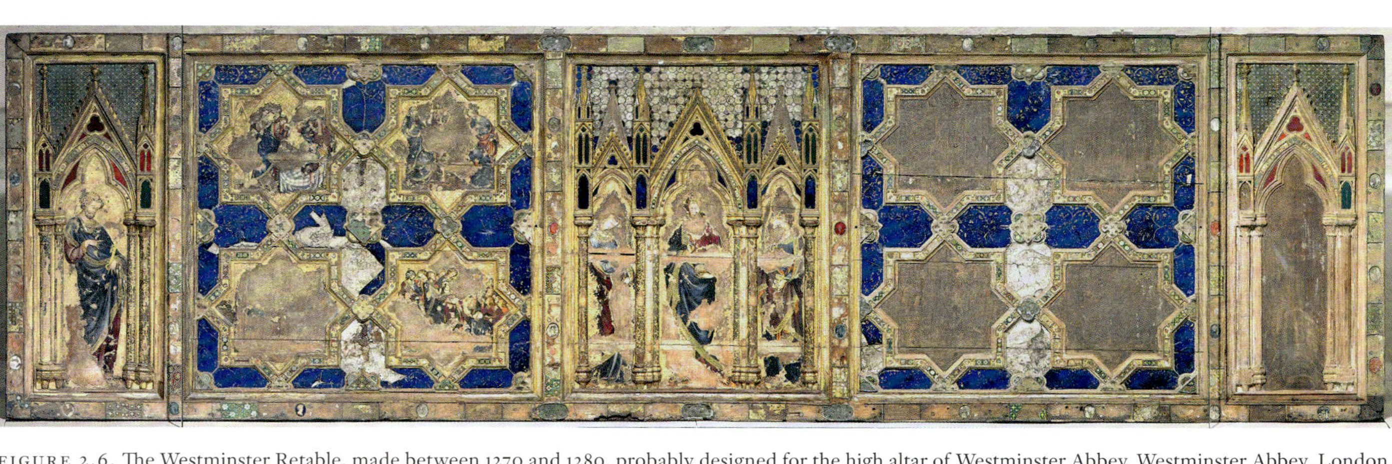

FIGURE 2.6. The Westminster Retable, made between 1270 and 1280, probably designed for the high altar of Westminster Abbey. Westminster Abbey, London. © Dean and Chapter of Westminster.

meter high and more than three meters long—was painted about 1260, for the high altar of Westminster Abbey in London. Original varnish was detected on the figures, as well as on the mordant gilding (gold applied to an underlayer that often contains oil or resin or both) of the columns and on the moldings framing the figures, but it could not be demonstrated on the water-gilded background.[91] Analysis of the varnish on the figures identified linseed oil with certainty, whereas the varnish on the mordant gildings contained linseed oil with a pine resin. In one of these coatings a more "oil-rich layer, containing partially heat-bodied oil and a somewhat lower proportion of resin" could be demonstrated. Gold that has been applied by "water gilding" can be polished, but gold applied to a mordant cannot. It is certainly interesting, therefore, that in the Westminster Retable only those areas with unpolished gold were found to have been covered with an oil-resin varnish. It is possible that on these unpolished areas this varnish was meant to imitate the gloss of polished metal.

Color from the Sun and Brilliance from the Tin

The fourteenth-century *De coloribus faciendis* of Peter of Saint-Omer, copied by Jehan LeBègue as part of his compilation, includes no fewer than eight recipes for making gold-colored varnishes and, notably, not a single recipe for picture varnish.[92] Most of these recipes for golden varnish are similar to those already discussed, but three stand out because they provide more insight into the preparation and special optics of this substance. The first recipe explains how to make a golden varnish that, like the recipe in the treatise of Eraclius, is called *auripetrum*.[93] In this case, however, *auripetrum* is used not to varnish gold leaf, but to transform tin foil into the appearance of gold. The recipe describes how to temper "Spanish saffron" (*crocus hispanicus*) with "very clear glue" (*lucidissimo glutine*) or with "liquid varnish" (*vernicio liquido*). These different binders suggest that the recipe describes the practice of making golden tin for application in books (with animal glue) and for application on wood (with oil-resin varnish).[94]

The golden liquid thus made had to be applied to polished tin foil to transform it into the appearance of gold: "Very bright and well-polished tin assumes the appearance of gold to those who look at it, for it receives its color from the sun and its brilliance from the tin" (*stanno limpidissimo, i.e. pene polito et claro, superpositas speciem auri intuentibus mentitur quod a sole colorem et stanno accipit fulgorem*). In this passage on the optical qualities of golden tin, *De coloribus faciendis* perfectly describes the art of material mimesis so characteristic of medieval art. The visible light passes through the yellow varnish and reflects from the polished tin foil through the varnish back to the eye; in this way the golden tin imitates both the color of gold and its reflective sheen (fig. 2.3).

The second varnish recipe from Peter of Saint-Omer's treatise that should be singled out here recommends polishing varnish to make it shine, much the same way metals would have been polished.[95] The *Mappae clavicula* manuscript recommended using an onyx stone to polish the gold-colored oil-resin varnish discussed earlier. *De coloribus faciendis* similarly recommends using an onyx stone, but this time for polishing a gold-colored varnish made with an egg-based medium by a method that closely resembles a recipe from Theophilus's *Schedula* for making golden tin for use in books.[96] Peter of Saint-Omer writes that the tin must be dipped in a mixture of saffron and the white of egg. This is repeated three times, allowing the tin to dry between them. When you are done, you polish this gold-colored egg white varnish with an onyx stone. After the golden tin has been glued on paper or on wood, it should be polished again when the glue has dried. This time it is first washed clean with a sponge "dipped in cold water" (*spongia intincta in aqua frigida*), then it is rubbed down with "a linen cloth well wrung out" (*panno lineo extersa optime*), and finally it is "polished" (*frica*) once more with the onyx stone. The recipe also mentions that "if you have no onyx" (*si non habes onchinum*), the tin varnished with egg white can also be greased with the "oil made from linseed" (*de oleo quod fit de lini semine*). These rather intricate polishing instructions show the importance of making your imitation gold as lustrous as possible, quite similar, in fact, to the way real gold leaf would have been made lustrous by polishing. Using oil rather than polishing with an onyx stone shows that the optical effect of applying oil to an object was thought to approximate the effect of a polished surface. According to *De coloribus faciendis* the "golden tin" thus made can be used as a substitute for gold to decorate "pages" (*carta*) and "wood" (*ligna*) "on account of [gold's] high price."[97]

Finally, I should discuss one last recipe from Peter of Saint-Omer's treatise, since it brings us back to something I mentioned briefly at the beginning of this chapter: that making varnish and making oil-resin ointments for medicinal purposes might have been related practices. The recipe in question explains how to boil linseed oil with the inner bark of black plum (*Nigripruni*).[98] When this is done you have to clean the resin called *glassam* and mix it in a jar with "alum" (*aluminis*), "dragon's blood" (*sanguinem draconis*) and "pine resin" (*picem*).[99] As soon as these ingredients have become liquid, you add the linseed oil that was previously boiled with the bark of plum. This substance is to be cooked "as if you were making a compound ointment" (*secundum unctionem confectionis*).

Finding a Name for Varnish

Theophilus refers to varnish as *glutine vernition* (which, in some manuscripts containing the *Schedula*, is also called *glutine vernice* or *glutine vernicion*).[100] The term *gluten* is not commonly used in medieval recipe books to describe transparent or translucent surface coatings; it usually describes a variety of adhesive materials. Our modern word "glue" is etymologically related to it. Theophilus also uses *gluten* in this way to refer to hide glue, which he calls *glutine corii* (*corii* means skin), and a glue made from casein (the main protein in milk or cheese), which he calls *glutine casei*.[101] When oils and resins are cooked together, the resulting liquid has strong adhesive properties as well. Medieval artisans therefore used oil-resin mixtures to glue materials together and to make a thick and tacky substance, the "mordant." Mordants were used, among other things, to stick thin leaves of gold, silver leaf, or tin foil to paintings or other art objects.[102]

Since Theophilus often uses adjectives to specify the particular material he is talking about, *vernition* most likely referred to the main ingredient of his varnish. The two recipes for making *glutine vernition* mention that *gummi fornis* and "*gummi fornis* that the Romans call *glassa*" are to be mixed with linseed oil.[103] Medieval treatises use the term *gummi* to denote both gums and resins because the botanical distinction between the two substances was not made until the twentieth century. We can nevertheless be fairly certain that in this case it implies a type of resin, because the *gummi* has to be made liquid by heating.[104] As in the case of glue, Theophilus uses *fornis* to specify the type of *gummi* he is talking about and not, as is sometimes suggested in the literature, to refer to varnish in general.[105]

Both *vernition* and *fornis* derive from the Latin *vernix*, an ingredient that also appears in the *lucide* recipes in the *Mappae clavicula* family.[106] In another recipe the *Mappae clavicula* shows that *vernix* (here called *vernice*) was indeed used to make glue. In this recipe, titled *Collam Grecam facere* ("How to Make Greek Glue," *collam*, from *colligo*, "keep together"), *vernice* is heated with linseed oil as in Theophilus's recipe for making *glutine vernition*; the recipe not only mentions the closed pot for melting *vernice*, but also discusses using an iron rod to test the consistency of the liquid.[107] Theophilus's *glutine vernition* was therefore named after one of the main uses of oil-resin mixtures as glue, and it was probably made with similar ingredients and by a comparable method. This also suggests that, during the time of Theophilus, the complicated *lucide* recipes from the *Mappae clavicula* tradition, mixing many materials, had been replaced by a "simpler" mixture of ingredients initially developed for use as an adhesive.[108]

Sweat and Tears: Vernix *in Medico-botanical Treatises*

Since many of the early recipes indicate that *vernix* was a common ingredient in surface coatings and glues from an early date, its identification with a particular resin has aroused much speculation.[109] My aim here is not to open up the old debate about the identification of specific resins in varnish recipes, but a discussion of *vernix* is too fundamental for our history of varnishing to leave it out altogether. All the more so because this mysterious ingredient eventually gave its name to the substance we today know as varnish in English, *vernis* in Dutch, *Firnis* in German, *vernice* in Italian, and *vernis* in French, to name a few.

In his 1847 edition of the *Schedula*, Robert Hendrie elaborately discusses the identity of *vernix* and *glassa* as they occur in Theophilus's varnish recipes. Hendrie offers a variety of ancient medical texts on plants to show that the term *vernix* was applied to the resin sandarac.[110] He explains that some Greek and Arab authors, and as a result many of the medieval writers who relied on them, confounded amber, sandarac, and the resin flowing from the juniper tree.[111] Hendrie concludes that *vernix* was a direct term for the resin sandarac and that *glassa* was used to describe sandarac as well because of its resemblance to amber.[112] In the same year, Charles Lock Eastlake came to a slightly different conclusion. Eastlake proposed that medieval variations of the word *vernix* might be etymologically related to the late Greek *beroníkē* and *bereníkē* (in Greek the *b* would have been pronounced as *v*), which originally meant resin or resin tree. He explains that in later centuries *beroníkē* came to designate amber.[113] Since amber was expensive, cheaper and softer resins were substituted for it. And, according to Eastlake, this is why the term *vernix* and its variations were later also used to designate these softer resins—and sandarac in particular.[114] Like Hendrie, Eastlake therefore argues that *vernix* and *glassa* could be used interchangeably to refer to sandarac. But unlike Hendrie, he concludes that when they appear together, such as in the two varnish recipes of Theophilus, *glassa* is more likely used to refer to amber, while *vernix* refers to sandarac. Eastlake's identification of *vernix* with sandarac or amber is generally accepted in today's literature about historical varnishes.

We will likely never be certain what resins the different appellations in recipes refer to, but a group of early medico-botanical treatises shed some additional light on the problem. They not only identify sandarac as *vernix* and amber as *glassa* but, significantly, also connect these resinous substances to their use in picture varnish. The first example is in a twelfth-century Latin manuscript on medicinal herbs attributed to Matthaeus Platearius

(d. 1161), a physician from the medical school at Salerno.[115] His *Circa instans* (ca. 1150–70), also known as the *Liber de simplici medicina* (*Book of Simple Medicines*), is an alphabetic textbook on the characteristics, preparation, and use of various medicines. It became an important handbook for physicians, apothecaries, and botanists and was translated into many vernaculars. By the end of the fifteenth century it started appearing in numerous printed editions as well. The British Library of London has in its collections an early copy of the *Circa instans* (MS Harley 270), dating to the thirteenth century.

This manuscript contains an entry about *bernice*, which besides its medicinal benefits is supposed to be of great virtue to painters.[116] We read that *bernix* is a *gummi* that flows from a tree that grows overseas (*ultramarinis*) and that dries and hardens through the action of the sun (*que actione solis exsiccatur et induratur*). There are three colors of *bernix*—*subcitrini* (yellow), *subrufi* (red), and *subalbidi* (white/clear)—of which the best is the brightest or the clearest (*lucidutio vel clara*). At the end of the entry on *bernix*, the *Circa instans* mentions that it is also called *cacabre* or *classa*. One of the main virtues of *bernix* is its ability to glue things together (*conglutinendi*) on account of its viscous nature/gumminess (*gummositate*). Its other virtues are that it can be used to make things clear (*clarificandi*) and that it can conserve as well (*conservandi*). As such, painters lay it over colors to make them shine/brighter, and they use it to preserve colors (*pictores enim super alios colores ponunt, ut lucescat et alios colores conservet*). And finally we learn that *bernix* can prevent metals from rusting. After listing a few of the resin's medicinal purposes, the *Circa instans* also mentions that when it is powdered, *bernix* can enhance a women's beauty/face (*pulverem eius apponunt mulieres faciei*). This was indeed one of the other main uses of this resin, and as I will discuss at the end of this chapter, face painting was closely related to varnishing pictures. The *Circa instans* perfectly sums up the three uses of varnish in the arts that we also encountered in the earliest recipes describing how to make this substance: as a glue, as a brightening substance, and as a protective coating. When we study a few of the many herbals that copy and edit the *Circa instans*, it becomes clear that this painter's varnish must have been quite similar to the varnishes mentioned in recipe treatises.

The first and earliest example can be found in one of the most popular encyclopedias of medieval times: *De proprietatibus rerum* (*The Properties of Things*, ca. 1240–50), written by Bartholomaeus Anglicus (before 1203–72). Like the *Circa instans*, *De proprietatibus rerum* was originally written in Latin and survives in extant manuscript copies and printed editions.[117]

Bartholomew's encyclopedia was also translated into many vernaculars, including French (Jean Corbechon, 1372), English (John Trevisa, 1397), and Middle Dutch (Jacobus Bellaert, 1485). Starting with a book about God (*De Deo*), *De proprietatibus rerum* describes the state of the art of knowledge about the "properties of things" in nineteen books. Bartholomew relied heavily on the *Circa instans* for book 17, on trees and herbs.

Bartholomew does not include a separate entry on the resin *bernix*, but he mentions it when he discusses *bdellium*, an oleo-gum resin that has been used for thousands of years for its medicinal properties.[118] He explains that *vernix* is the name of a substance used by painters, which is also known by the name *bernix*. He continues that *vernix* is a type of *gummi* that oozes from a tree different from the tree that exudes *bdellium*. *Vernix* is of great use to painters (*pictores*); they use it in a "very liquid" (*multum liquidem*) state to glue things together (*glutenendi*), to makes things clear (*clarificandi*), and to conserve their work (*conservandi*).[119] Bartholomew's remark that painters use *vernix* in a liquid state, which is not included in the *Circa instans*, suggests that this resin was generally available in a solid state. This not only corresponds with the instructions in recipe books, where a hard resin must be heated so that it becomes liquid enough to be mixed with linseed oil, but also with the name that varnish eventually got in Italy, *vernice liquida*.[120] It is clear that in the work of Bartholomew *vernix* is not amber, because Bartholomew uses the term *electrum* to describe this particular resin.

To finally solve our resinous riddle, we have to turn to another popular encyclopedia on the medicinal uses of herbs and plants. That work is part of an early series of printed books published by the end of the fifteenth century. It first appeared in German in 1485 under the title *Gart der Gesundheit*, printed in Mainz in the workshop of Peter Schöffer. Although it is often attributed to Johannes de Cuba, his authorship cannot be established with certainty.[121] In 1491 the work was translated into Latin as *Ortus sanitatis* and printed by Jacob von Meydenbach, again in Mainz. Like Bartholomew of England, the author of this last edition made substantial use of the *Circa instans*. It is interesting, therefore, that it not only contains the account of painter's varnish (which is not present in the 1485 German edition) but adds a few facts that are not present in any of the earlier works. The painter's varnish is discussed in the entry on *vernix* (fig. 2.7).[122]

We read that if you mix *vernix* with linseed oil it is called *vernice*, and that with this substance you can "illuminate" (*illuminantur*) or "consolidate" (*consolidantur*) the colors of your painting. This may be one of the earliest examples where the word *vernix* is transformed into its modern meaning

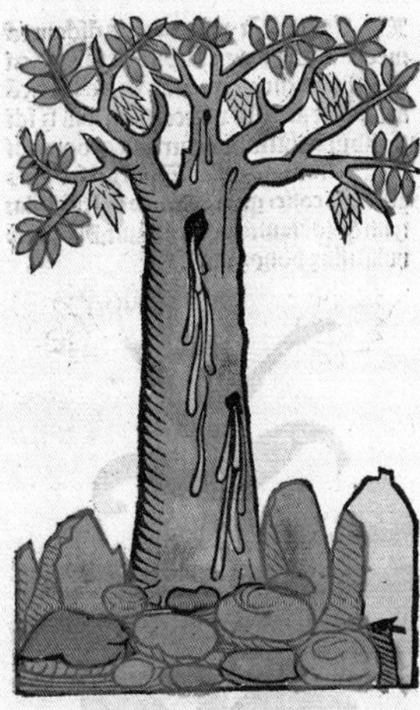

FIGURE 2.7. Folio from *Ortus sanitatis* (Mainz: Jakob Meydenbach, 1491), a woodcut showing resin flowing from a tree. Cambridge University Library (Inc.3.A.1.8[37]), fol. 232v. Reproduced by kind permission of the Syndics of Cambridge University Library.

of varnish. According to the *Ortus sanitatis*, the resin *vernix* is also known as *sandaros*, *baratin*, *gressa*, or *sandaraca*. This shows that, as Eastlake and Hendrie suggested, *vernix* was indeed the common word for the resin sandarac. The entry on *vernix* continues with the information from the *Circa instans* about the three quality grades of *vernix*, but in addition it explains that *vernix*—sandarac—flows from a tree called the juniper (*iuniper*). This last is illustrated by a beautiful woodcut showing resin flowing from cuts in a tree (fig. 2.7).

Today sandarac is known as the resin that flows from *Tetraclinis articulata* (Vahl) Mast., also known as the sandarac tree, and not from the juniper tree (*Juniperus* spp.) as the previous treatise suggests.[123] During the period studied in this chapter, however, the botanical distinction between the varieties of resin-producing North African coniferous plants was not yet established.[124] This means that when recipes or encyclopedic texts discuss the resin *vernix*—sandarac—they refer more generally to a resin flowing from the cypress family of trees (Cupressaceae), which includes both the juniper tree and the sandarac. This last fact is further shown by two premodern glossaries: the fourteenth-century *Breviarium Bartholomei* (1387), composed by the medical writer John Mirfeld, and a related, untitled glossary of the mid-fifteenth century. These glossaries include various entries for all the names the resin from the juniper tree (*gummi juniperi*) was known by: *bernix*, *vernix*, *classa*, *glassa*, *glossa*, *carabre*, *karabe*, *kacabre*, and *cacabre* (*vulgo dicitur lambra*).[125] This list includes the names—variations of *vernix* and *glassa*—that we have encountered in the varnish recipes studied in this chapter. And what is more, one of the entries on *bernix* that appears in both glossaries explains that when *classa* and linseed oil are mixed together the substance is called *vernix*: "*Bernix*, *classa* and *gummi juniperi* are the same: Also called *vernix*, which is made from oil of flax [*oleo seminis lini*] and *classa*, and this illuminates [*illinentur*] and consolidates [*consolidantur*] the colors of paintings [*colores picturarum*]."[126]

The glossary also indicates that "amber" (*cacabre vulgo dicitur lambra*) could be used to describe the resin from the juniper tree as well, and that it was the same as *classa*.[127] The entry on *classa* in its turn explains that *classa genus est lapidis. respice in bernix* ("*classa* is of the stone variety; compare *bernix*"). It is therefore significant that only the entry on *classa*/*glassa* says this resin is of the stone variety; all the other resins are defined as of either the *gummi* or the *resina* variety. Since ancient times, amber had been considered a precious material that was mounted in jewelry and valued, like precious stones, for its color, rarity, and hardness.[128] Therefore *glassa*, meaning resin from the juniper tree, might also have been used for a much harder (fossilized) resin that was thought to have once flowed from the juniper tree. Even though it was known that amber, like the other resins harvested from plants, originated as a tree sap, of course nobody could have found it in that form and established what tree it came from. The optical resemblance between these two resins may be the reason amber was classified with the resin flowing from the tree identified as juniper (fig. 2.1). What is more, since ancient times a yellow to red amber resin had been valued precisely for its resemblance to gold, while both amber and gold in their turn were likened to the radiance of the sun.[129] This may be why, on another level, amber resin, and its optical equivalent sandarac—which produce a reddish varnish (figs. 2.8 and 2.9)—were considered the most suitable resins for varnishes that were meant to increase the luster of gold or to create a material mimesis of gold on other metals. Once again, this material shows the deep connection between

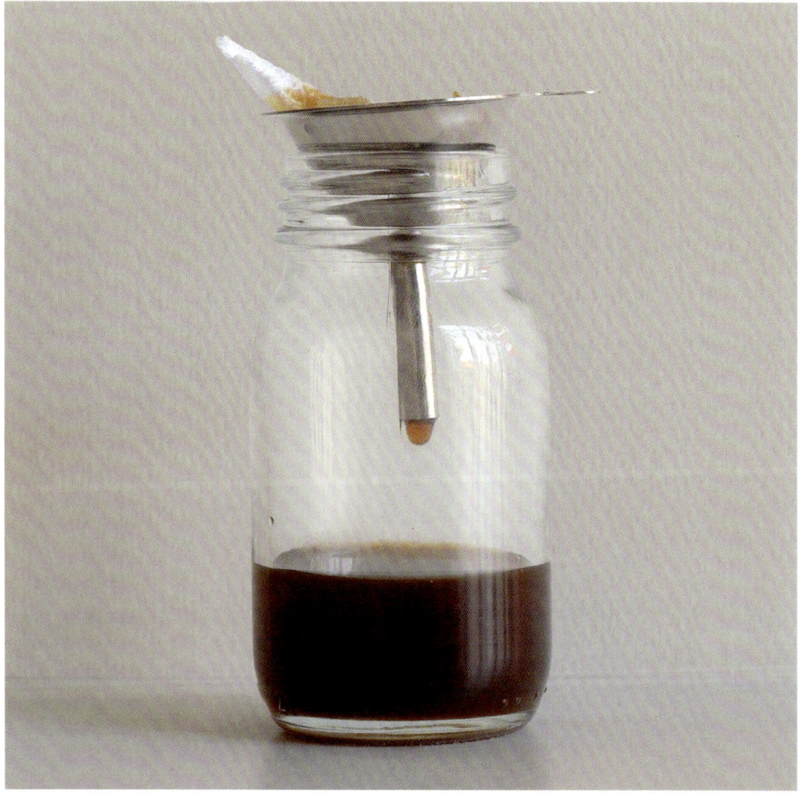

FIGURE 2.8. Linseed oil-sandarac varnish (3:1) being filtered through linen cloth, after being heated at about 245°C for two hours. Although not all sandarac was incorporated in the oil, the varnish was extremely viscous and pulled a thread; see figure 2.9. Photo by Marjolijn Bol.

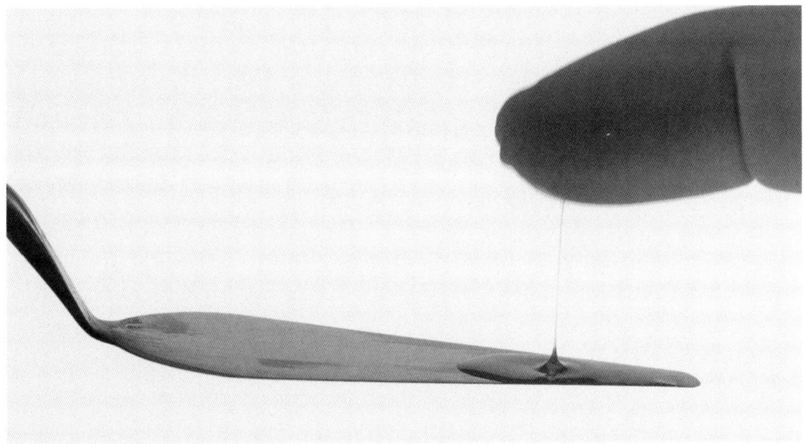

FIGURE 2.9. Sandarac varnish pulling a thread. Photo by Marjolijn Bol.

the early history of varnish and its ingredients and the history of enhancing and imitating gold.

Vernix *as Varnish Proper*

As the *Ortus sanitatis* showed, by the end of the fourteenth century the mixture of linseed oil and the resin *vernix* became known as *vernice*, and in Italy as *vernice liquida*. Another early instance where *vernix* is not just mentioned as an ingredient in varnish mixtures but is used to denote the varnish proper can be found in Eraclius's recipe for making golden varnish discussed above: *Quomodo vernicietur aurum ne perdat colorem* ("How to varnish gold so that it will not lose its color").[130] If this recipe can indeed be dated to the twelfth century, it would be one of the earliest examples in which *vernix* is used as a verb to describe the act of varnishing (*vernicietur*). But because the manuscript this varnish terminology derives from was transcribed in the fifteenth century, it cannot be excluded that LeBègue or an earlier copyist might have "updated" some of the vocabulary or added recipes, or both.[131] Whatever the case, there is plenty of evidence that, from the thirteenth century onward, *vernix* and its variations were used to denote varnishes and the act of applying them. At the same time, the vernacular became increasingly popular for recording artisanal recipes, and here too *vernix* became the general name for varnish.[132]

There are many examples of this, but I will mention only two relatively unknown expense accounts from The Hague that show how, in fourteenth-century Dutch, *vernix* was transformed to mean both varnish and the act of varnishing. The first expense account records a payment to the painter Dirk Janszoon to restore a painting depicting Mount Calvary. We read that the work was well painted but in need of restoration because—owing to its age—it had become so dark and damaged that one could no longer recognize the picture (*van ouderdom verduystert ende vergaen, dan met die picture langer niet wel bekennen en mochte*).[133] Dirk Janszoon was asked to make the colors bright again and to varnish the painting so that the good work could be admired for longer (*weder is doen verlichten ende vernissen, opdat dat goede werck te langer soude mogen bekennet bliven*). The same archive contains another expense account that mentions the painter Dirk Janszoon in relation to the restoration of a painting. In this case the artwork had to be brightened and varnished because it had suffered too much from the stove (*vernist ende verlicht heeft, wanttet van den stove zeer vergaen was*).[134]

The term *vernissen*, used to describe Janszoon's restorations, is derived from the name for the resin known in Middle Dutch as *vernijs*.[135] The use of *verlichten* in the same context is also of interest here because it is related to the Latin *illumino* (*in*, "upon," and *lumen, lumin-*, "light"). This term was used to describe the general act of making something bright and the act of brightening something with colors, which is of course where the term manuscript illumination comes from. It is therefore possible that *verlichten* may be interpreted as Dirk Janszoon's being asked to repaint (parts of) the painting. But since we have seen that both the earliest recipes and the medico-botanical treatises studied in this chapter consistently describe the varnish as a substance that can be used to illuminate the colors it is applied to, it is more likely that in this context the term *verlichten* indicates that Dirk Janszoon was making the painting "bright" again by applying a new varnish after removing the old, damaged varnish.

Vernis, *Women, and Make-Believe*

This chapter has argued that varnishing was born as an art of material mimesis, devised to imitate the optical appearance of one material with another. As such, varnishes were not just used on medieval panel paintings but also applied to many other art objects ranging from polychrome sculptures to gilt leather wall coverings and from reliquary shrines to various other kinds of woodwork. Medieval artisans used this viscous liquid to evoke the gleam of gold and to suggest the appearance of gold on silver and tin. And when applied over colors, the varnish helped make panel paintings shine like polished enamel. The relation between the art of panel painting and the art of enameling will be discussed further in chapters 4 and 5. Here I will conclude with one last case in which the practice of varnishing is connected to what at the time was considered a deceitful form of imitation: its history as face paint.

One of the most striking examples can be found in a fourteenth-century poem by Jan van Boendale (ca. 1280 to ca. 1351), a writer from Brabant who moved to Antwerp to become secretary of the city (where he also became known as Jan de Clerk). Van Boendale was the author of various narrative and moralistic verses. In one of these works, written about 1330–34 (Oxford, Bodleian Library, MS Marshall 29), he let his alter ego Jan Teesteye conduct dialogue with his friend Wouter to demonstrate that their time is just as good as the past (or even better). In the section of the poem titled "More about Women" (*Noch meer vanden wiven*) Jan Teesteye is not entirely positive about the behavior of the ladies of his day and age. He complains that they are not satisfied with the appearance God gave them and try to look like something they are not. To illustrate his point, Van Boendale describes how women put creams and ointments on their faces to make themselves appear more beautiful, the way the master varnishes an image embellished with deceiving decoration:

> They [the women] grease and anoint their faces to appear beautiful and admired by many; but as a master varnishes [*vernist*] an image with all its deceiving decoration, which shines beautifully as though it were gold, on the inside it remains wood, the same way a woman has varnished [*vernist*] her body so that it looks beautiful and shining, but it is still a futile thing. What there is still stays the same. That she can never change.[136]

Van Boendale thus compares the practice of varnishing and make-believe in the arts of his time to another cultural practice: the way women create false beauty with face paint. It is not entirely clear whether Van Boendale considered a wooden object imitating gold and gems as morally reprehensible as a woman's applying face paint or whether he mainly intended to elucidate the nature of face painting. But what makes his comparison especially interesting here is that the Dutch word *vernis* is used to refer both to the act of varnishing an artwork and to the cosmetics ladies used to whiten and smooth their faces. Another example of the word varnish used in the sense of maquillage, albeit a more favorable one, is in the fifteenth-century biography of the Dutch Saint Lidwina (1380–1433; *Tleven van Liedwy, die maghet van Sciedam*, ca. 1435).[137] Lidwina fell ill after she broke a rib by falling on the ice while skating. She died after thirty-eight years as an invalid, which, according to legend, she bore without a single complaint—considered a miracle. Various other miracles happened during her life, but the one of interest here occurred after Lidwina's death. Her biography reports that as soon as Lidwina passed away, her face was not like that of a dead person but shone with a "lovely white shining color and glistened beautifully" (*scoon wit blinckende verwe ende glinsterde also betamelic scoon*), as if she were anointed with "oils or with varnish" (*olyën of met vernis*).[138] Lidwina's body therefore shone divinely as if covered with the materials that women use to improve their complexions, which again are referred to by the same terminology used for the substances coating artworks.

Other sources confirm that the varnishing and face painting might have had more in common than Van Boendale's association of outward beautification with falsification. They show that the varnish used as maquillage and the varnish applied to artworks were made from similar materials. We have already seen that the *Circa instans*, in its entry on the resin *vernix*, explained that in powdered form it can enhance the beauty of a woman's face. And it is well known that the resin mastic was sometimes used for the same purpose.[139] But a most interesting connection between painter's varnish and face paint can be found in Cennino Cennini's *Il libro dell'arte*. Cennini writes that painters are sometimes asked to coat the skins of both men and women.[140] He explains that the colors for the face can be tempered with egg yolk, but that if you want to "liven them up" (*vuoi per caleffare*), "oil" (*olio*) or "liquid varnish" (*vernicie liquida*) can be used as well. The latter, so Cennini writes, is "the most powerful of the temperas" (*la quale è più forte tempera*).[141] Like Van Boendale, Cennini does not have a favorable opinion of this practice. He points out that women who use such artificial preparations grow old before their time:

> But I tell you definitely that if you want your face to keep its color for a long time, make a habit of washing yourself with water from a spring, from a well, or from a river, keeping in mind that if you use another, man-made concoction your face quickly becomes wizened and your teeth black, and in the end women grow old before their time. They turn into the most revolting old women there could be. And this is enough said to you on this explanation.[142]

Indeed, almost ironically, the varnish for faces, often containing toxic ingredients such as lead white or vermilion, "brightened" the face of a lady for only a short time; unlike the varnish for artworks, it did nothing for the long-term "preservation" of her countenance.

3

Crystal Clear

Changes in Varnishing Practice, 1450–1500

In chapter 2 we saw that recipes written before the fifteenth century mention the role of varnish in giving brightness to paintings and to gold or describe its role in making silver or tin look like gold. For this purpose artists used a viscous substance, applied in the sun in no fewer than three coats using the hands or a sponge. Even though varnish was used to brighten paintings, there is no evidence that the clarity or particular hue of the substance itself was of special concern to artisans at this time (except, of course, for the varnishes used to imitate gold, which had to be yellow). On the contrary, the processes and materials used to make oil-resin varnishes would have turned them relatively dark brown (fig. 2.8).[1] This changed in the course of the fifteenth century. We have already seen that Cennino Cennini starts his varnish instructions with the remark that you should select "the most liquid and clear and pale [varnish] that you can find" (*la tua vernicie liquida e lucida e cchiara la più che trovare menti*).[2] Unfortunately, *Il libro dell'arte* does not give further details about how to obtain or make such a clear and liquid varnish, but other recipes written or compiled in the fifteenth and sixteenth centuries show that Cennini's remark indeed appears to reflect a new concern of painters at the time.

To obtain a relatively liquid, clear, and colorless varnish, you must select certain types of resins and oils. Resins, for instance, come in various hues that range from yellowish white to dark brown (fig. 2.1). And nut oils are paler and display less yellowing than linseed oil (fig. 3.1). The way the varnish and its ingredients are processed also affects its final viscosity, color, and

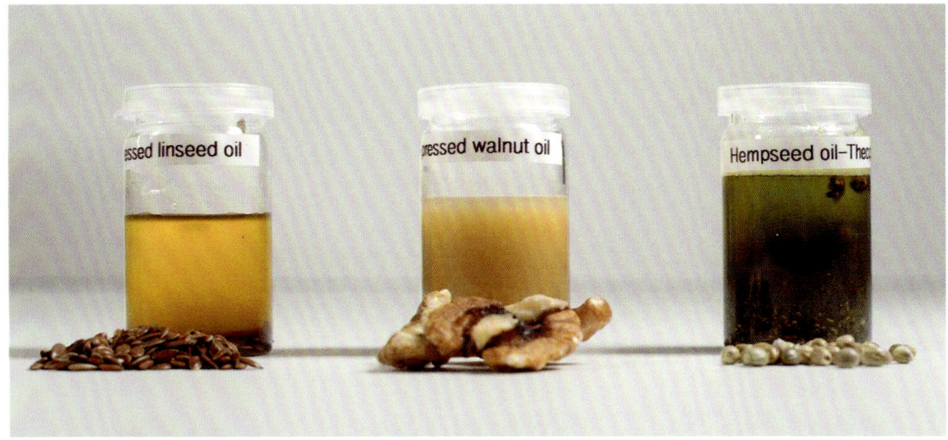

FIGURE 3.1. *Left to right:* Cold-pressed linseed oil, walnut oil, and hemp oil, all expressed by the Theophilus method. The oils are shown immediately after pressing and before filtering. Photo by Marjolijn Bol.

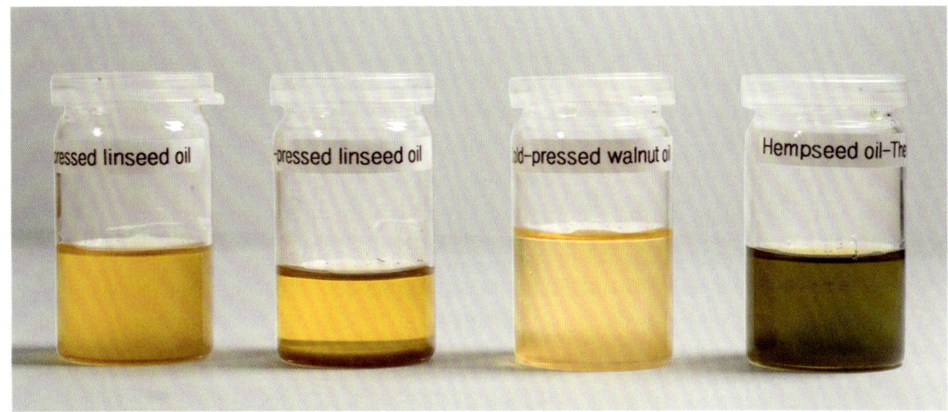

FIGURE 3.2. *Left to right:* Cold-pressed linseed oil (first two flasks), walnut oil, and hemp oil. The oils were filtered and allowed to stand for a day after pressing. The second flask from the left contains (unfiltered) linseed oil allowed to stand for about three weeks, thus clarifying significantly. Photo by Marjolijn Bol.

clarity. Freshly extracted oil, for example, contains mucilaginous materials and other impurities (fig. 3.2).

These impurities impede oil's ability to dry, make it look turbid, and may even influence its discoloration as it ages.[3] Further, when drying oils used to make varnish are cooked at high temperatures, as some of the harder resins

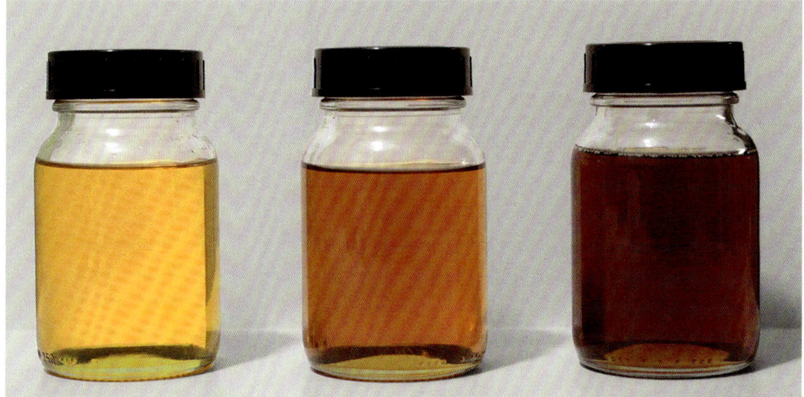

FIGURE 3.3. *Left to right:* cold-pressed linseed oil with no further treatment; cold-pressed linseed oil heated to 165°C; cold-pressed linseed oil heated to 269°C (both for five hours). Photo by Marjolijn Bol.

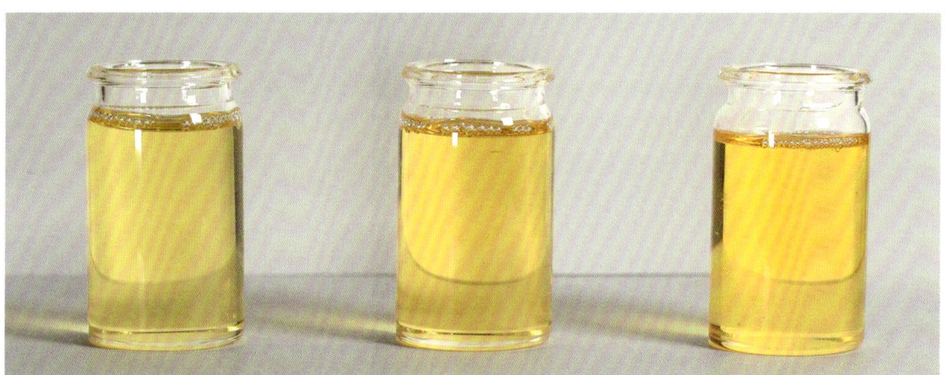

FIGURE 3.4. The same oils after being kept in the dark for almost two years. *Left to right:* cold-pressed linseed oil, no further treatment; cold-pressed linseed oil heated to 165°C; cold-pressed linseed heated to 269°C (both for five hours). There are minute differences, but the cooked oils have lightened dramatically in two years. Photo by Marjolijn Bol.

require, they become significantly darker and more viscous as well (fig. 3.3). When such a dark yellowish brown varnish is applied to a painting, it may influence the optics of the paints, making whites look yellow or turning blues green. The recipes discussed in this chapter show how, since the mid-fifteenth century, painters actively searched for new materials and methods to make a clear varnish of the palest tone possible that still had good drying qualities. In this search, as we will see, it was not just the color of the varnish that changed. The thick varnish medieval artisans wanted made way for thinner varnishes as they explored various new ingredients and procedures. It is no coincidence that these changes in varnishing practice are documented in fifteenth-century recipes. Even though these new explorations and experiments with the clarity and viscosity of varnishes were not due to a special invention by Van Eyck, they were driven by the changes in panel painters' ambition instigated by Van Eyck's art and that of his contemporaries. This chapter will show that as the principal aim of panel painting changed from the mimesis of materials to imitating the visible world, painters adapted the material characteristics of varnish to suit its new purpose.

The Strasbourg Recipe Family: Varnish Like Crystal

The first evidence of changes in varnish practice can be found in the so-called Strasbourg family of recipes, named after the oldest manuscript, which was lost in a fire at the Strasbourg Library in 1870. Fortunately its contents have been preserved through a nineteenth-century copy made by Charles Lock Eastlake, now kept in the National Gallery, London (shelf mark 75.023 STR).[4] The manuscripts belonging to the Strasbourg tradition can be dated to the fifteenth and sixteenth centuries and for the most part are written in German. Most of the recipes in these manuscripts pertain to manuscript illumination, but they also contain several recipes for making oil-resin varnishes that cannot be found in the earlier recipe compilations studied in chapter 2. Also part of the Strasbourg family are various printed recipe compilations, the earliest three representing the first printed recipes on art technology: The *Artliche kunste—mancherley weyse Dinten und allerhand farben zubereyten*, first printed in 1531 by at least three publishers; the 1532 *Schreyberey*; and the 1533 *Allerhand Farben* (all three have similar subtitles).[5] The recipes in these popular pamphlets range from how to make pigments and how to write in gold or silver to the method for etching iron and the way to color parchment and feathers. The three treatises are almost identical; only the *Schreyberey* and *Allerhand Farben* contain a number of recipes, including one for painter's varnish, that cannot be found in the 1531 treatise.

A more comprehensive collection of published recipes, also organized around the art of illumination, is in Valentin Boltz von Ruffach's (1515–60) *Illuminierbuch* (1549).[6] Valentin Boltz was not a professional illuminator or artisan, but he was active as a pastor in Württemberg and Basel and was the author of various plays. Boltz acknowledges his inexperience with the matter at hand in the introduction to his *Illuminierbuch* and remarks that because of this work there will be "many resentful and angry artists" (*etliche mißgünstigenn nydige künstler*).[7] They believe that "one should not make these things common knowledge, since it diminishes the art" (*man solt die ding nit gmein machen / zů verkleinerung der kunst*). Boltz writes that he nevertheless did not start his work to anger anyone but expected that those experienced in the art of illumination would improve and add to his work to make it better. He also believes that many "virtuous people" (*tugentreiche menschen*) will find joy in reading about the professions of others, in this way addressing his work to an amateur audience as well. Much of the content of Boltz's *Illuminierbuch*, including his three recipes for making oil-resin varnish, can be traced back to the various recipes in the manuscripts belonging to the Strasbourg tradition.[8] Boltz's emendations and additions to these varnish recipes are especially interesting here, because he provides more information about the function of some of the new materials mentioned in the manuscripts. The *Illuminierbuch* and the earlier recipe pamphlets proved to be in popular demand and were reprinted, translated into other vernaculars, and incorporated into numerous printed recipe collections in the course of the sixteenth and seventeenth centuries. The wide distribution and availability of these treatises, as indeed many of their authors themselves pointed out, made the specialized knowledge of the workshop available to a wider literate public while giving artisans the opportunity to learn about new and established procedures.[9] That such printed recipe compilations became popular precisely when the art of panel painting had just changed its ambition from the mimesis of materials to the mimesis of the visible world makes it all the more interesting to study which varnish recipes were selected for publication and thus eventually became part of the written canon of painterly technique.

Linseed, Hemp, or Nut Oil

The lost Strasbourg manuscript, which dates to the beginning of the fifteenth century, contains two recipes showing that a clear varnish was desirable, perhaps even precious, at this time. The first recipe teaches us how to make varnish by mixing either one part *gemeinen virnis glas* or one part "mastic" (*matik*) with three parts "linseed oil or hemp oil or old nut oil" (*lin ölis oder hanf ölis oder alt nut ölis*).[10] Hemp oil has a slight greenish hue, while nut oil, as I noted at the beginning of this chapter, is much less yellow than linseed oil (fig. 3.1). Also, compared with sandarac or amber, the resins most likely indicated by *gemeinen virnis glas*, the resin mastic is paler and becomes liquid at a relatively low temperature (95 to 120°C for mastic as opposed to 135 to 150°C for sandarac and 250 to 325°C for amber). That the author of this recipe mentions that the nut oil must be old is also important. When oil is allowed to stand for a long time, the impurities present in freshly extracted oil form a sediment, making the oil less turbid. The walnut and linseed oils shown in figure 3.1, for instance, were extracted using a mechanical oil press that expels the oil much the way a traditional oil press does. Immediately after pressing, the oils are extremely turbid, with many small impurities floating around. In a few hours many of these impurities have settled and the oil slowly starts to become clearer. The longer the oil stands, the clearer it will become until eventually it completely loses its initial turbidity (fig. 3.2). In addition, drying oils or varnishes that have darkened from being heated may in time regain most of their original hue. The linseed oil-mastic varnish shown in figure 3.6, for instance, lightened significantly after being stored in the dark for almost two years (compare figs. 3.6 and 3.7 and figs. 3.3 and 3.4).[11]

When you have chosen the oils and resins you want to use, the recipe from the Strasbourg manuscript continues with instructions similar to the methods for preparing varnish discussed in chapter 2, but with one significant difference.[12] First your resin has to be crushed to a fine powder and mixed into the hot oil, which has previously been heated and "skimmed" (*geschumet*, i.e., remove the scum). This "heating" and "skimming" of the oil is similar to the "reduction" of oil or varnish by two-thirds of the original amount that we saw in most other varnish recipes in chapter 2.[13] But adding the resin directly to the oil is different from the procedures used to incorporate resin into oil described by the earlier recipes, where the resin had to be heated and melted before it could be added to the already hot oil. This difference is likely because the Strasbourg manuscript recommends using mastic, which, having a lower melting temperature, would have been easier to incorporate into oil than, for instance, sandarac. The Strasbourg recipe continues that as soon as the resin has dissolved you should keep cooking the varnish, using a "smaller heat" (*kleiner hitze*) while stirring so that it does not burn (*das es nüt an brünne*). When the varnish starts to become "reddish thick like liquid honey" (*gerotted dikelecht werden als zerlossen honig*) it can be tested to see if it is ready. To test the varnish, you take a drop on a knife blade and let it cool (fig. 2.12):

> Grasp with a finger on the drop and slowly pull away the finger and let the varnish form a thread with the finger [*griff mit einem finger uff den troph züch den finger langsam uff und lat der virnis fedemlin mit dem finger uff ziechen*]. Then the varnish is well cooked.[14]

After the varnish has cooled, you strain it, as in the varnish recipes in the previous chapter, through a "thick linen cloth into a clean glazed container." The Strasbourg recipe concludes that this way you have made a "precious and clear varnish, the best one" (*guten edelen lutern virnis den besten*). Prepared at a lower temperature, a varnish made with mastic resin would indeed have been lighter in color than a varnish made with, for instance, sandarac (compare figs. 3.6 and 2.8).

The second recipe in the Strasbourg manuscript that expresses an interest in the clarity of varnish explains how to make a "good varnish that is clear and bright like a crystal" (*guten virnis machen der luter und glantz ist als ein cristalle*). You mix a material called *gloriat*, which can be found "in the apothecary" (*in der apotheken*) with "twice as much oil" (*zwürent als vil ölis*).[15] We are told that this *gloriat*-oil varnish then is cooked and processed in the same manner as the varnish in the previous recipe and also is ready when it pulls up a thread. Boltz gives us more information about the nature of *gloriat* in an almost identical recipe that he included in his *Illuminierbuch*, but that in this case has to be mixed with "linseed oil" (*lynsot ölls*). *Gloriat*, we learn, is the same as turpentine (*Nim Glorien / das ist Terpetin*).[16]

Turpentine is an oleoresin exuded by the different genera of Pinaceae trees. This includes the pines (*Pinus* spp.), which produce "common turpentine"; the European larch tree (*Larix decidua* L.), which produces what became known as turpentine of Venice; and the various species of fir (*Abies*), which produce what later became known as Strasbourg turpentine.[17] These turpentine oleoresins all have slightly different characteristics, and we cannot be entirely certain which was implied by the term *gloriat*.[18] What they have in common, however, is that unlike sandarac or mastic, which quickly harden into brittle tears after exudation and are collected as such, the various turpentine oleoresins are tapped in a semiliquid state owing to their relatively high ratio of volatile components.[19] To blend such oleoresins with oils requires much less heat than blending oils with hard and solid resins (fig. 3.5).

As a result, oleoresin-oil varnishes do not become as dark as the varnishes that have to be prepared at much higher temperatures, such as those made with amber or sandarac, or even mastic (compare figs. 3.5 and 3.7). A downside to these varnishes is that oleoresins take much longer to harden than solid resins do. Other materials were sometimes added to make these varnishes form a hard layer faster. Such ingredients are not mentioned in the recipe for *gloriat*-oil varnish in the Strasbourg manuscript; but, significantly, a closely related varnish recipe in Boltz's *Illuminierbuch* does include two materials that would have been suited for this purpose. To make Boltz's *gloriat*-oil varnish, turpentine and linseed oil have to be heated together and "thoroughly skimmed" (*schumbs wol*).[20] When this is done you add "mastic" (*mastix*) and burned bone (*gebrenten beins*). Several other varnish recipes

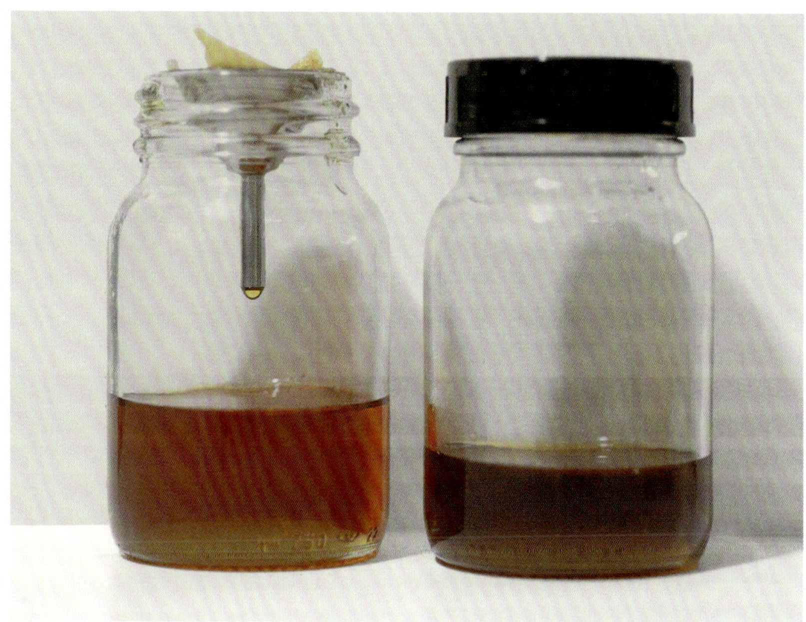

FIGURE 3.5. Linseed oil-mastic varnish (3:1). *Left:* varnish cooked at 170°C; *right:* varnish cooked at 220°C (both for three hours). Photo by Marjolijn Bol.

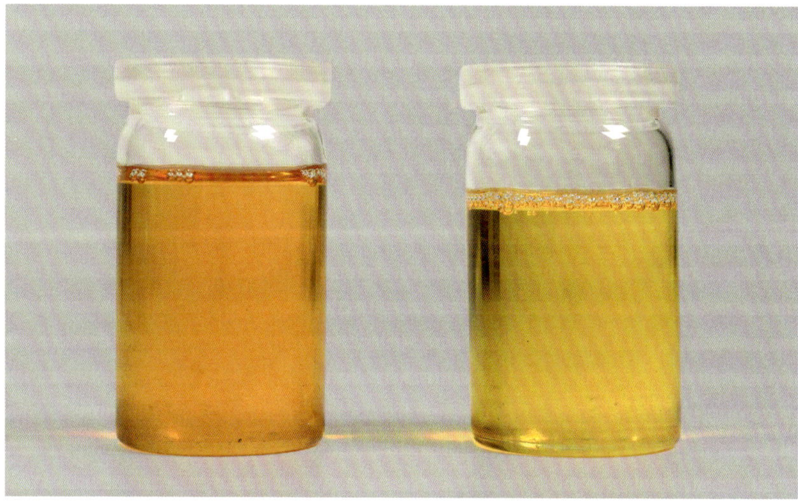

FIGURE 3.6. The same linseed oil-mastic varnishes after being kept in the dark for two years. Like the drying oils and the sandarac varnish, they became significantly lighter. Photo by Marjolijn Bol.

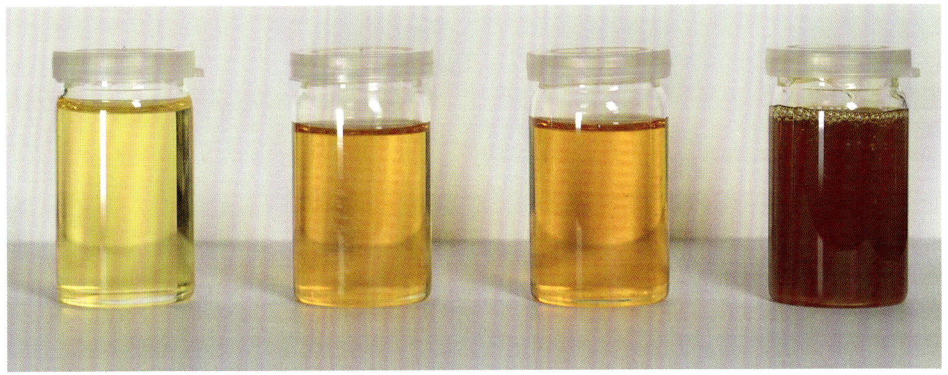

FIGURE 3.7. *Left to right:* cold-pressed linseed oil; cold-pressed linseed oil-oleoresin varnish cooked at 130°C; turpentine oil-oleoresin varnish cooked at 130°C; and linseed oil-sandarac varnish cooked at 245°C (which has become significantly lighter after two years of storage; compare fig. 2.8). Photo by Marjolijn Bol.

in Boltz's *Illuminierbuch* and the Strasbourg manuscript supply more details about the supposed function of mastic and bone ash in oil-resin varnishes.

Bone Ash and Mastic

The Strasbourg manuscript explains that you have two choices in preparing a varnish with bone ash: it can be made with "old hemp oil" (*alt hanf öli*) or, "if you want to have it strong" (*wiltu in aber stark*), it can be prepared with hemp oil and "mastic" (*mastik, 4 lot*) or "turpentine" (*terpentuum, 6 lot*).[21] Hemp oil is expelled from the seeds of varieties of *Cannabis sativa* L., and it has a deep green hue (fig. 3.1). Compared with linseed oil, hemp oil dries relatively slowly.

The Strasbourg recipe explains that first you heat and "skim" (*schum es*) the oil, then you take equal parts "pumice" (*bimses*)[22] and "calcined bone from an old knuckle-bone" (*gebrentes beines eines alten knorren*). These ingredients are added to the hot oil while continuously "skimming it carefully" (*schum es recht wol*). If you decide to make the stronger version of the varnish, either mastic or turpentine must be stirred in when the oil is "boiling hot" (*siedendig heisse*) until the substance "becomes sticky like a thread" (*gerat zech werden als ein faden*). After being cooked this way, the varnish should be put in the sun for two days. This last step would have helped the impurities in the varnish settle out, aided also by adding pumice stone (porous volcanic rock) and bone ash. Pumice and bone ash attract the water component in the oil. If the oil or varnish is allowed to stand after these ingredients have been added, they form a sediment at the bottom of the vessel. The oil poured off from this sediment is clearer and has better drying properties.[23] Finally, exposure to sunlight also makes the oil component in the varnish significantly paler (fig. 5.2). Recipes show that exposing oil to sunlight was common practice when preparing oils for grinding with pigments—and I will discuss this further in chapter 5.

Boltz's *Illuminierbuch* also contains a recipe for making a varnish with bone ash that, according to its title, was used "on parchment or leather" (*Fürnüss off perment oder låder*).[24] Just like the recipe in the Strasbourg manuscript, it is divided into two parts, but instead of instructing readers how to make varnishes of different strengths by making either a pure oil varnish or one mixed with resin, Boltz explains how to make a relatively slow-drying oil-resin varnish or a relatively quick-drying one.

The first part of the recipe explains how to make the slow-drying varnish with three parts "linseed oil that is clear" (*leinöle dz luter syg*) to one part pulverized mastic; if you do not have linseed oil you can use the same amount of "old nut oil" (*alt nussöl*) or "hemp oil that is very clear and beautiful" (*hanfföl das gar luter und schön syg*). Boltz first explains that the oil has to be skimmed (*schums*) while cooking it, as in the Strasbourg recipe. When your oil has thus been properly "skimmed and boiled" (*verschumt und verwallet*), the pulverized mastic can be added, and the varnish is then heated on a lower fire until the resin has melted. Again, the varnish is ready when a small drop that has cooled allows you to pull up a thread with your finger. The second part of Boltz's recipe goes on to explain how to prepare this varnish so "that it dries quickly" (*daß er bald trucknet*). To give the varnish better drying properties, you take the "bone of a sheep" (*schaafsbein*) and calcine it in a closed pot over a strong fire for two hours. When this is done you "grind it like flour" (*stoß es wie ein meel*), sieve it, and "stir the size of a nut into the hot varnish" (*rühr es einer nüß groß in dem heissen firnis*). According to Boltz this produces a varnish that "dries very quickly on everything you apply it to" (*so trucknet es gar bald / worauf du ihn streichelt*).

Calcined bone ash (a compound of calcium, including calcium oxide) is a secondary dryer. This means that on its own it does not have much of a drying function in oil but in the presence of a primary dryer, such as the various lead compounds, calcium compounds function as a catalyst.[25] In this recipe, however, no primary dryers are mentioned and, as I suggested above, the siccative effect of calcined bone ash may likely be attributed to its ability to bind the water present in unrefined painter's oils.[26] That painters explored a variety of materials for this purpose becomes clear from another German manuscript, which belongs to the same recipe tradition: the *Liber illuministarum.*

Alum and Bread

The *Liber illuministarum* (Cgm 821, Bayerische Staatsbibliothek, Munich) was compiled at the monastery at Tegernsee between the second half of the fifteenth century and the first decades of the sixteenth.[27] Next to various recipes for making oil-resin mixtures that record the earlier varnishing practices discussed in chapter 2, the *Liber illuministarum* contains two unique varnish recipes that cannot be found in any of the earlier treatises. These recipes also point toward a new search for clearer, lighter, and better-drying varnishes.

The first recipe introduces a new method for cooking the oil before it is mixed with a resin called *glas*. The oil must be cooked with nine slices of "house bread" (*hausprot*) and a material called *alaun* (alum) until the bread turns brown.[28] In another recipe the *Liber illuministarum* explains the role of bread in preparing this oil-resin varnish. The recipe describes how to make a gold-colored varnish for tin by mixing together the following ingredients: *fuerneis* (just as in the earlier recipes—most likely sandarac or juniper resin); *mini* (minium, red lead-based pigment); and *leinoell* (linseed oil) or *magoell* (pale, slow-drying oil made from the seeds of the poppy family—Papaveraceae Juss.). While mixing these materials, a slice of bread is added because it "attracts impurities present in the oil" (*ain schnitel prot quia attrahit immundiciam olei*).[29] Alum would have been added for a similar reason. The appellation "alum" was used in the past to denote the various chemical compounds of aluminum. Of these, it most commonly referred to the double sulfate of aluminum and potassium, today also known as "potash alum."[30] Alum can bind suspended mucilages in the oil.[31] Adding alum and bread thus helped bind the impurities, and after they formed a sediment, the clarified oil could be poured off. As with the calcined bone ash and ground pumice stone mentioned in the recipes from the Strasbourg manuscript, cooking the oil with alum and bread not only makes it clearer but, by removing suspended mucilaginous materials, helps it dry better.

The second recipe in the *Liber illuministarum* that is of interest here explains how to make a varnish that does not need the sun to dry. Unfortunately the recipe is incomplete because a folio is missing from the manuscript. The part of the recipe that did come down to us shows that the lost text must have explained how to prepare a drying oil before it can be mixed with "mastic" (*masticem*). The first remaining lines of the recipe read:

"When it has settled at the bottom, you sieve it through a linen cloth, and to this oil you add some of the mastic and cook it" (*ad fundum decenderi tunc cola perpannum lineum et de oleo illo funde ad masticem et mitte bullire*). The purified oil and mastic are cooked until they have "united into one mass" (*una materia simul mixta*).

Since the first part of the recipe is missing, we can only speculate how the oil used to make this varnish was prepared. But its mentioning that the oil is strained through a cloth after a substance has settled at the bottom of its container does suggest some additional method of purification. The author of the recipe finally observes that when this oil-mastic varnish has cooled "it becomes a very good varnish that dries without the sun and is twice as good as any other" (*erit optima vernisa que siccatur absque sole et in duplo plus valet quam aliud*). As we will see later on in this chapter, having varnish that can dry without the sun becomes increasingly important in the course of the sixteenth century.

Painter's Varnish for Scribes and Illuminators

Unlike the earliest varnish recipes discussed in chapter 2, the Strasbourg family of recipes does not contain any information about how, or even if, oil-resin varnishes ought to be applied to panel paintings or other wooden art objects.[32] The most obvious reason may be that the recipes belonging to the Strasbourg tradition principally record the practice of scribes and illuminators. The recipes for viscous, slow-drying oil-resin varnishes discussed above were of course not suitable for varnishing the delicate paintings in manuscripts. Instead, to give luster to paintings on paper and parchment, recipes in the Strasbourg tradition explain how to make varnishes from egg white, gum, and honey.[33] So why did these illuminator's recipe compilations include so many instructions for making oil-resin varnishes? To solve this problem, we have to study another group of recipes that discuss the role of varnish in making colored parchment see-through.

The Strasbourg manuscript contains two recipes that explain how to use varnish when making parchment translucent, one of which follows immediately after the three recipes for making oil-resin varnishes.[34] These recipes explain how to make green parchment so that "one can see through it what one wants as if through glass" (*das man da dur sicht was man wil als dur ein schön glas*).[35] To make such a green-glass substitute, one should first search for "the clearest parchment" (*des lutersten megdenbermenten*). If you then want to color the parchment "a beautiful, fine green" (*schön fin grun*) so that "you can see through it as you can through beautiful glass" (*man da *dur sicht was man wil als du schön glas*), you should use vinegar to grind verdigris and a dye used to color leather. The parchment is left in this liquid overnight to take on a green color, then it is stretched on a frame and coated on both sides with a drying oil or a "clear varnish made of mastic" (*luter virnies das usser mastikel gemachet sÿ*) until it looks quite "shiny" (*glantz*).[36] Like varnished panel paintings, the varnished parchment had to be dried in the sun: "Place it in the hot sun and let the parchment dry thoroughly" (*so setz es in eine heisse sunne und las das bermit wol truken werden*).[37] Once dry, the sheets can be removed from the frame, "cut into three or four parts, and kept in a book or press to keep them flat."[38] The clear mastic varnish this recipe refers to is likely one of the mastic-varnish recipes discussed above, which were indeed described as *luter*.[39] That these translucent parchment windows were commonly colored green may also explain why the Strasbourg oil-mastic recipes recommend hemp oil as an alternative to linseed oil and nut oil (fig. 3.2). With its green hue, hemp oil would have enhanced the color of the parchment, which had to resemble contemporary green, translucent glass (compare figs. 3.8 and 3.9).[40]

The mystery of how scribes might have used these pieces of see-through parchment is solved by another group of recipes that, somewhat confusingly, are not always included in the treatises that explain how to prepare the green parchment. One of these recipes is in the so-called *Amberger Malerbüchlein* (Amberg, Staatlichen Provinzial Bibliothek, MS 77, ca. 1492). It explains that parchment made green and see-through according to the methods above will soothe readers' tired and strained eyes.[41] Perhaps even more spectacular, it was believed to magnify text:

> When you lay the parchment over some writing, that which is small will appear to be large, and what one reads or studies through it does not harm the eyes and it keeps the sight fresh and secures the eyes [*Wer daz permait legt auff ein schrifft wÿ chlain sÿ ist so scheint sÿ groz und wÿ vil man dar durch list oder studirt so schat es den augen nicht und behat das gesicht in schoner frist und sichert dy augen*].[42]

To ease tired eyes by reading through pieces of green, translucent parchment, the varnish used to clarify such parchment, as the recipes in the Strasbourg tradition indeed suggest, has to be perfectly clear as well. Besides these interesting contrivances that use the translucent green parchment to magnify text and protect sight, the search for clear varnishes might have also been related to one of its other uses: as a substitute for window glass. These recipes represent a separate group that again first appears in the Strasbourg

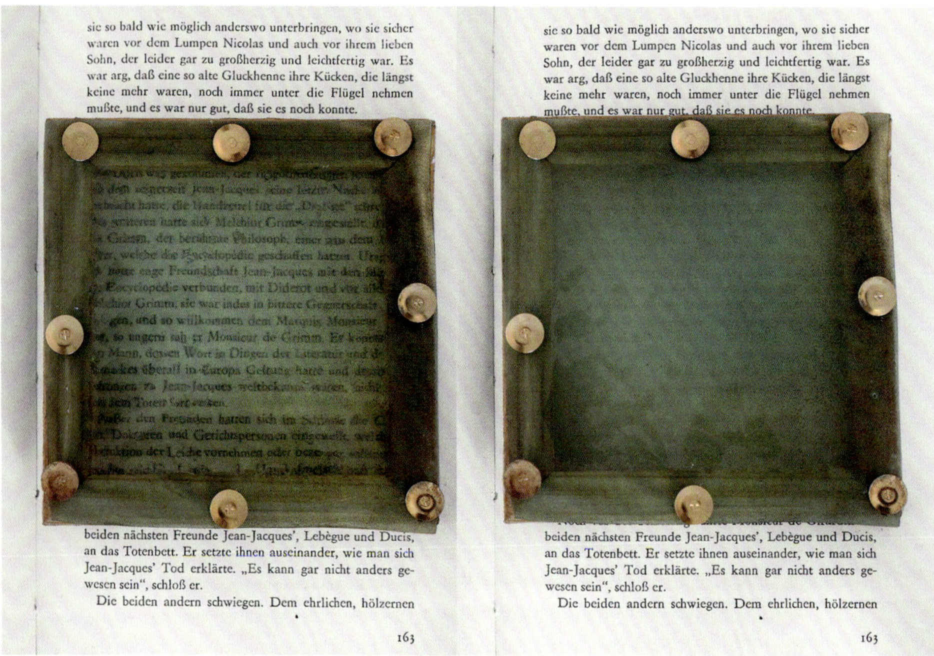

FIGURE 3.8. Reconstructions of see-through parchment meant to look like green glass: *left:* colored parchment before oiling with linseed oil; *right:* colored parchment after oiling with linseed oil (parchment was colored green by soaking it overnight in a bath of copper green dissolved in vinegar). Photo by Marjolijn Bol.

tradition. The procedures for making parchment windows are similar to the techniques for making eye screens. That this concerns a different series of recipes becomes clear only from the terminology used to describe the end product. Instead of the word *schirmm*, this translucent parchment is consistently described as *venster glaz* ("window glass"). Recipes show that this parchment *venster glaz* was sometimes also made in colors other than green or was decorated with colorful designs.[43] The Elizabethan scholar Hugh Platt (1552–1608) elaborates on this practice in his *Jewell House of Art and Nature* (1594).[44] From a recipe on how "to make Parchment clear and transparent to serve for divers purposes," we learn that we should procure the finest and thinnest parchment and oil it "with a pensill, with the oile of sweet Almonds, oile of turpentine, or oile of spike; some content themselves with linseed oile." When this is dry, it will "shew very cleere, & serve in windowse in steade of Glasse, especially in such roomes as are subject to overseers." Before being varnished, the transparent, colorless parchment

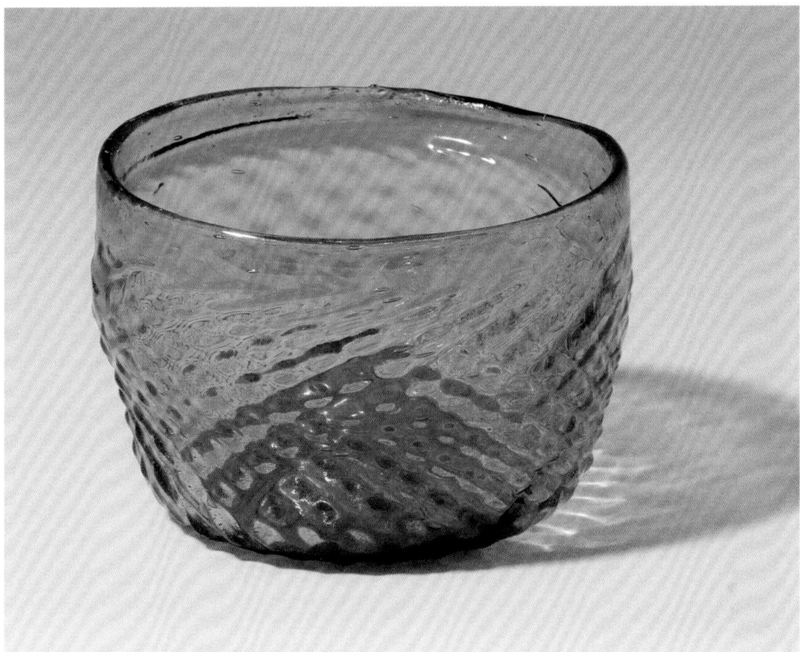

FIGURE 3.9. Beaker (*Maigelein*), late fifteenth century. Mold-blown glass. Metropolitan Museum of Art, New York. Public domain.

can also be decorated with a design. Cut into square panes and stretched onto a frame, the parchment windows will "keep the room very warm" and "will indure the blustring and stormy weather much better then [*sic*] paper." Platt's rather derogatory remark about using linseed oil, which "some content themselves with," shows that it was indeed not the preferred binding medium for this practice.

That the varnish recipes included in the Strasbourg manuscript may represent a special category, closely tied to the practice of varnishing parchment to make it see-through, is further supported by the three printed pamphlets I mentioned at the beginning of this chapter. The 1531 *Artliche Künste* explains how to make colored, see-through parchment by covering it with a thin coat of "painter's varnish" (*maler vierneß*).[45] After this is done, the varnished parchment has to "dry in a place where there is no dust" (*laß es trucken werden an einer stedt da es nit steübet*). The 1531 *Artliche Künste* does not explain how to make this painter's varnish, but—directly after their recipe for making see-through parchment—the 1532 *Schreyberey* and the 1533 *Allerhand Farben* do include a recipe titled "How to Make Painter's Varnish"

(*Maler firnis wird also gemacht*).[46] To make painter's varnish you cook five pounds of "linseed oil" (*lein öl*) with half a quarter pound (*ein halben vierdung*) of finely ground and sieved *atgstein* (a German term for *Bernstein*, i.e., "amber") and the same amount of *venediisch glaß*. These ingredients are cooked for "three hours, or until it is strong enough according to your liking" (*und laß es drey stund syden / oder so lange biß es starck genug wird nach deinem gefallen*). Variations of the word *glassa*, as we saw in chapter 2, have been used since the time of Theophilus to refer to resins used to make varnish.[47] But because Venetian glass is known to have been used since the sixteenth century as a dryer and optical filler in oil paint, references to *venediisch glaß* in these early varnish recipes are also sometimes interpreted to refer to crushed, colorless glass.[48] If *venediisch glaß* was added as an optical filler or dryer in this recipe, its author does not show any awareness of it. Rather, the recipe mentions another material that may be used as a dryer. To prepare the painter's varnish so that it "dries faster" (*bald drucken*), you add *sylbergledt*. *Sylbergledt*—or litharge (lead oxide) as it is known today is a lead compound that acts as a siccative in drying oils by accelerating their uptake of oxygen and thus their autoxidation.[49] In addition to including lead as a dryer, this recipe stands out for another reason. Most of the medieval varnish recipes mention fairly high ratios of resin to oil—three parts oil to one or two parts resin being most common. Even if we interpret both *atgstein* and *venediisch glaß* to refer to resin—and it seems likely that we should—the ratio of oil to resin is nothing less than twenty to one, implying a relatively thin varnish.

The recipes contained in the Strasbourg family of manuscripts are the earliest indication that the medieval practice of varnishing with dark and viscous substances might be slowly changing. The varnish recipes contained in the various related manuscripts and published treatises seem to indicate that artisans were looking for paler resins and oils, sought methods for clarifying drying oils of impurities, and actively searched for ways to give their varnishes better drying properties. But since the Strasbourg varnish recipes were likely recorded as part of the practice of producing see-through parchment—and might therefore record a particular varnishing practice developed to this end—we cannot be certain that they convey changes in the practice of easel painters. It is nevertheless quite likely that the two artisanal traditions developed alongside each other. That this may indeed have been so is suggested by another group of varnish recipes, dating to the end of the sixteenth century. These recipes show that many of the materials and procedures mentioned in the Strasbourg family of recipes were used by painters as well.

Secreti Diversi: Clear and Clarified Oils and Resins

The first changes in the practice of varnishing easel paintings or, at the very least, a newly expressed interest in relatively clear and quick-drying varnishes, are evinced in two late sixteenth-century manuscripts and a printed treatise dating to the same period. The first is an Italian manuscript that bears the title *Segreti d'arti diverse* but is better known as the Marciana manuscript (Venice, Biblioteca Marciana, MS It. III.10). Mary Philadelphia Merrifield, who edited the Marciana recipes as part of her *Original Treatises*, dates the manuscript to the first quarter of the sixteenth century, but in their more recent edition Fabio Frezzato and Claudio Seccaroni place it in the second half of the sixteenth century.[50] The Marciana manuscript contains no fewer than twelve varnish recipes. Some discuss the new methods for making varnish described above, while others tell how to make solvent-based varnishes that have no precedent in earlier compilations and treatises.

The second manuscript of interest, now kept in the National Library of France, Paris (MS Fr. 640), contains various recipes that bear a strong resemblance to those included in the Marciana manuscript, but it also gives more details on the why and how of these new varnishing practices. This work, written in 1580s Toulouse, stands out for the practical nature of many of the processes it describes, which the author declares he himself has experimented with.[51] Internal evidence in the manuscript suggests that, rather unusually, the author really does appear to have engaged with the recipes he recorded. Equally interesting remarks on Italian varnishing practice are found in a third related work: Armenini's 1586 *De' veri precetti della pittura*, dedicated to "beginners, to scholars, and to lovers of fine arts" (*i principianti, a gli studiosi, et alli amatori delle belle arti*).[52] The varnish recipes contained in these works do not just describe how to procure liquid, clear, pale varnishes; they also explain why this had become the primary concern of oil painters in the sixteenth century.

Mastic, Greek Pitch, and Alum

Of the twelve varnish recipes in the Marciana manuscript, only one specifies that it is suitable for varnishing oil paintings: "A most excellent clear and drying varnish proper for colors, both in oil-painting and in other kinds of painting" (*vernice ottima chiara et diseccativa ben per colori et a olio et per ogni dipintura*).[53] The recipe recommends taking the best and clearest ingredients: two parts "clear and good nut oil" (*dolio di noce chiaro et bello*), one part "clear and good Greek pitch" (*pece greca chiara et bella*), and one part "clear and good mastic" (*mastice chiaro et bello*).[54] You first separately grind the Greek pitch and the mastic to a fine powder. Then, when the nut oil has been cooked "until one-third has evaporated" (*tanto che scemi el terzo*), you put in your powdered Greek pitch "a little at a time." MS Fr. 640 gives a detailed explanation of what the appellation *pece greca* in this recipe most likely referred to. We learn that *pix græca* is also known as "colophony" (*colophoine*), and that this is the solid residue that remains when distilling "turpentine" (*tourmentine*) to make "turpentine oil" (*dhuile de tourmentine*).[55]

From left to right, figure 3.10 shows the three materials the recipe is talking about; the turpentine resin obtained after tapping the tree; its distillate, known as oil of turpentine; and the residue after distillation, known as Greek pitch or colophony (sometimes also called rosin). The varnish recipe

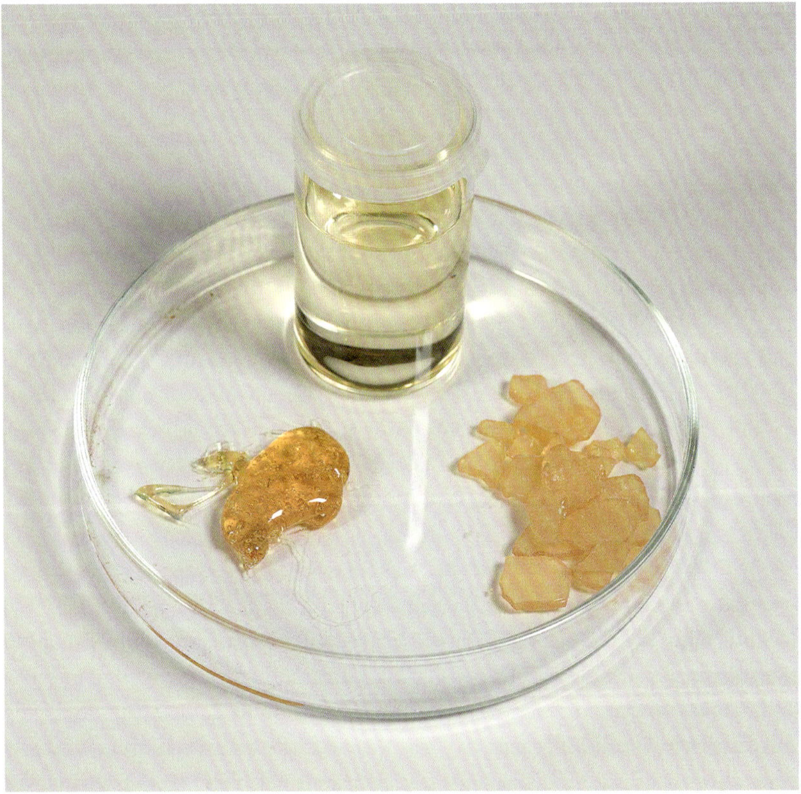

FIGURE 3.10. *Left to right:* oleoresin, distilled turpentine oil, and colophony. Photo by Marjolijn Bol.

in the Marciana manuscript continues by warning that only when your Greek pitch has been fully incorporated into the nut oil can powdered mastic be added in the same manner. As soon as all of this has dissolved, you take the varnish from the fire and "strain it through a fine and old linen cloth." Having thus carefully prepared the varnish with the clearest of ingredients, you can further improve its clarity by cleaning and selecting your mastic and by cleaning your varnish with alum:

> If you wish your varnish to be still clearer, prepare the mastic with tepid water in the following manner: Take the largest and clearest tears of mastic [*grani del mastice più grossi et chiari*] you can find and soak them in tepid water so that they become tender; then select the best pieces, dry them, and pound them.
>
> You may also try the effect of adding a little burned and pulverized rock alum [*allume di roccha arso polverizzato*] when the other ingredients are dissolved, so that the whole may virtually be seasoned with it, straining it afterward. This is done to purify it better [*Questo si fa per farla più purgata*].[56]

Alum, we have seen, was also mentioned in the sixteenth-century century *Liber illuministarum*, but the Marciana manuscript details for the first time its supposed role in removing impurities in oil-resin varnishes. In his *Veri precetti della pittura*, Armenini includes a similar varnish recipe in which he also observes that a little fine dust of alum is used to improve clarity when it is boiled with the varnish (*più lustra se vi si getta che bolle un poco di allume di roccha abbrugiato*).[57] The varnish also has to be made from the clearest of materials: "white, lustrous mastic" (*mastice che sia bianco, & lustro*) and "clear oil of walnut" (*oglio di noce chiaro*). Armenini explains that besides its use as a coating for paintings, this varnish can be used with some colors, such as blues and lakes, and "will make them dry faster." A comparable recipe in the Marciana manuscript provides an interesting list of the various qualities that make up the previous oil-mastic varnish, which is here made with linseed oil instead of nut oil: "A varnish which spreads like oil, dries quickly, and is very lustrous and beautiful, appearing like a glass mirror, and which is admirable for varnishing lutes and similar things" (*Vernice che distende come olio et seccha presto et è molto lustrate et bella et pare uno specchio di vetro et per stare a la cosa et sopra liuti et simile cose è mirabile*).[58] According to the Marciana manuscript, this quick-drying, lustrous varnish that spreads like oil is made this way: Mix two parts linseed oil (the recipe mentions taking one *libra*, meaning one pound) that has previously been cooked "in the proper manner" (most likely meaning reduced on the fire) to one part "pulverized clear and fine Greek pitch" (*pece grecha chiara et bella polverizzata*). When the Greek pitch has dissolved completely, one part "powdered mastic" (*mastice macinato*) is added and dissolved as well. When all of this has united, "some burned and pounded rock alum the size of a nut" (*quanto una noce di allume di roccha arso pesto*) should be added and mixed until it is entirely dissolved and incorporated. After the varnish has been strained "through an old linen cloth," it is ready for use. This varnish is recommended for many things, such as wood, iron, paper, leather, and all kinds of painting, and it also "withstands water" (*et per stare alla aqua*). We learn that if you find this varnish too viscous, you can dilute it with linseed oil "in the proper manner." These recipes, in contrast to the earlier ones, suggest that by the end of the sixteenth century artisans sought less viscous varnishes to cover their work.

The Marciana manuscript includes another varnish recipe that gives more details about how an oil-resin varnish should be strained from its sediment after it has been cooked. The recipe calls on the expertise of "Master Jacopo de Monte San Savino, the Sculptor," who has tried this varnish himself.[59] Jacopo Sansovino (1486–1570) was an influential sculptor and architect who is best known for bringing the style of the High Renaissance to Venice, one of his masterpieces being the library at the Piazzo San Marco, today known as the Biblioteca Marciana.[60] It is therefore interesting that this recipe records the famous sculptor as an authority on varnish, which it says is good "for every kind of work and on all materials." To make this varnish, you add one part *vernici in grana* (likely sandarac)—ground to a fine powder—to three parts "clear nut oil" (*olio di noce chiaro*). The nut oil is to be cooked "over a slow fire in the same way linseed oil is boiled" (*quoci l'olio lento igne in pignattino invetriato, come si cuoce l'olio di lino*). The method of cooking linseed oil for this purpose can be found in another recipe in the Marciana manuscript that explains how to make a mordant for applying gold leaf to glass.[61] We learn that linseed oil must be purified this way: "Boil it over the fire with water for three or four hours, then let it settle and separate it from the water" (*l'olio di lino si purifica cosi: fallo bollire al fuoco tre o quattro hore con l'acqua, poi lascialo riposare poi separalo dall'acqua*). After your oil has been purified in this manner, the varnish recipe in the Marciana manuscript continues, noting that sandarac must be added, and "so much clear incense finely powdered as will impart a pleasant savor" (*modo tanto incenso chiaro polverizato sottile, che condischa discretamente tutta la materia*).[62] It concludes that adding alum and exposing the varnish to the sun helps form the sediment that is necessary to clarify it:

> If you please, you may also add a sufficient quantity of burned and pounded roche alum to have a sensible effect on the whole composition; and adding alum will improve the varnish if you stir it until it is dissolved. It should then be strained through a linen cloth and afterward exposed to the sun and open air [*al sole, et al sereno*[63]] until a sediment is formed, which should be separated by pouring off the varnish, after which it will be ready for use.[64]

Water Washing

The nut oil used to make the varnish in the previous recipe was said to be purified by cooking it with water. When oil and water are mixed and allowed to separate, the water draws out some of the aqueous matter present in the oil. Recipes suggest that using water to clean and clarify oil became popular in the seventeenth century, but scattered evidence from earlier treatises such as the Marciana manuscript suggests that it must have been known earlier. One of the earliest examples in this respect is in a varnish recipe from the *Mappae clavicula* tradition that, as we saw in chapter 2, recommended adding rainwater to the varnish while cooking it.[65] But one of the most elaborate descriptions of the process of water washing oils to clarify them, here without the use of heat, is in the 1557 edition of Alessio Piemontese's *Secreti.*

First published in 1555, this immensely popular work saw over ninety editions and was translated into many languages, including Latin, French, English, Dutch, German, Spanish, Danish, and Polish.[66] William Eamon argued that "Alexis of Piedmont" can likely be identified with the humanist Girolamo Ruscelli (1504–66), who claimed authorship of the *Secreti* in his 1567 *Secreti nuovi.*[67] In the introduction to the previous work, Ruscelli explained that a secret academy, of which he was a member, collected the *secreti* from books, from manuscripts, and from practitioners by word of mouth. One of the goals of the academy was to try out all the recipes they had found: "First, during all those years, we continually experimented on all the secrets that we could recover from printed books or from ancient and modern manuscripts."[68] The 1555 edition was sold out within a year, and the popular work was reprinted by five different Italian publishers in 1557. These 1557 editions first include the instructions for clarifying linseed oil to make the costly blue pigment ultramarine out of the semiprecious stone lapis lazuli (*purificare l'oglio di semelino per l'azurro oltramarino*). The clarified linseed oil is an ingredient of the "pastilles" used to extract the pigment from lapis lazuli, which, in addition to linseed oil, were made with various other ingredients typically used to make varnish as well. As with the varnish recipe in the Marciana manuscript, these ingredients must be the clearest you can obtain (here rendered side by side in the original Italian and in a 1595 English translation): "clear and neat turpentine" (*trementina chiara & netta*), *ragia di pino bella*,[69] "faire Pix Graeca" (*pece greca bella*), "faire masticke" (*mastice bello e netto*), and "cleare newe waxe" (*cera nuova lustrante*).[70] The recipe thus suggests that to keep the hue of this expensive blue pigment pristine, artisans considered it important to use clear and colorless materials. As soon as these ingredients have been melted and well incorporated, the purified linseed oil—previously prepared—is added. The linseed oil is purified as follows:

> Take what quantities of oyle of line seede you shall thinke good, so that it be faire and cleare, of a yellow colour like Golde [*bello e chiaro del color croceo, cioe color d'oro*], and put it in a horne of glasse, or in an ore horne that hath a hole in the bottome [*e quella quantita' che a te pare, e mettilo in un corno di vetro over die buc*], and put upon it some fresh water, and stur it well with a stick, then let it stand still a little while, and open the hole underneath, and let out the water, doing so seven to eight times, or until the water come out as cleare, as when you did put it in: and in this manner men purifie the said oyle, then keep it in some bessell or glasse, against you have neede of it. . . . And note well, that when you heare speak of oyle, it is of this purified oyle.[71]

Reconstructions of Alessio Piemontese's recipe show what happens when unrefined cold-pressed linseed oil is mixed with water. After being stirred with the oil, the suspended proteinaceous solid matter settled out from it (figs. 3.11 and 3.12). This helps clean the oil and, depending on the amount of water used, also makes it lighter.[72] Mixing oil with water, like adding bread, alum, or bone ash, was therefore yet another method, perhaps even the most ancient one, for creating a purer, clearer oil.

The First Solvent-Based Varnishes

MS Fr. 640 stands out from the previous treatises because, for varnishing easel paintings, it completely does away with the traditional oil-resin varnishes. One of these new varnish recipes for painters explains how to make "mastic varnish dry in a half hour" (*Vernis de mastic sec en demye heure*).[73] The recipe resembles those in the Marciana manuscript and the treatise of Armenini discussed above, but here nut oil has been replaced with turpentine oil and wine spirits. The author of MS Fr. 640 explains that you heat

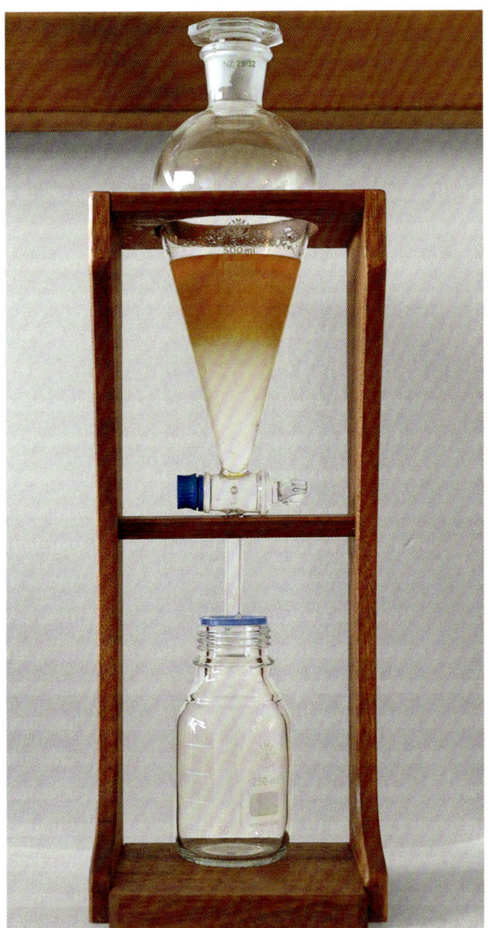

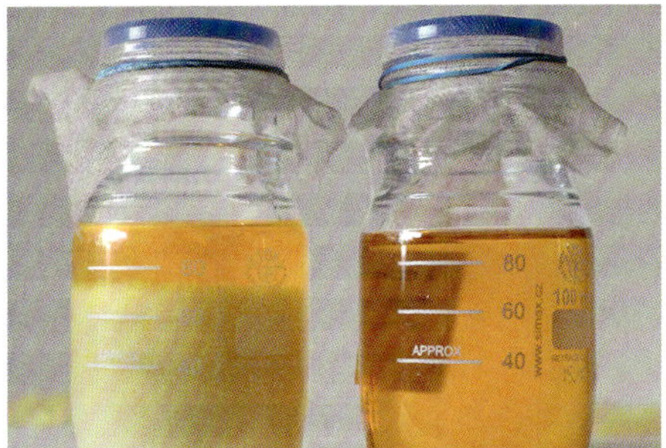

FIGURES 3.11 AND 3.12. Alessio Piemontese's method for water washing oil. Figure 3.11 (*left*) shows a separating funnel with the oil separated from the water after first being mixed with it. Figure 3.12 (*right*) shows (*left to right*) regular cold-pressed linseed oil and linseed oil that has separated from its mucilaginous materials after being mixed with water that was siphoned off nine times. After several days, more oil comes floating to the top. Photos by Marjolijn Bol.

together turpentine, turpentine oil, and *eau de vye*, and as soon as this liquid has become "very hot" you put in some "subtly ground mastic." When all of this has united, you try the liquid on a knife and add turpentine oil if it has "too much body" (*trop de corps*) or mastic if it is too thin. In a marginal note that appears on the same folio, the author of MS Fr. 640 describes how to determine whether your varnish has the right consistency: "You know this varnish does not have body enough when it does not take well on a panel in oil, for it is like water" (*On cognoist que ce vernis nha point de corps asses quand il ne prend pas bien sur le tableau a huile car il est co[mm]e eau*). Replacing linseed oil with turpentine oil and wine spirits—both solvents that decrease the viscosity of resins when liquid—would indeed produce a varnish much thinner than traditional oil-resin varnishes.

Having thus explained how to cook a quick-drying mastic varnish for paintings, the recipe continues with an elaborate description of how to procure the clearest mastic, which has to be "white and purified of any dirt and dust and black dross" (*blanc & purifie de toute terre & poulsiere & mache noire*). Like the instructions in the Marciana manuscript, MS Fr. 640 explains that you must choose your mastic carefully or, to make it even better, wash and dry it before using it in the varnish. The reason, we learn, is that impurities in the mastic "will remain within the varnish, and when you set it on white or flesh color, it will look like fleas and blemishes" (*demeureront aulx vernis & co[mm]e tu lassoyras sur quelque blanc ou carnation il semblera que ce soient pulces & taches*). Most of the mastic that can be purchased today has already been sorted and is therefore relatively clean, but on close inspection the small dark impurities the author of MS Fr. 640 talks about can still be seen in some of the mastic tears, and especially in the sandarac resin depicted in figure 2.1. As soon as the resin dissolves, these impurities would indeed have ended up in the varnish liquid. Applied to a painting they would have shown up as small black specks on the lighter parts of the painting or, in the words of the author of MS Fr. 640, look like fleas.[74] Mastic is not soluble in water, but it does swell up when immersed in it. Soaking the resin may therefore have helped release some of the impurities at its surface.

The recipe in MS Fr. 640 continues that once you have selected clear (or purified) mastic, it has to be pulverized very finely and passed through a sieve.[75] The finest tears of purified mastic are particularly suited for "small works" (*petits ouvrages*). Various other marginal notes on the same folio explain that this varnish "almost dries as you work" (*Il deseiche presque en travaillant*), has to be "laid down cold on the panel with a very clean fingertip, and should be spread vigorously" (*Ce vernis se pose a froid sur le tableau avecq le bout du doigt bien net & le fault estendre bien vivement*). "One lays the varnish with a finger so as make a thin layer; because when thick, it yellows" (*On couche le vernis avecq le doigt affin de le coucher maigre Car estant espes il jaulnist*).[76] The turpentine oil-mastic varnish was not perfect for all purposes, however. Because mastic resin reacts to moisture, it is prone to "blooming," meaning the varnish may develop a cloudy, whitish surface. The varnish described here, therefore, does not resist moisture very well. This fact is noted on various occasions by the author of MS Fr. 640, and a rainproof varnish, we learn, should instead be made from a drying oil and *rousine*: "Mastic varnish does not resist rain, whereas that of oil and *rousine* does."[77]

Thin and Lustrous

Like the author of MS Fr. 640, Armenini believed "the thinnest and most lustrous" (*la più sutile, e più lustra*) varnish had to be made without drying oils.[78] To make it, painters melted *oglio d'abezzo chiaro* in a little pot over a "slow fire" (*lento fuoco*). When this had thoroughly liquefied, an equal quantity of *oglio di sasso* was added. *Oglio di abezzo* refers to a turpentine oleoresin and may be the same as Strasbourg turpentine (resin from fir). The appellation *Oglio di sasso* refers to the crude oil, also known as petroleum, found beneath the earth's surface—such as the famous oil of Modena in the seventeenth century—which in some places can be found seeping from the earth.[79] Sometimes the volatile component in such oils is so high that they might be a natural distillate.[80] Petroleum had been found in Europe since the Middle Ages and was used for lighting or for medicinal purposes and, in the case of varnishing, was used to thin varnishes, as was oil of turpentine. Armenini explains that the mixture of *oglio d'abezzo* and *oglio di sasso* is blended by hand while it is still warm and then spread uniformly over the work, which has previously been heated in the sun (*mesticanodo con la mano così caldo stendevano sopra il lavoro prima posto al Sole*).[81] He greatly admires the skill necessary to varnish a painting by what he considers this traditional method: "I have seen it used in this manner by the most skillful artists throughout all of Lombardy, and I was told that Correggio and Parmigianino used this type of varnish in their works, if one believes their disciples."[82]

But however much Armenini may have still esteemed the method of varnishing in the sun, some recipe authors start by cautioning against this practice. As early as the first decades of the fifteenth century, Cennino Cennini wrote that "if you want to have the varnish dry without the sun, cook it thoroughly in the first place; for the panel will be very well off not to be strained by the sun too severely."[83] The author of MS Fr. 640 also warns against this practice when he observes that "it is better to heat the varnish a little bit, rather than to put it out in the sun," because "this makes the panel warp" (*cela faict enveler le tableau*).[84] And Vasari, as I discussed in chapter 1, based his story of the invention of oil paint on Van Eyck's alleged frustration with having to varnish his panels in the sun, which made them crack at the joints.

Unlike MS Fr. 640, Armenini also mentions the use of various oil-resin varnishes for paintings. It is therefore significant that here too he shows a concern for not making these varnishes too viscous. In one of these recipes Armenini explains how to make a varnish with equal quantities of mastic (*mastice*) and sandarac (*sandaracha*) crushed together.[85] Your crushed sandarac and mastic should be covered with oil of walnut (*oglio di noce*) and then cooked over a fire. Accordingly, the mixture is strained, and one-third part *oglio di abezzo* is added. This is done, so we learn, "with little boiling, since that would make the varnish viscous" (*bollir poco, perche la vernice verrebbe viscosa*). Armenini concludes the passage on varnishing with another remark about the viscosity of varnishes: he writes that "covered in their pots, [the varnishes] may be kept for a long time, becoming purer and thinner with age" (*le quali poi coperte nel suo vasetto si conservano lungo tempo, con si più purgate, & sottili*). Significantly, the mastic and sandarac varnishes I made based on some of the recipes discussed in this chapter did indeed lighten dramatically after they had been stored in the dark for almost two years (compare figs. 3.5, 3.6, and 3.7).

Varnishing: A Lost Art or a New Beginning?

One of Armenini's most remarkable comments about varnishing easel paintings is that he considers it a tradition almost lost to his own time: "Very few pay attention to varnishes in our times, perhaps more through avarice and neglect than for good reasons. Nevertheless, since varnishes are necessary, we shall discuss how they were made and used by the best craftsmen who are no longer with us."[86]

The author of MS Fr. 640 observes something similar about Italian varnishing practice—or the lack thereof: "The Italians scarcely varnish their paintings because they lay their paintings very thick & they are a long time drying on the inside, though on top they make a dry skin & crust."[87]

Whether or not there is a connection between Armenini's observation that the art of varnishing is no longer practiced in Italy and MS Fr. 640's technical explanation for omitting this final surface coating on works that had been thickly painted remains to be seen. It is nevertheless well known that painters often waited to apply varnish until their paintings had thoroughly dried. Cennini details this practice in an elaborate recipe:

> Know that the most lovely and the best varnishing that exists comes about because the longer you put it off after the painting of your panel, the better it is; I say definitely waiting several years, and at least one . . . varnishing at the point when the pigments with their binders have run their course, they

then become really fresh and lovely, also staying in the same state forever [*nono poi freschissimi ebegli e stando in edenti forma senpre*].[88]

The practice of postponing the varnishing of a painting for some years is also described by Albrecht Dürer (1471–1528) in his correspondence with his patron Jakob Heller (ca. 1460–1522), a rich merchant from Frankfurt. After Dürer finished the altarpiece Heller had commissioned from him, he wrote his patron a letter on August 26, 1508. Dürer explains that if Heller keeps the painting clean and does not sprinkle it with holy water, "it will stay fresh and clean for five hundred years" (*das sie 500 jahr sauber und frisch sein wirdt*).[89] Dürer further informs his patron that he will come back "in about 1 or 2 or 3 years" (*etwa uber 1 jahr, 2 oder 3*) to see "if the painting has become really dry" (*ob sie woll dürr were worden*). If it has, he will apply "some excellent varnish that you cannot apply to the painting when it has just been finished" (*so wolt ich sie von neuem mit ainem besondern fürneis, den man sonst nit kan machen, uf ain neues uberfierneisen*). With this special varnish, the work will last "100 years longer than it would before" (*so wirdt sie aber 100 jahre lenger stehen den vor*). Dürer stresses that Heller should not allow anyone else to varnish the work, "because all the other varnishes are yellow and you would ruin your painting" (*denn alle andere furneiß sind gelb, und man wurde euch die Taffel verderben*).

For Dürer and for Cennini, the act of varnishing had everything to do with preserving their paintings for a long time and, in Dürer's case, without yellowing because of the special varnish only he knows how to make. Similar remarks about how varnishes ensured the durability of one's painting and united and preserved its colors can be found in other sixteenth-century treatises as well. Armenini, for instance, writes that varnishes protect the beauty of colors for a long time, bring out their luster, and are useful for bringing out small details in a painting. The author of MS Fr. 640 explains that "one does not varnish to make paintings shine, for it just takes the light out of them" (*On ne vernist pas pour faire luire les tableaulx car cela ne faict quoster leur jour*).[90] Instead, the varnish "heightens some colors which have soaked in" (*rehaulser les couleurs qui sont imbeues*) and is used to keep them from dust. In another recipe the author of MS Fr. 640 describes in more detail what he means by this. The varnish made sure the painting was not matte in one place while shiny in another:

Colors in oil that are imbibed [*Couleurs a huile qui semboivent*]

It is best that colors in oil are imbibed, that is to say they do not remain shiny [*luisantes*] after they are dry, for they do not die [*ne meurent*]. But if

> in some places they are shiny, it is because the fattiness of the oil has remained in that part, which would make the colors die. The varnish mends all this and unites and renders it similar in one place as in another [*Le vernis racoustre tout cela & lunist & rend semblable aulta(n)t en un endroit quen laultre*].[91]

The previous examples all stress how varnish serves the painting underneath. Indeed, the varnish for easel paintings did not just undergo a transformation in ingredients between the fifteenth and sixteenth centuries; its function changed as well. Unlike the case with medieval artworks, which required the thick and glossy *lucide* varnishes to create the material mimesis of gold and enamel, fifteenth- and sixteenth-century varnishes were meant to disappear as much as possible on the surface of the painting, protecting it from dust and unifying and preserving the complexion of its colors. A similar desire to bring out colors can be found in earlier varnish recipes as well, but in the fifteenth and sixteenth centuries the brown and yellow medieval varnishes became less desirable because of easel painters' new ambition to imitate the visible world: whites had to remain pristine, and blues should not turn green. To achieve this, painters believed the varnish itself should have as little color as possible. As a result of this search for a pale and clear painter's varnish, linseed oil slowly made way for less yellow alternatives.

Solvent-Based Varnishes against Green Skies

We have already seen that the recipes contained in MS Fr. 640 show that varnishes for easel paintings were preferably made from a mixture of turpentine oleoresin and turpentine oil, with or without a solid resin such as mastic. Another varnish recipe in this manuscript explains how to make such a clear "varnish for panels" (*Vernis pour tableaulx*) with turpentine and oil of turpentine.[92] First you heat "venice turpentine" (*tourmentine de venise*), and as soon as it has the "right" temperature you add its distillate: "turpentine oil, the whitest you can procure" (*de tormentine de la plus blanche que tu pourras*). Once these two ingredients have been properly united, the substance is immediately taken from the fire. The recipe continues that this varnish can also be made without the fire, but this makes it dry more slowly (fig. 3.7). The varnish thus made is "appropriate for panel paintings and other painted things without corrupting the colors or yellowing [*sans corro(m)pre les couleurs ne jaulnir*]. And it dries both in the shade and in the sun, and overnight and during the winter as well as in the summer. It is commonly sold 15 sous a lb."[93]

The French manuscript includes various other pertinent remarks about why oil-resin varnishes should be avoided for easel paintings. One can be found in a recipe for making a varnish for frames and guitars with spike lavender oil (*Vernis dhuile daspic*)—the distillate from *Lavandula latifiola* Medik.[94] We learn that "framemakers, to avoid the trouble of polishing their ebony, varnish it with this, as do guitarmakers" (*Cestuy cy est promptem[ent] sec Les quadraniers pour eviter la peyne de polir leur hebene le vernissent de cestuy cy Les quiterniers aussi*). The recipe warns that except for their moldings, this varnish is not as suitable for panel paintings because for this, "fine, good turpentine" (*tourmentine fine bien*) has to be used instead. On the same folio, in the margins, the author of MS Fr. 640 explains that spike lavender varnish is not as appropriate, because it "eats away at the colors since it is too penetrating" (*mange les couleurs pour estre trop penetra[n]t*). The recipe concludes with a most interesting observation about the history of the use of linseed oil varnishes on landscape paintings:

> One did not use to varnish the landscape of a panel when linseed varnish was in use, because it yellowed the landscape [*On ne souloit poinct vernir les paisages dun tableau quand le vernis de lin estoict en usage pource quil faisoict jaulnir le paisage*]. But with varnish made with turpentine [*tourmentine*], one varnishes everywhere. You can put in pulverized mastic [*mastic pulverise*] extracted in tears or otherwise, and it will be more desiccative [*desiccatif*], than with sandarac [*sandaraque*].[95]

Not only the author of MS Fr. 640 notes the drying agency of mastic resin over sandarac; it was hinted at in numerous recipes since the *Mappae clavicula* compilation. But even more interesting, we learn here that varnishes were avoided for landscapes when linseed oil was still their main ingredient because it would yellow them too much. Figure 3.13 shows that a traditional varnish made with linseed oil (cooked with sandarac and allowed to become lighter for two years) does indeed impart a yellow hue to the blue layer of azurite underneath, turning it green.

A remark about this exact effect of traditional oil-resin varnishes on blue paints occurs in a recipe from MS Fr. 640 for "thick varnish for planks" (*Gros vernis pour le plancher*). The varnish for planks is made with two parts "common thick turpentine" (*grosse tourmentine commune*) mixed with one part "fine turpentine oil" (*huile de tourmentine fine*). According to the recipe, this varnish is very different from the substance used to varnish floorboards in the past:

FIGURE 3.13. *Left to right:* Linseed oil-sandarac varnish; linseed oil-mastic varnish (both varnishes were two years old and thus much lighter than when first made); and turpentine oil-oleoresin varnish over azurite. Photo by Marjolijn Bol.

> There is a varnish that takes a long time to dry and drips more than two months after it has been applied to the planks [*Il y a du vernis qui est long a seicher & degoute plus de deux moys apres quil a este aplique aulx planchers*]. But this one [the turpentine–turpentine oil varnish] does not drip like that of times past, which was made of linseed oil, garlic boiled in it to extinguish it and rid it of grease, and with wheat. And this one yellowed and rendered greenish the blue color of paintings [*Et cestuy cy jaulnissoict & rendoit verdastre la couleur bleue des tableaulx*].[96]

This method of preparing MS Fr. 640's yellow varnish of the past closely resembles some of the procedures described in the varnish recipes recorded in the Strasbourg tradition. The previous passage from MS Fr. 640 therefore seems to confirm what the scattered evidence from recipes that started appearing in the first decades of the fifteenth century already suggested; painters and other artisans were looking to replace traditional linseed oil-resin varnishes with materials that were less yellow. (Compare the different varnishes applied over azurite; see fig. 3.13.) Indeed, whereas the recipes discussed in chapter 2 show that linseed oil was used in almost every varnish recipe, in the course of the fifteenth century paler oils, especially nut oil, start making their appearance in varnish recipes as well.[97] In some of these examples the varnish recipes appear to reflect practices that evolved out of the desire to make green, see-through parchment. But recipes dating to the end of the sixteenth century show that in the art of easel painting important changes had taken place in varnishing as well.

Easel painters sought to produce a varnish that differed from earlier ones in two fundamental ways: they searched for a less viscous and faster-drying varnish so that it could be applied without the sun's heat. And most important, they looked for a varnish that would not change the colors of their paintings. The sources suggest that at first this search centered on procuring paler resins and clearer and paler oils; recipes discuss new methods for clarifying oil-resin varnishes of their mucilaginous materials using bone ash, alum, bread, and water and by cooking them over a slower fire so they would darken less. That aged oils were sometimes preferred may indicate that oils were also allowed to stand for a long time, making them clearer and lighter. Eventually, however, drying oils appear to have been replaced by volatile oils mixed with clear turpentine oleoresins and mastic. And even more, according to MS Fr. 640 this had become the only appropriate substance a painter should use to varnish his easel painting. It is therefore almost ironic that in the course of the sixteenth century, as painters were increasingly looking for a way to varnish in the shade, precisely the invention that Vasari attributed to Jan van Eyck—who allegedly discovered an oil varnish out of frustration at having to varnish in the sun—came to be used less and less as a final surface coating on easel paintings. Recipes suggest that drying oils—especially linseed oil—were avoided because they could impart a yellow hue to lighter colors or could turn blues into greens. Certainly it was not one painter's technical discovery that changed varnishing practice between the fifteenth and sixteenth centuries. Rather, the new ambition of painters striving for the mimesis of the visible world inspired a more general search for new methods and materials better suited to this purpose. Varnishes were no longer developed to imitate the splendor of the goldsmith's work but assumed a new role in the art of painting: the varnish became a colorless protective coating meant to protect and bring out all the colors and details of the visible world meticulously conveyed with paint alone.

4

In Search of Splendor

Gems and Their Imitations before 1400

Whoever thou art, if thou seekest to extol the glory of these
doors,
Marvel not at the gold and the expense but at the craftsmanship
of the work.
Bright is the noble work; but being nobly bright, the work
Should brighten the minds so that they may travel, through the
true lights [*lumina vera*]
To the True Light where Christ is the True Door.[1]

ABBOT SUGER OF SAINT-DENIS,
De administratione (ca. 1144–48)

Abbot Suger (ca. 1080/81–1151), the famous patron of Gothic art and architecture, filled his abbey church at Saint-Denis with material splendor. From the gold and silver metalwork embellished with numerous precious stones and enamel to the sumptuous stained glass windows, Suger's church was a spectacle of refracted and reflected light (fig. 4.1). In his various writings, Suger detailed the reasons behind this sumptuous program of decoration. He explains that "the loveliness of the many-colored gems" calls him away from his "inferior world" to the "higher one." And that this "worthy meditation" induces him to reflect, "transferring that which is material to that which is immaterial" (*de materialibus ad immaterialia transferendo*).[2] One of Suger's poems recorded in *De administratione* (cited in

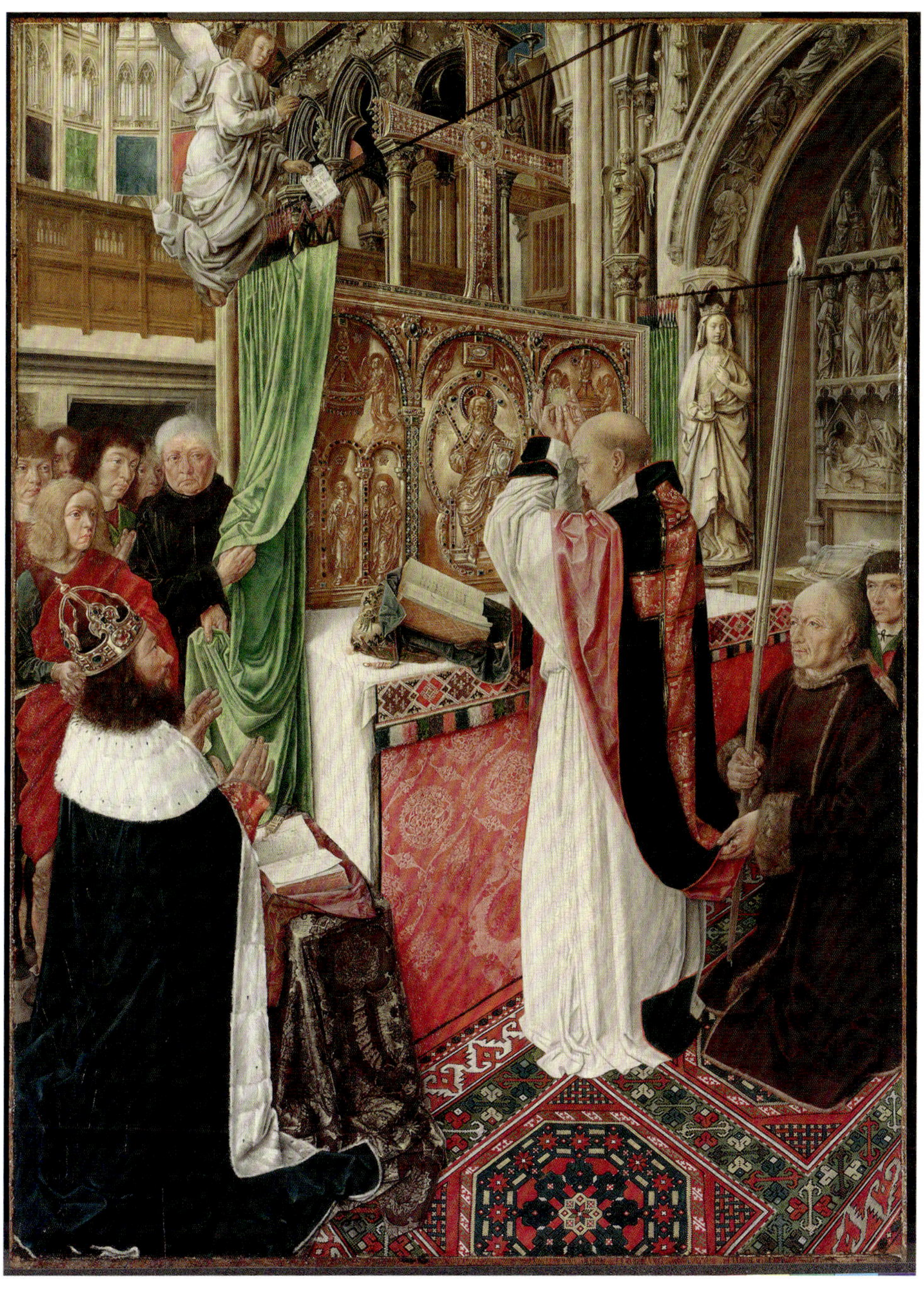

FIGURE 4.1. Master of Saint Giles (active ca. 1500), *The Mass of Saint Giles* (showing the royal mausoleum of Saint-Denis in Paris), ca. 1500. Panel from polyptych, oil on panel. National Gallery, London. © The National Gallery, London.

part above) perfectly sums up the essence of this idea. The verses, which were appropriately placed above the two gilded entry doors of the church, explain how the "true lights" (*lumina vera*) of polished gold open the "true door" (*janua vera*), thereby evoking the "true light" (*verum luminem*) of the Divine.

The significance of precious stones and precious metals for the aesthetics and meaning of medieval art is well studied. The prominent role of gems and gold in numerous biblical stories turned these natural materials, found in and on the earth, into symbols of the Divine. And since gold and gems were often thought to have a divine origin, many believed they were imbued with special powers. With its streets of gold and the twelve rows of precious stones making up the foundations of the heavenly city, Saint John's description of the New Jerusalem (Revelation 21:19–20) became an important inspiration, and even justification, for the splendor of medieval metalwork and many other artistic practices.[3]

Yet gold and precious stones did not become symbols of the Divine just because of their prominent role in scripture. As Suger's verses show, God the Father was himself considered to be "light," and this is why the ability of the purest and most precious of earthly materials to reflect and refract the visible light came to be identified with the Divine on another level as well.[4] The deep fascination with the optical powers of the most precious materials produced by nature led medieval artisans to explore a wealth of substances and techniques to achieve a similar optical resplendence with manmade materials. In some cases these materials produced by craft were thought to be imbued with powers like those of their natural counterparts. For Suger this meant that the subtle glow of the colored glass windows designed for his abbey urged him "onward from the material to the immaterial" (*una quarum de materialibus ad materialia excitans*), as did metalwork set with precious stones. This chapter investigates ideas about the nature and genesis of refraction in precious stones in relation to the arts designed to imitate them. These ideas and practices regarding gems and their imitation set the stage for the early history of glazing discussed in chapter 5. Indeed, even though the art of glazing would eventually become known as specific to the toolbox of the easel painter, in its early beginnings, as this chapter will show, the materials and methods used to make glazes and other translucent layers were explored across a range of arts by a variety of artisans interested in imitating precious minerals. Since much of the medieval thinking about gems goes back to ancient mineralogical texts and theories about light and sense perception, this chapter takes a long-term approach to investigating these ideas.

Translucidus and *Perspicuus*: Describing the Optics of Precious Stones

Since the earliest treatises on mineralogy, precious stones have been characterized as rare, hard, of beautiful color, and, particularly in the West, translucent or transparent: that is, they can transmit visible light.[5] In modern gemology, this last property of gems is called clarity. Of the previous characteristics, only rarity cannot be influenced by human artifice. The stone's hardness, however, governs how far color and clarity, the other two characteristics that determine a stone's value, can be enhanced.

When precious stones are found, their surface is rough and irregular, so their potential clarity and saturated color is diminished by the scattering of light. To reduce this scattering, precious stones must be made smooth at the microscopic level. To smooth a gem, the lapidary needs an abrasive at least as hard as the mineral he wants to work. During the early medieval period, gemstones were shaped, smoothed, and polished with various abrasives by rubbing them against flat, smooth fixed plates made from stone or metal.[6] This method allowed the lapidary to give his gem a domed, polished top surface, often quite irregular in shape, and usually a flat or hollowed-out base: the *cabochon*. Figure 4.2 shows a thirteenth-century gold ring set with precious stones that were shaped and polished by this method; in this example we see blue and purple minerals of the corundum variety, known today as sapphires.

The hardness of a gemstone determines the amount of polish it can receive. The harder a gem, the better it resists scratching and abrasion and the more highly it can be polished. This also means that the hardest precious stones are best at retaining their polish. When relatively hard minerals have been polished to perfect smoothness, not only can we perceive their potential clarity and saturated, glowing color, but their surface glints with the "specular reflection." The human eye perceives specular reflection as a bright white sparkle, such as can be seen on the polished cabochons in the finger ring shown in figure 4.2.[7]

Of the thirty-seven books that make up Pliny the Elder's *Historia naturalis,* the books on stones were among the most influential during the post-classical period, shaping the content and vocabulary of numerous medieval encyclopedias and lapidaries either directly or, more commonly, through the many adaptations of Pliny's work.[8] Pliny uses two terms to describe the clarity of precious stones. He used *translucidus,* a compound from *trans,* "through," and *lucere,* "to shine" (from *lux,* "light") to denote all light-transmitting minerals ranging from colored, translucent gems to

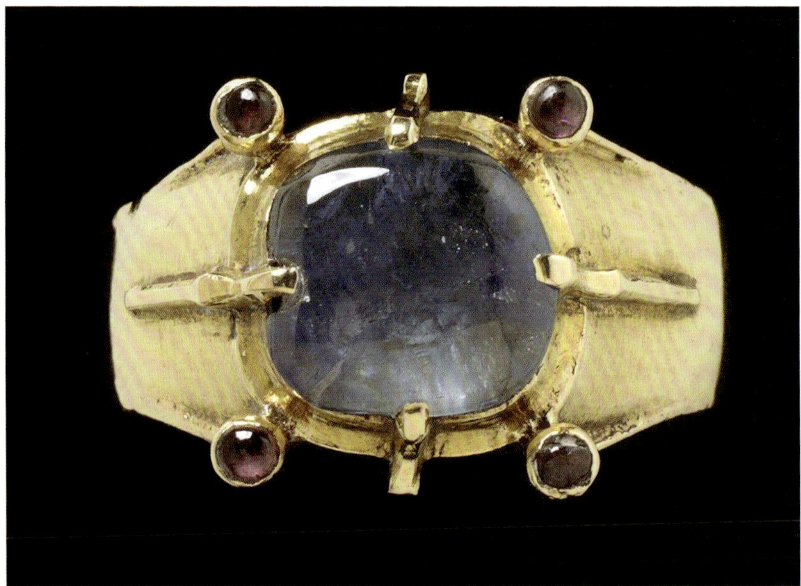

FIGURE 4.2. Ring, ca. 1250–1300, England (made). Gold, blue sapphire, and purple sapphires. Victoria and Albert Museum, London. © Victoria and Albert Museum, London.

colorless, transparent ones.[9] Pliny also used terms derived from *perspicuus*, a compound of *per*, "through," and *spicere*, "to look," to explain that colored or colorless minerals can be penetrated by the sight: they are "see-through." To clarify the optical characteristics of precious stones beyond these two terms, Pliny compared them to substances that are visually similar. He explains, for instance, that the most highly prized *berulli* (possibly beryl varieties of various hues) imitate the pure green of the sea, while others look like olive oil; *carbunculi* (used to refer to various translucent red stones such as garnet) are named from their resemblance to fire; and while the best *amethysti* (amethyst, purple quartz variety) have the perfect shade of Tyrian purple, the lesser ones may have the color of red wine or "degenerate" nearly into *crystalli* (rock crystal, a transparent variety of quartz).

An influential early author who relied on Pliny and his legacy was Isidore, archbishop of Seville (Isidorus Hispalensis, ca. 560–636). Isidore of Seville wrote what would become one of the most popular post-classical encyclopedias, the *Etymologiae*.[10] By tracing the etymology of the Latin language, he presents the state of knowledge until his time. "Etymology," Isidore explains, "is the origin of words, when the force of a verb or a noun is inferred through interpretation." For example, *flumen* (river) is so called from *fluen-dum* (flowing) because it has grown by flowing.[11]

Besides Pliny, Isidore relied on numerous other ancient and early Christian authors. Since many of these texts are now lost, the *Etymologiae* became pivotal for transmitting this knowledge to later centuries. That the *Etymologiae* survives in nearly a thousand manuscript copies testifies to its enduring popularity in the premodern period.[12]

Isidore explains that "all of the gems [*gemmas*] that are not light transmitting [*non translucidas*] are called 'blind' [*caecas*] because they are made opaque [*obscurentur*] by their density [*densitate*]."[13] In his etymology of precious stone—*gemma*—Isidore points out that this Latin term derives from the optical resemblance of gems to resin: "Precious stones [*gemmae*] are so called because they are *transluceant* like resin [*gummi*]."[14] Even though this is likely not the true etymology of the word *gemma*, it shows that Isidore considered the ability to transmit light one of the essential properties that makes a stone precious.[15] He concludes that gems add great beauty to gold with their lovely colors, and through his explanation of why some gems are called "precious" (*pretiosi*), Isidore introduces a religious element to his etymology of *gemma* as well. The rarity of precious stones is one of the reasons for their esteem, something they have in common with the word of God:

> They are called precious stones because they are valued dearly, or because they can be distinguished from base stones, or because they are rare [*rari*]—for everything that is rare is called great and precious, as may be read in the book of Samuel [I Kings 3:1 (*Vulgate*)]: And the word of the Lord was precious [*pretiosus*] in Israel; that is, it was rare [*rarus*].[16]

Like Pliny, Isidore uses the term *perspicuus* to explain the special optics of materials that allow us to see through them. About the sky, for instance, he relates that in Greek it "is called *ouranos* from *orasthai*, that is, from seeing [*videndo*], from the fact that the air is *perspicuus* and very pure [*purior*] for looking out [*speculandum*]."[17] And about the *iacinthus*, one of the purple (*purpureis*) precious stones that Isidore describes, we learn that in fair weather it is *perspicuus* and pleasing, but in cloudy weather it loses its strength and fades before one's eyes. Since Isidore explains that the *iacinthus* is so hard that it can be incised and engraved only with an *adamante* (diamond), it might be identified with a purple variety of corundum, which is indeed the hardest natural mineral before the diamond (fig. 4.2). Finally, Isidore also uses *perspicuus* to explain the special optics of glass (*vitrum*), because he believes its name derives from its transmitting light in such a way that it allows us to "see" (*videre*) "through" it:

> Glass [*vitrum*] is so called because due to its perspicuous quality [*perspicuitate*] it transmits light [*transluceat*] to one's sight [*visui*]. Anything contained inside other minerals [*metallis*] is hidden, but any sort of liquid or visible thing contained in glass is displayed to the outside; it is somehow both enclosed and opened up [*et quodammodo clausus patet*].[18]

This brief excursion into the optical terminology used by two of the most influential early writers on stones will help us understand how medieval writers used these same terms to describe and theorize about the transparency and translucency of gems and the arts imitating them—in particular the art of enameling and glazing with oil paint. Because the historical use of *translucidus* and *perspicuus* cannot always be translated by "translucent" and "transparent" as this book defines the terms, I will cite the original Latin (and later vernacular equivalents) throughout this and subsequent chapters. Beyond avoiding possible confusion of terminology, this decision also demonstrates the specificity with which optical effects were described in the past. To medieval authors, the term *translucidus* ("light through") became a descriptor for materials that are illuminated from within by setting them on a reflective surface, such as polished precious stones set in metalwork, the art of enameling, and the art of glazing with oil paint, whereas terms derived from *perspicuus* were used to describe materials that allow light to pass through to the other side so we can "see through" them. As such, this last term would become common in descriptions of glass that was not fused to a base material (e.g., window glass or glass tableware).

Gems in the Water of Paradise: Ideas about the Genesis of Precious Stones

The idea that refractivity was one of the defining characteristics of precious stones became intertwined with ideas about the ways nature produces gems and, more specifically, with ideas about their genesis in the waters of the Garden of Eden.[19] In the Old Testament, chapter 2 of Genesis gives an account of God's creation of man and woman and tells how God placed Adam and Eve in the garden to cultivate it. The paradisal garden is watered by a river that, on flowing out of Eden, is divided into four large rivers called Pishon, Gihon, Hiddekel, and Euphrates (Genesis 2:10–14). Saint Augustine of Hippo (354–430), one of the great church fathers of the Latin West, explains this passage in his two influential commentaries on Genesis, *A Refutation of the Manichees* and *The Literal Meaning of Genesis*. Saint Augustine writes that the river Pishon "is the river which goes round the whole land of Havilah; there is gold there, indeed the gold of that land is the best; there is *carbunculus* [a translucent red stone] there and *lapis prasinus* [literally leek-green stone, perhaps emerald]."[20]

Saint Augustine comes back to these paradisal gems when he explains that the four rivers flowing out of the Garden of Eden are connected to the four cardinal virtues.[21] The Pishon is associated with "prudence" (*prudentia*):

> This prudence [the Pishon] then goes round the land [Havilah] which has gold and carbuncle and leek-green stone, that is, a discipline of life that glistens brightly as if refined from all earthly dross, like the best gold; and truth which no falsehood can overcome, like the brilliance [*fulgor*] of the *carbunculum* which is not overcome by night; and eternal life, which is signified by the greenness [*viriditate*] of the *lapidis prasini*, because its vigor never withers.[22]

By connecting stories from the Old and the New Testaments, Saint Augustine became the first theologian to write about human salvation in terms of historical progression.[23] Through his historical reading of the Bible, he observed parallels between the Garden of Eden and the splendorous New Jerusalem in Saint John's book of Revelation. His explanation of the meaning of the river Pishon must be understood in this light. Not only is the Pishon, overflowing with gold and precious stones, symbolic of the pure state of mankind before the Fall, Saint Augustine believes its presence in the Garden of Eden portends the restoration of this purity with Christ's Second Coming in the New Jerusalem of the Revelation.[24]

Saint Augustine's historical reading of the Bible led to the idea that the four paradisal rivers are real rivers that can still be found on earth. Because the four rivers were still known to man, it was thought that the Garden of Eden could still be found on earth as well.[25] It was here, in the divine waters of paradise on earth, so it was believed, that the most highly prized and most powerful precious stones were born. Influenced by the works of Saint Augustine and various other early Christian texts, medieval lore describes how the rivers of paradise distribute precious stones from the Garden of Eden to other parts of the world. Again, it was Isidore of Seville's encyclopedic dictionary that was largely responsible for the early transmission of this idea. In his etymology of "Asia," Isidore explains that the terrestrial paradise is in the East and is irrigated by a spring that is divided into the headwaters of four rivers. We learn that after the fall of humankind paradise was blocked off by a wall of fire and a garrison of angels. According to Isidore, the gems of paradise are carried by the river Euphrates: "The Euphrates is a river in Mesopotamia rising from paradise [*de paradiso exoriens*], well supplied with gemstones [*copiosissimus gemmis*]; it flows through the middle of Babylonia."[26]

Born from the Elements

Ideas about gems in the waters of terrestrial paradise can be found in numerous encyclopedias, travel chronicles, maps, religious texts, and various other forms of medieval lore.[27] In the course of the twelfth century, these ideas about where nature produced its most rare and precious gems became integrated with theories about their natural formation. Medieval thinkers believed that stones, like all natural materials, were born from the four elements: earth, air, water, and fire. The idea that these four elements are the essence of all matter has its origin in pre-Socratic philosophy. These theories were transmitted to the Latin West through a translation of Plato's *Timaeus* by a certain Calcidius (fourth century AD), about whom very little is otherwise known.[28]

In the preface to her book on the healing properties of minerals, Hildegard von Bingen (ca. 1098–1179), Benedictine abbess of the Rhineland, provides a rather ingenious account of how the elements generate precious stones in paradise. Hildegard's lapidary is part of the nine books that make up her *Physica* (1151–58), a work treating of the scientific and medicinal properties of natural materials, which, besides stones, includes books on plants, the elements, trees, fish, birds, other animals, reptiles, and metals. Hildegard attributes the healing properties of precious stones to their divine genesis. What is more, unlike minerals found elsewhere, she argues that only paradisal gems have the power to do good.

According to Hildegard, "precious stones are born from fire and water; whence they have fire and moisture in them" (*et sic pretiosi lapides ab igne et ab aqua gignuntur; unde etiam ignem et humiditatem in se habent*).[29] Implicitly referring to paradise, she claims that gems are made in the East, in areas where the warmth of the sun causes the mountains to have heat as powerful as fire. When the rivers in this area "boil from the sun's heat" (*ubi aqua ignem tangit*) and touch the mountains, a "froth" (*spumam*) appears. Hildegard compares this to the froth "produced by hot iron, or a hot stone when water is poured over it" (*velut ignitum ferrum seu ignitus lapis facit, si aqua super eum funditur*). This foam then hardens into gemstones "in the space of about three to four days." Hildegard explains that when the rivers flood again, the gems are carried to other countries, where they are later discovered by humans: "the mountains where so many and such large stones have sprung up this way, shine like the light of day" (*prefati autem montes tot et tantis gemmis sibi hoc modo innatis, velut quedam lux diei ibi elucent*). Hildegard thus gives a resourceful explanation of how gemstones are formed from water and fire, helped by earth and air, while not neglecting their divine origin in the paradisal rivers springing from the Garden of Eden in the East.

The Dominican clergyman Thomas of Cantimpré (1201–72) provides more details about the role of the elements in creating refractivity in gems, something he believes explains why the clearest and rarest precious stones are found "in the East," close to where paradise on earth was thought to be. In his *Liber de natura rerum* (between 1228 and 1244), in book 13 on fountains Thomas describes how, at the center of the earthly paradise, there is a fountain of the clearest water (*fons limpidissimus*).[30] This fountain has so much water that it divides into four streams that, starting in India, irrigate the earth. One of the four rivers, the Pishon, falls in a miraculous manner from a mountain in India.[31] Book 14 of the *Liber de natura rerum* continues with Thomas of Cantimpré's mineralogical treatise. It is introduced with a general theory about how gems are generated in veins in the earth from variations of the four elements. If a lot of earth enters the mix, bigger, "opaque" (*obscurus*) stones are produced, and if a stone contains a lot of water it becomes "clear" (*lucidus*).[32] We learn that blue stones contain a lot of air, and those exposed to fire turn red. Thomas continues that some regions produce stones that are precious and have powerful virtues, the best of which come from the East. They are more "clear" (*claritate*), "bright" (*fulgoris*), and "highly prized" (*cariores*) than others because in this region elemental vapors are purer (*uia ibi sunt vapores elementorum a sordibus puriores*).[33] This is why the stones carried to us by the four rivers from terrestrial paradise are rarest and most precious.

In the mid-fourteenth century, the Bavarian scholar Conrad of Megenberg (1309–74) made a German translation and adaptation of Thomas of Cantimpré's encyclopedia, using it as the basis for his *Buch der Natur* (ca. 1349). A fifteenth-century illuminated copy of Conrad of Megenberg's work contains an interesting frontispiece miniature to the book "On Precious Stones" (*Von den Edeln Stainen*, fig. 4.3).

The painted page shows two men pointing toward a stream of gemstones pouring from a large mountain. A church in the background connects this "river of gems" to their divine origin. The miniature therefore combines the idea that gems flood from paradise in the four rivers with Thomas of Cantimpré's claim that the Pishon falls from a huge mountain in India.

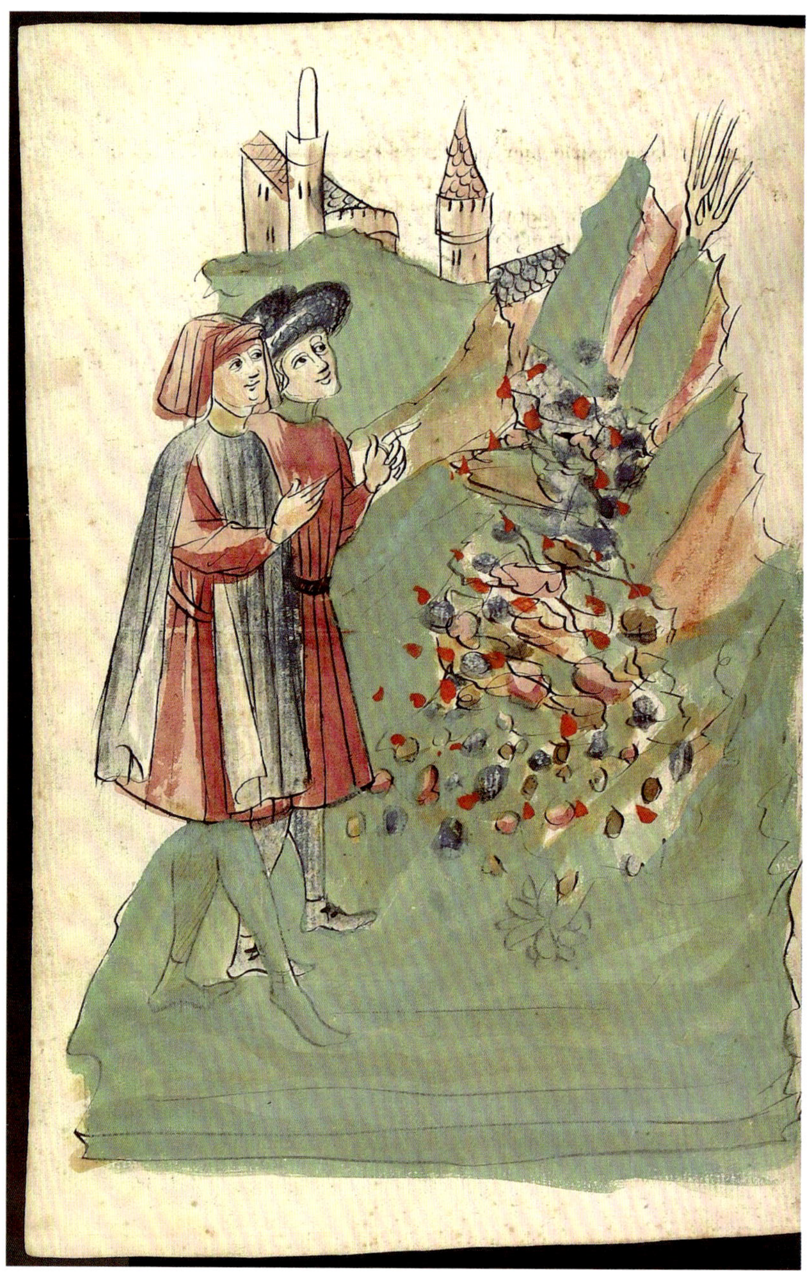

FIGURE 4.3. Workshop of Diebold Lauber (attr.), frontispiece miniature, *Buch VI: Von den Edeln Stainen* of *Konrad von Megenberg, Das Buch der Natur* (*The Book of Nature*, ca. 1442–48), showing two men pointing at a stream of precious stones flowing from a mountainous landscape. Universitätsbibliothek Heidelberg (Cod. Pal. germ. 300, 320v).

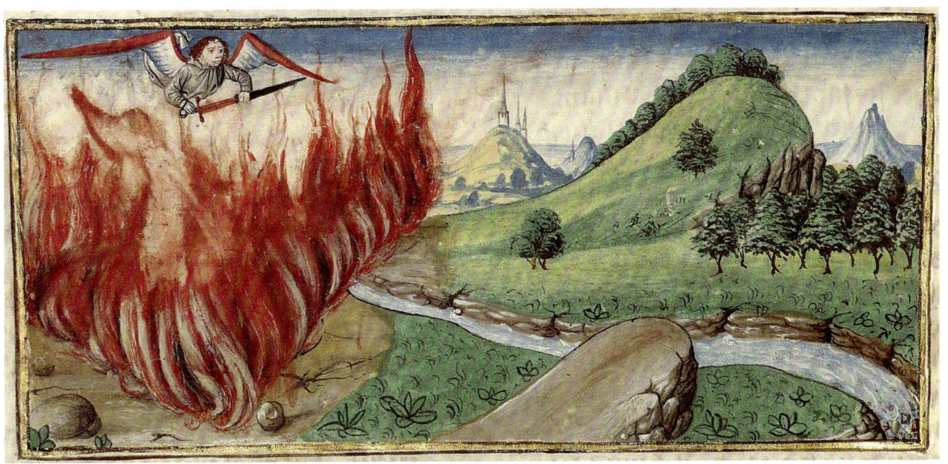

FIGURE 4.4. *The Angel of Paradise with a Sword*, Ghent, Belgium (made), ca. 1475. Tempera colors, gold leaf, and gold paint on parchment: leaf 43.8 cm × 30.5 cm (17¼ in. × 12 in.). MS Ludwig XIII 5, v1, fol. 54v, J. Paul Getty Museum, Los Angeles. Image made available under Getty's Open Content Program.

A similar depiction of the paradisal gems can be found on another fifteenth-century illuminated page, here illustrating a French translation by Jean de Vignay (before 1283 to after 1340) of the *Speculum historiale* (Mirror of History, between 1230 and 1250) by the French scholar Vincent of Beauvais (d. 1264). The *Speculum historiale* describes the history of the world from the creation of Adam and Eve to Vincent's lifetime. In a late fifteenth-century Flemish copy of this work, kept in the J. Paul Getty Museum, Los Angeles, a beautiful landscape illustrates Vincent's account of the regions that are close to the terrestrial paradise. The landscape depicts the Garden of Eden fenced off by a wall of flames and an angel drawing a red-hot sword just as we have seen that Isidore described. One of the paradisal rivers emerges from this divine barrier with an assortment of precious stones that have gathered in the lower right corner of the miniature (fig. 4.4).

Water and Precious Stones

We have to turn to *De mineralibus* (ca. 1261–63), the lapidary written by the great theologian and philosopher Albertus Magnus (ca. 1200–1280), for a more elaborate theory about the causes of refractivity in minerals. Most medieval writers on stones were indebted to the legacy of Pliny. The lapidary of Albertus Magnus stands out because it is embedded in Aristotelian thought instead.[34] Albert even distances himself from the work of Pliny with the remark that the *Natural History*'s discussion of minerals is unsatisfactory because it "does not offer an intelligent explanation of the causes common to all stones."[35] Precisely this was the ambition of Albertus Magnus. In his own words, he wants to show how stones "are mixed from the elements, and how each is constituted in its own specific form." The corpus of Aristotelian literature contains very little on mineralogy that Albert could rely on, which explains much of the originality of *De mineralibus*.[36] Albertus Magnus also argues the novelty of his treatise in another way; contrary to most lapidaries of his time, which describe only precious stones, his treats of "stones in general."[37]

Like the earlier sources studied in this chapter, Albert considers clarity the defining characteristic of precious stones: "Stones are called precious because they are see-through [*perspicui*] to a greater or lesser degree."[38] We learn that regular stones, on the other hand, are those minerals that are not see-through, are not limited in size, and are not precious.[39] This is also why Albert considers marble the most noble among the regular stones; marbles have something "see-through" (*perspicuum*) in them, causing them to "gleam" (*fulget*) or "sparkle" (*micat*).

It is noteworthy that the term *translucidus*, popular with Pliny and his followers, occurs only four times throughout *De mineralibus*.[40] Instead, Albertus Magnus uses terms derived from *perspicuus* to describe the optics of precious stones. Albert's reason for choosing this term lies in his discussing the causes for "perspicuousness" in minerals in terms of Aristotle's theory of vision as set out in the *De sensu et sensato* (*On Sense and Sensible Objects*).[41]

In *De sensu*, Aristotle theorizes that the *diaphanēs* (a Greek compound from *dia*, "through," and *phainein*, "to show") is not something we "see" but something "through" which things can be "seen" or "shown."[42] Aristotle explains that because the eye is made from the diaphanous element water, humans have vision. In Aristotelian theory of vision, therefore, the eye functions as an extension of the other diaphanous element air, allowing us to see through both diaphanous media. It is important to understand that the diaphanous merely constitutes the potential for seeing; it can be made visible only through the activity of light. Albert explains that if eyes had been made from an element characterized by opaqueness, such as earth, humans would have been blind. Isidore's remark about "blind" precious stones discussed earlier likely comes from the same theoretical framework.

Like the eye, the diaphanous in precious stones is water, and the more water a gem contains, the clearer it will be. Yet Albert explains that the common material from "perspicuous stones" (*lapidibus perspicuis*) is not made up from water alone: "evidence of this is that nearly all kinds of stones sink in water."[43] The reason is that the water that makes up a large part of the material of precious stones is congealed by having been acted on by earth in a process called *congelare* (from *con*, "together," and *gelare*, "to freeze").[44] The stones that are not precious, and therefore not perspicuous, are made through a different process known as *conglutinare* (from *con*, "together," and *glutinare*, "to glue"). In this case the separate particles of the element earth are "glued together," as it were, by water, producing an opaque stone.

Albertus Magnus uses another principle from Aristotle—the idea that art imitates nature—to prove his theories about the genesis of minerals by analogy to the ingenuity of craftsmen who produce visually identical substances. Aristotle first introduced the idea that art imitates nature in his *Meteorology* (4.381b3–9) to justify his use of technical terms such as "boiling" and "broiling" when explaining natural phenomena: "The affections produced are similar though they lack a name; for art imitates nature."[45] This influential concept was picked up by Albert, who substantiates his ideas about the genesis of precious stones by analogy to the art of making glass:

> And evidence of what we have said—namely that water acted upon by dryness, either hot or cold, is the common material of these [perspicuous] stones—is that glass is made from a moisture of this sort [*est quod vitra fiunt ab hujusmodi humiditate*], which is melted out of various ashes [*quae a cineribus diversis*], either lead [*sive plumbi*] or flint [*sive silicis*] or iron [*sive alicujus*] or anything else, by the strongest fire. That this moisture is water is shown in that it is solidified by cold and is melted and liquefied by intense dry heat.[46]

But because natural mixtures are always purer than those made by man, precious stones are always of a more perfect clarity than glass:

> For these stones are a sort of glass produced by the operations of nature [*Illi enim lapides sunt quredam vitra per naturae opus producta*]; and therefore they are of a more subtle mixture [*subtiliores in commixtione*] and a clearer [*clarioris*] perspicuousness [*perspicuitatis*] than glass made artificially. For although art may imitate nature, nevertheless it cannot reach the full perfection of nature [*quia ars licet imitetur naturam, tamen ad plenum opus naturae attingere non potest*].[47]

Albert continues with a detailed explanation of how color is generated in precious stones. He introduces his discussion with an interesting observation about the nature of "white" and "black." We learn that white is nothing else than "many perspicuous [*perspicuis*] parts distributed in something else," whereas black is caused by "opaque parts predominating over the perspicuous parts in the same body."[48] Black, in fact, is caused by the mere lack of "perspicuousness."[49] What Albert describes here is the effect of the scattering of white light, which turns finely powdered transparent materials into the appearance of a white mass, as I explained in the introduction. Albert explains this in more detail in his treatise on meteorology. Here he explains why snow is white by comparing it to powdered glass and rock crystal:

> For snow must necessarily be white, since it is itself composed of parts of transparent material [*materialiter de perspicuis*] with clotted air dispersed among them [*aere concreto in partibus divaricato*]. All such parts take in some light between them [*recipiunt de lumine inter se*], and so they are white [*albae*]; but nevertheless they limit the sight, as is seen when rock crystal or glass or any other transparent material is powdered [*amen visum terminantes, sicut in crystallo contrita patet vel vitro contrito vel quolibet alio perspicuo non colorato*].[50]

The comparison of snow to rock crystal and glass is connected to Albert's theories about the ways nature produces clear and colorless stones, among which he lists *crystallus*, *beryllus*, and *adamas*. In line with the ancient idea that transparent minerals were petrified ice, Albert argues that transparent gems are made up from a large amount of air and water, compacted by the "attack of earthy material" and extreme cold.[51]

Color, on the other hand, is generated in precious stones when certain elemental vapors act on them during formation. Red and clear stones are thus produced from smoky, earthy vapor acting on air and water, while clear blue stones are produced from predominantly watery vapors with a bit of subtle, burned earthy vapor. The colors between red and blue—that is, yellow or green—are produced from different proportions of the same earthy and watery vapors. To explain why some green stones are clearer than others, Albert once more brings up the art of glassmaking. We learn that when glass made from a "mixture of lead" (*commixtione fit plumbi*) is fired many times, it becomes clearer: "By repeated firing the perspicuousness [*perspicuum*] is purified and rendered subtle, and the clear brightness of fire [*claritas luminis ignei*] is imposed on the nature of water; and so it becomes clear [*et ideo clarificatur*]."[52]

Adding lead oxides produces a yellow color in glass, while green glass was generally produced by adding copper. It is likely, however, that here Albertus Magnus refers not to lead as a producer of color in glass but to lead as a base for colored glasses, which, as R. J. Charleston argued, was common practice at this time.[53] Such colored lead glass could be painted on glass or metal (enamel), used to make mosaic stones, and used to imitate precious stones as well (fig. 4.5). Albert points out this last fact when he writes that

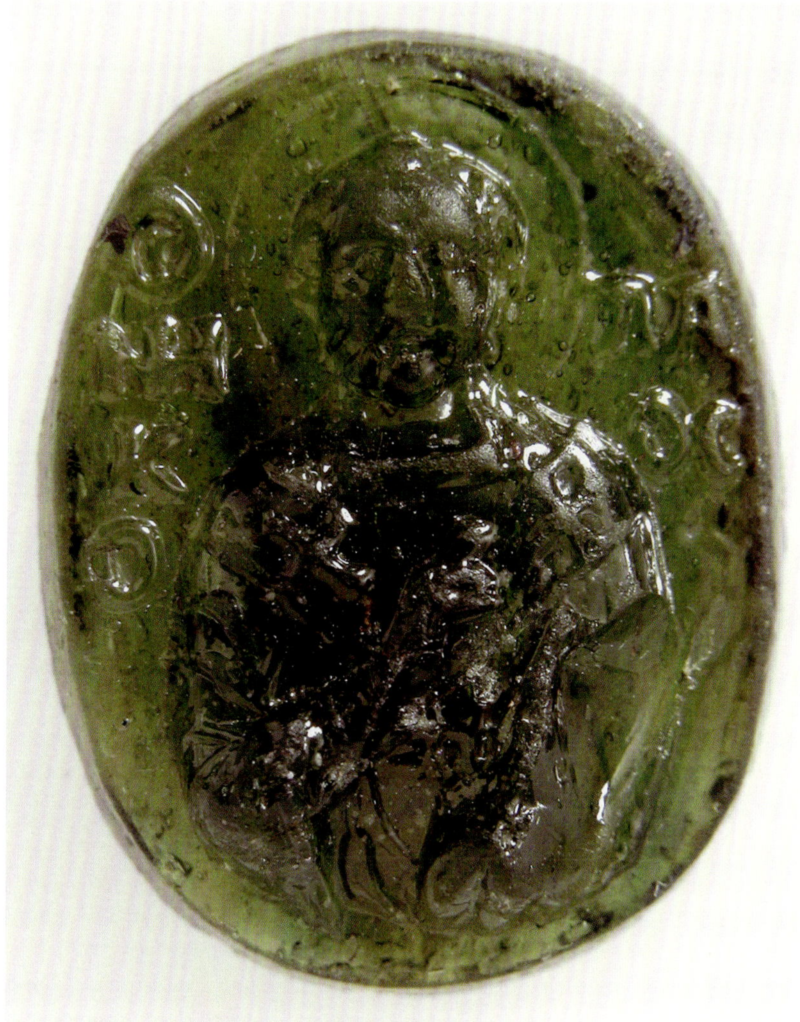

FIGURE 4.5. *Cameo with Saint Nicholas*, twelfth to thirteenth century, Venice? (made), Byzantine or Italian. Molded glass paste. Metropolitan Museum of Art, New York. Public domain.

"sometimes colors of this sort are made in mere glass, and likewise images; and the ignorant common people think that they are stone."[54]

That Albert may have been speaking about lead glass as a base for colored glass is also supported by various medieval recipes that discuss the art of glassmaking. Theophilus mentions lead glass in a recipe for making finger rings (*De anulis*) by mixing ashes, salt, powdered copper, lead (oxides), and the "colors of glass that you want."[55] And Eraclius, in the third book of *De coloribus*, includes a similar recipe that explains how to make lead glass from "good and shining lead" that has to be burned until it is reduced to powder, cooled, and mixed with sand.[56] To make this glass green you add "copper filings." Elsewhere in his treatise, Eraclius includes a passage from Isidore of Seville's *Etymologies* (16.16.3) about the different uses of glass. These include using colored glass to make imitations of various gems and using transparent glass, which looks like crystal:

> Glass is also tinged in many ways so as to imitate *jacinctos* and green *saphiros* and onyx stones [*onichinos*] and other colours of gems. And there is no material fitter for mirrors, or for pictures especially, than the white glass, and particularly that which is made like crystal [*proximaque in cristalli similitudine*]; so that for drinking cups it has driven gold and silver quite out of use.[57]

Coloring Crystals: Oil and the Imitation of Precious Stones

Medieval scholars studied the genesis of gems through craft practices resembling the natural processes or through practices of material mimesis designed to create an optical substitute for precious stones. We have seen that Hildegard's gems were made like foam on hot metal and Albert's clear gems were made in a manner similar to the way artisans make glass. Another method used since ancient times to imitate gems is closely related to the glazing technique that is the subject of this book. It may, in part, even explain the origin of the technique.

The earliest surviving reference to this practice is about making an imitation emerald.[58] The passage can be found in the *Epistles*, or *Moral Letters to Lucilius* (*Ad Lucilium epistulae morales*) of the Roman philosopher Seneca the Younger (1 BC to AD 65). Seneca writes that the pre-Socratic philosopher Democritus of Abdera (460–370 BC) discovered how a *calculus* (pebble) could be transformed into *zmaragdum* (emerald) by boiling it. This method, Seneca points out, was still practiced in his own time on colored stones, which are "amenable to this treatment."[59] A similar practice used for imitating precious stones is mentioned in Pliny's encyclopedia. He writes that the Indians have found a way to stain rock crystals to make them look like more expensive jewels.[60] Such methods for making imitation gemstones were allegedly so convincing that Pliny complains that "there is considerable difficulty in distinguishing genuine stones from false; the more so, as there has been discovered a method of transforming genuine stones of one kind into false stones of another."[61] Isidore of Seville similarly writes about how convincing fake *smaragdi* could be and laments the abundance of such practices of deception:

> As a substitute for that most precious stone, the emerald [*smaragdus*], some people dye glass with skill [*vitrum arte inficiun*], and its false greenness deceives [*falsa viriditas*] the eyes with a certain subtlety, to the point that there is no one who may test it and demonstrate that it is false [*quoadusque non est qui probet simulatum et arguat*]. It is the same with other matters in one way or another, for there is no mortal life free from deception.[62]

Considering the negative light in which the encyclopedic writers regard the practice of imitating gemstones, it is perhaps not surprising that most lapidaries do not discuss the specifics involved in making these imitations. Pliny does not consider it appropriate even to mention the names of the authorities on the matter:

> There are treatises by authorities, whom I at least shall not deign to mention by name, describing how by means of dyestuffs emeralds [*smaragdum*] and other translucent [*tralucentes*] colored gems are made from rock crystal. . . . And there is no other trickery that is practiced against society with greater profit [*neque enim est ulla fraus vitae lucrosior*].[63]

It has been suggested that an echo of the contents of the treatises on gemstone imitation that Pliny refers to here may have been preserved in the so-called Stockholm Papyrus (*Papyrus Graecus Holmiensis*).[64] This document, among the earliest testaments to the Greek alchemical tradition, may have been copied as a funerary gift about AD 200–300, but it is believed to have had a much older core.[65] The papyrus contains a variety of recipes for making alloys and dyeing textiles, and more than seventy recipes for imitating precious stones using less precious materials, including emerald (more than twenty recipes), ruby, beryl, amethyst, and sunstone.[66] The recipes are various, but most involve "corroding" or "opening up" a transparent mineral so that it becomes receptive to the color added in the next step. For the emerald, the coloring substance is always the pigment verdigris, finely ground and mixed with oil, resin, or vinegar.[67] The Stockholm Papyrus includes a separate instruction on how to make this pigment. In a recipe titled "Preparing Verdigris for Emerald," the green pigment is made by suspending clean copper plates covered with oil above strong vinegar (fig. 4.6):

> **Preparing Verdigris for Emerald**
>
> Clean a well-made sheet of Cyprian copper by means of pumice stone and water, dry, and smear it very lightly with a very little oil. Spread it out and tie a cord around it. Then hang it in cask with sharp vinegar so that it does not touch the vinegar, and carefully close the cask so no evaporation takes place. Now, if you put it in in the morning, then scrape off the verdigris carefully in the evening; but if you put it in in the evening, then scrape it off in the morning and suspend it again until the sheet becomes used up. However, as often as you scrape it off again, smear the sheet with oil as explained previously. The vinegar is [thus rendered] unfit for use.[68]

The pigment verdigris, a copper salt known today as "copper green," has special material and optical properties that make it particularly suitable for imitating translucent green stones. Ground with oil or resin, verdigris not only has the potential to approach the deep green hues of the emerald, it also can transmit light so as to approach its clarity. Verdigris is the only relatively stable green pigment known at the time that has a refractive index closely matching that of oil, resin, and crystal. This means that when verdigris is ground with such a binding medium and then used to coat the crystal, light can pass through while reflecting back a green hue similar to that of the emerald (compare figs. 4.6–4.8).

In addition, when verdigris is mixed with hot oil or resin or both, it dissolves. When a pigment is dissolved in its medium its refractive index becomes irrelevant (because it is now a "solution" without particles). The result is a perfectly translucent paint with great color saturation. That verdigris dissolves in resin (and in vinegar), makes it suitable for another method of producing imitation gems described in the papyrus: quench-cracking crystals to give them a green color. In one of the recipes for preparing such a quench-cracked emerald, the Stockholm Papyrus explains how a mixture of verdigris, celandine (*Chelidonium majus* L., a yellow to orange dye made the from roots, stems, and leaves of this plant, also used to color textiles),[69] and "scythian black" (?), have to be pulverized and mixed with "liquid resin":

Preparation of Emerald

Take pure pyrites or rock crystal and make the composition in the following way:

Verdigris, 2 drachmas; celandine, 1 drachma; Scythian black, 3 drachmas; liquid resin, which one holds in the mouth,[70] as much as necessary. Pulverize the dry materials, mix the resin with them, and set it aside. Take liquid alum, pour water on it so it becomes very watery, and preserve it in a clay vessel. Heat the stone in an earthen vessel and cool it in the alum. Heat the stone and put it in the above-named composition. However, if you desire that it should be greener, then again mix pulverized verdigris with it.[71]

The second mixture described in the recipe, made of "liquid alum" with water, is used to quench the minerals after they have been heated in an earthen vessel.[72] This causes a quick change in temperature that would have cracked the stone without breaking it, causing internal fractures reaching up to its surface (fig. 4.9).

According to the recipe, the stone should then be heated again and submerged in the first verdigris-resin solution. It is not entirely clear whether this green liquid is heated with the stones or whether they are again quenched in it, but when we consider that the stones have to take up the resin, which is quite viscous when cold, it seems likely that the resin would at least have been kept warm. When heat is applied the now fractured mineral opens up again, and through capillary action the solution of resin and copper green can penetrate the stone (fig. 4.8). If copper green would not dissolve in resin, the stone would have enclosed the oil while filtering out the pigment particles before they could enter the microscopic cracks.[73]

Verdigris is indeed the only pigment for coloring transparent minerals mentioned in the papyrus; the other colors are all true dyestuffs; that is, they will form a solution with the liquid they are immersed in.[74] The papyrus mentions the red dye alkanet derived from the root of *Alkanna lehmannii* Tineo (formerly known as *Alkanna tinctoria* (L.) Tausch) and the red dye kermes derived from the bodies of female *Kermes vermilio* Planchon (a scale insect) for imitating translucent red stones. For imitating translucent blue stones on crystal the papyrus mentions the blue dye indigo obtained from various plants such as dyer's woad, *Isatis tinctoria* L. (Brassicaceae), and from indigo, *Indigofera tinctoria* L.[75] These same dyes are also mentioned in the papyrus's recipes for dyeing textiles. It is therefore possible that the practice of coloring transparent minerals was related to, or even derived

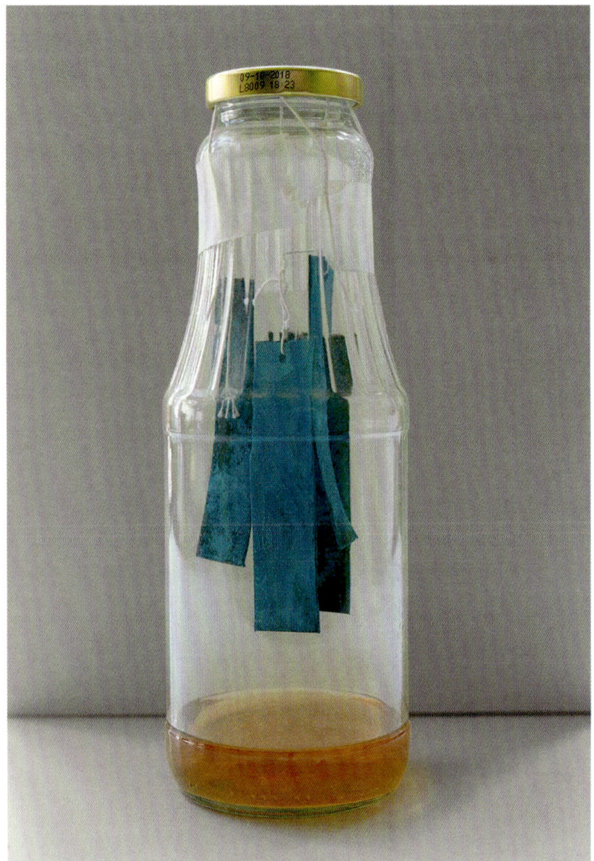

FIGURE 4.6. Exposing copper to vinegar in a closed environment to produce the green pigment verdigris. Photo by Marjolijn Bol.

from, early technologies for making fibers receptive to dyes. The recipes for dyeing cloth are various as well, but the basic procedure involves cleaning the wool of grease and dirt before mordanting it with salts of aluminum (alum), iron, or copper. Acting as a bridge, these metallic salts ensure that the dyestuff can form a permanent chemical bond with the textile fiber.[76] This mordanting step in dyeing cloth is therefore quite similar to "opening" up or "etching" precious stones with alum, as we have seen in the recipes for making imitation gems out of rock crystal. Even though it is not entirely clear in these recipes how alum would have helped with coloring quartz, it may have been used because it was known for its ability to "lock" in color in other materials, especially textiles.

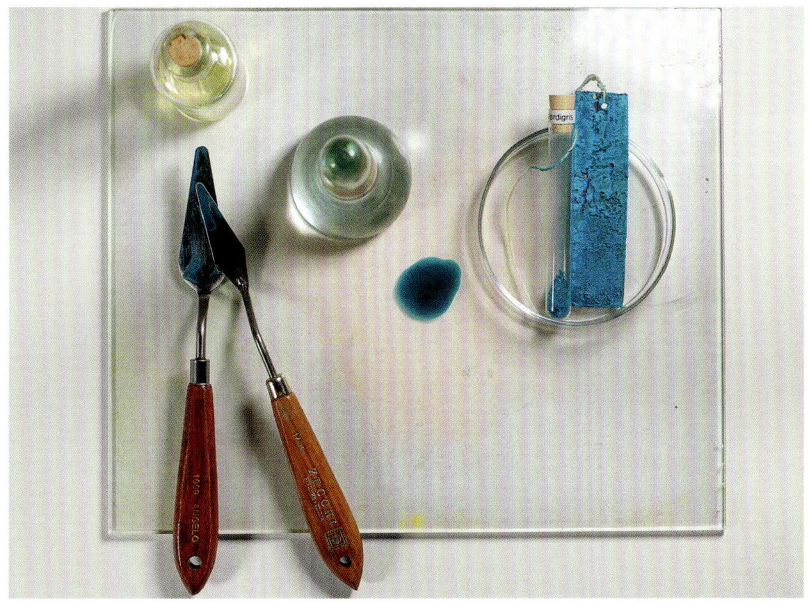

FIGURE 4.7. Copper with verdigris crust; the scraped pigment; and verdigris ground into paint. Photo by Marjolijn Bol.

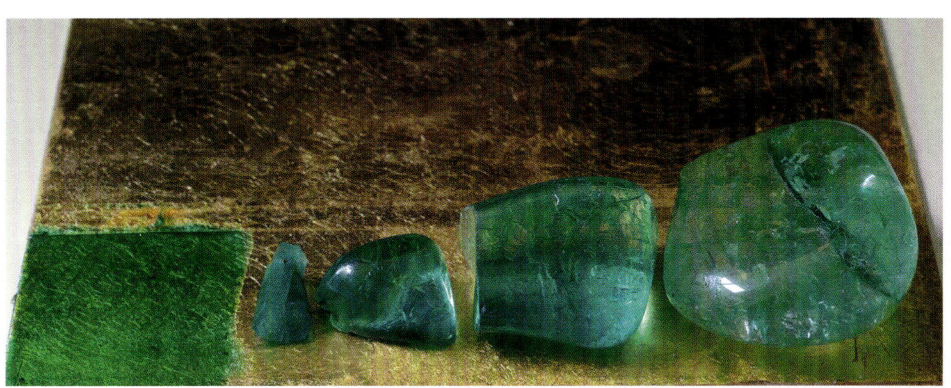

FIGURE 4.8. *Left to right*: Verdigris glaze; raw, unpolished emerald; three quench-cracked crystals dyed green with verdigris dissolved in hot oleoresin. Photo by Marjolijn Bol.

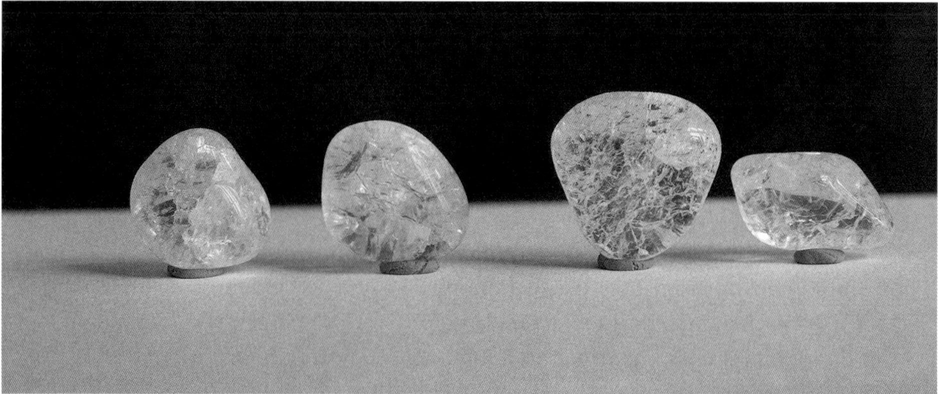

FIGURE 4.9. Quench-cracked quartz crystals. Pieces of quartz were heated and afterward quenched in cold water. The photo shows the numerous small fissures that result from this treatment. This network of fissures (reaching up to the surface of the crystal) allows the dye to penetrate the stone when it is heated again and quenched in the dye liquid, resulting in a colored stone (fig. 4.8). Photo by Marjolijn Bol.

Ultimately, the role of verdigris in imitating emeralds led natural philosophers to believe that emeralds and copper green were related substances. Albertus Magnus, for example, explains the genesis of emeralds from the pigment verdigris: "Emeralds occur in veins of copper, which gives the gems their perspicuousness [*perspicuo*] because it has not yet actually become copper—for the rust of copper is green" (*quia rubiginis aeris habet viriditatem quod in venis aeris*).[77] The translucent emerald was thus thought to grow from verdigris or, as Albertus Magnus describes the pigment, from the "rust of copper." An earlier Arabic source, the treatise on stones by the Persian Muslim scholar and polymath Al-Beruni (AD 973–1048), shows that Albert's theory was already in existence before he wrote his mineralogical treatise. Al-Beruni includes a similar yet more critical stance on the idea that emeralds and copper are related substances. He writes that according to the Persian physician Rhazes (AD 854–925) emeralds and verdigris must be related because the color of the emerald is like that of verdigris. However, Al-Beruni adds the note that this can be accepted only if emeralds are found in copper mines, not in gold mines.[78] Al-Beruni was right; only those gems that were considered the lesser species of emeralds, malachite in particular, were found in copper mines.[79] In the West, the theory that emeralds were generated from copper was nevertheless persistent and can be found in numerous mineralogical treatises in subsequent centuries. More than two hundred years after Albert wrote his book on minerals, the Italian metallurgist Vannoccio Biringuccio (ca. 1480 to ca. 1539), while questioning various other ancient theories about the emerald, is still certain about its origin in copper: "It must be believed that all these strange things [emeralds], wherever they may come from, are colored by the virtue and power of minerals of copper [*tenti per virtu & potētia dele miniere del rame*]."[80]

The previous theory likely persisted because of the various artisanal practices that used verdigris and copper to imitate the color and clarity of emeralds. We have seen that besides coloring crystals, copper was also used to produce a green color in glass to imitate emeralds, and in chapter 3 above, verdigris ground with oil was used to give parchment windows the appearance of green glass. These interrelations between the materials copper, verdigris, and glass in artisanal practice came to play a prominent role in theories about the genesis of precious minerals in nature.

Material Mimesis and the Arts of Refraction and Reflection

The sources discussed in the previous section show that imitating precious stones was a deceitful practice. Surviving medieval art objects nevertheless suggest that imitation gemstones also served a variety of decorative purposes and were knowingly ordered by their patrons. The materials used to make such imitation gems share many similarities with the materials used to make the more fraudulent imitations. The social practices leading artists to make imitations were various: they could be economic, for instance when the patron could not afford a more costly stone or solid gold or silver; imitations could be made because certain precious stones were scarce; or they could be inspired by ritual practices, such as substituting grave goods for the "originals" that may have existed during the lifetime of the deceased. But whether it used real or imitation gems, medieval jewelry invariably shows that artisans focused on bringing out the play of light between reflecting metals and the colored, translucent materials applied to them.

From the fifth to the eighth century, the most popular setting for precious stones in European metalwork was cloisonné, a practice transmitted from the late Roman period by Byzantine craftsmen. In some cases early medieval jewelry might have been made in workshops in Byzantium and only afterward distributed across Europe.[81] This means that a piece of jewelry used by the Franks in the sixth century might not have been made by the Franks.

To make cloisonné decoration on metalwork, small cells (*cloisons* in French) are created by soldering or otherwise affixing metal strips or wire to make cells into which the gems or their imitations can be set. Red garnet (a neosilicate mineral) was by far the most popular mineral used in early medieval cloisonné ornaments. Garnets are one of the most common precious stones found on earth and are a beautiful translucent red that enhances and is enhanced by the glistening yellow splendor of the gold underneath.[82] Another advantage of garnets is that, owing to their crystalline structure, they can be split into thin sheets, making them particularly suited for filling up the decorative patterns of the small cells.

Among the more spectacular examples of the many jewels of this type are two gold shoulder clasps found in a burial mound at Sutton Hoo, England. The grave, attributed to King Redwald of the East Anglians (d. ca. 625), contained a ship twenty-seven meters long filled with treasures, including a variety of cloisonné ornaments made from the fifth to the seventh century AD.[83] The two shoulder clasps found in this ship are among the most refined examples of early medieval cloisonné jewelry (fig. 4.10). The center piece of the gold clasp is inlaid with garnets and millefiori glass, surrounded

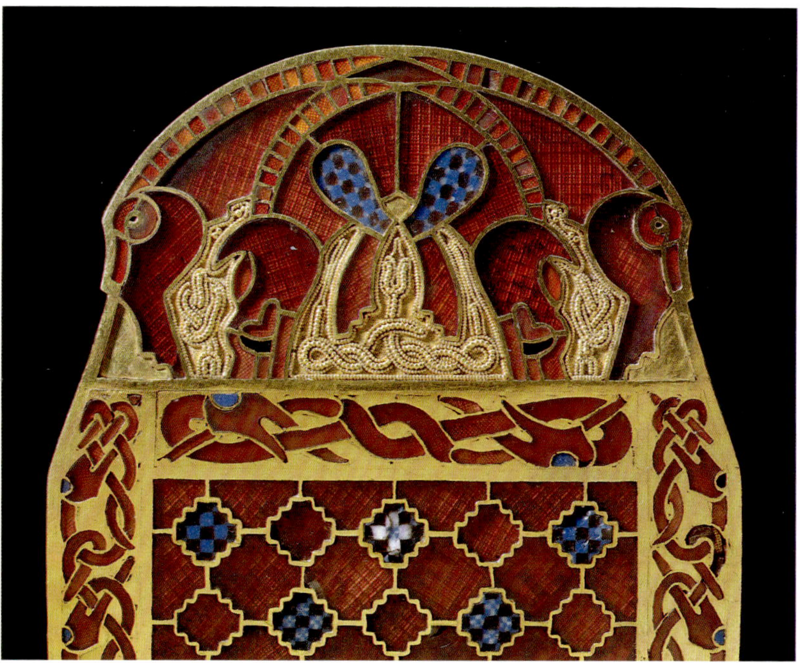

FIGURE 4.10. Gold and garnet shoulder clasp from the Sutton Hoo ship burial, ca. 625. British Museum, London. © The Trustees of the British Museum.

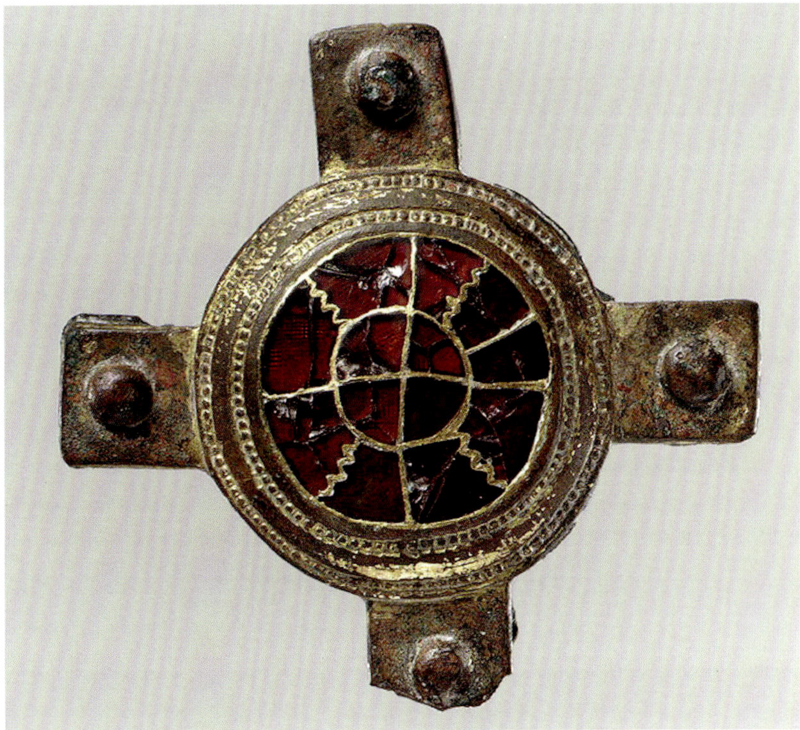

FIGURE 4.11. Cloisonné strap mount, first half of the sixth century, Frankish. Copper alloy, gilt, central medallion with gold and garnets. Metropolitan Museum of Art, New York. Public domain.

by interlacing beasts with translucent blue glass eyes. At the rounded end, the ornament is crowned by two boars, similarly designed from inlaid garnets and millefiori glass. The garnets are placed on gold foil embellished with a decorative waffle pattern that delicately reflects through the translucent red stones.

Garnets were so commonly available that they were used to adorn not only pieces of solid gold, but also those jewels and ornaments that were made to look like gold. Numerous such objects survive, but I will single out a sixth-century Frankish strap mount, now kept in the Metropolitan Museum of Art, New York (fig. 4.11).

This ornament is made from a copper alloy that has been gilded so it looks like gold. The cells are filled with garnets cut into thin slices. The thicker gold foil underneath the garnets has been given a delicate waffle pattern that glistens as light is reflected back through the stones. Thus the gilded copper set with gems and glass on reflective gold foil was meant to imitate the material appearance of jewelry made of solid gold like the shoulder clasps in figure 4.10.

Translucidus *and Enamel*

Because the cloisonné setting had been used since ancient times to fit precious stones into metalwork, it is believed that cloisonné, one of the oldest enameling techniques, was invented out of the desire to imitate precious stones with glass in these settings. Many of the ornaments set with garnets have pieces of glass set in their cells as well, sometimes even in decorative patterns such as in the millefiori glass in the shoulder clasp in figure 4.10. However, to make true cloisonné enamel, powdered glass mixed with the appropriate metal oxides is fused to a metal base—usually gold, silver, or copper—in a furnace at high temperature. The result is an extremely durable, lightfast vitreous coating that can range from opaque to translucent or transparent.[84] Cloisonné enamel became especially popular in Byzantine art by the middle of the ninth century. David Buckton argued that the popularity of enamel may have risen at this time because the supply of garnets had dried up. As a result, according to Buckton, enamel was no longer considered "a poor man's substitute" and came to be seen as a technique worthy of embellishing even the most precious work of the goldsmith.[85]

An early example of such a precious object decorated entirely with cloisonné enamel is the Fieschi Morgan Staurotheke, dated to the beginning of the ninth century and now kept in the Metropolitan Museum of Art (fig. 4.12 depicts the cover of the staurotheke [reliquary]).[86]

This box, possibly made in Constantinople, was meant to hold a relic of the true cross. It is made from gold, gilded silver, and enamel worked in cloisonné. The inside of the lid is worked in the niello technique, which uses a black mixture—usually sulfur, copper, silver, and lead—as an inlay on engraved or etched metal. On the outside the staurotheke is decorated with various colors of opaque and translucent enamel. The translucent red, blue, and green enamels are fused to a "bed" of polished gold foil that can be seen shimmering through (fig. 4.12).

In dealing with the art of the goldsmith, book 3 of Theophilus's *Schedula diversarum artium* contains a detailed description of how to make such cloisonné enamel on gold. The *Schedula*'s instructions are part of a larger group of recipes that detail how to make and decorate a gold chalice.[87] Theophilus first explains how to make the cells by soldering or otherwise adhering thin gold strips or wires to the chalice. These compartments have to be carefully

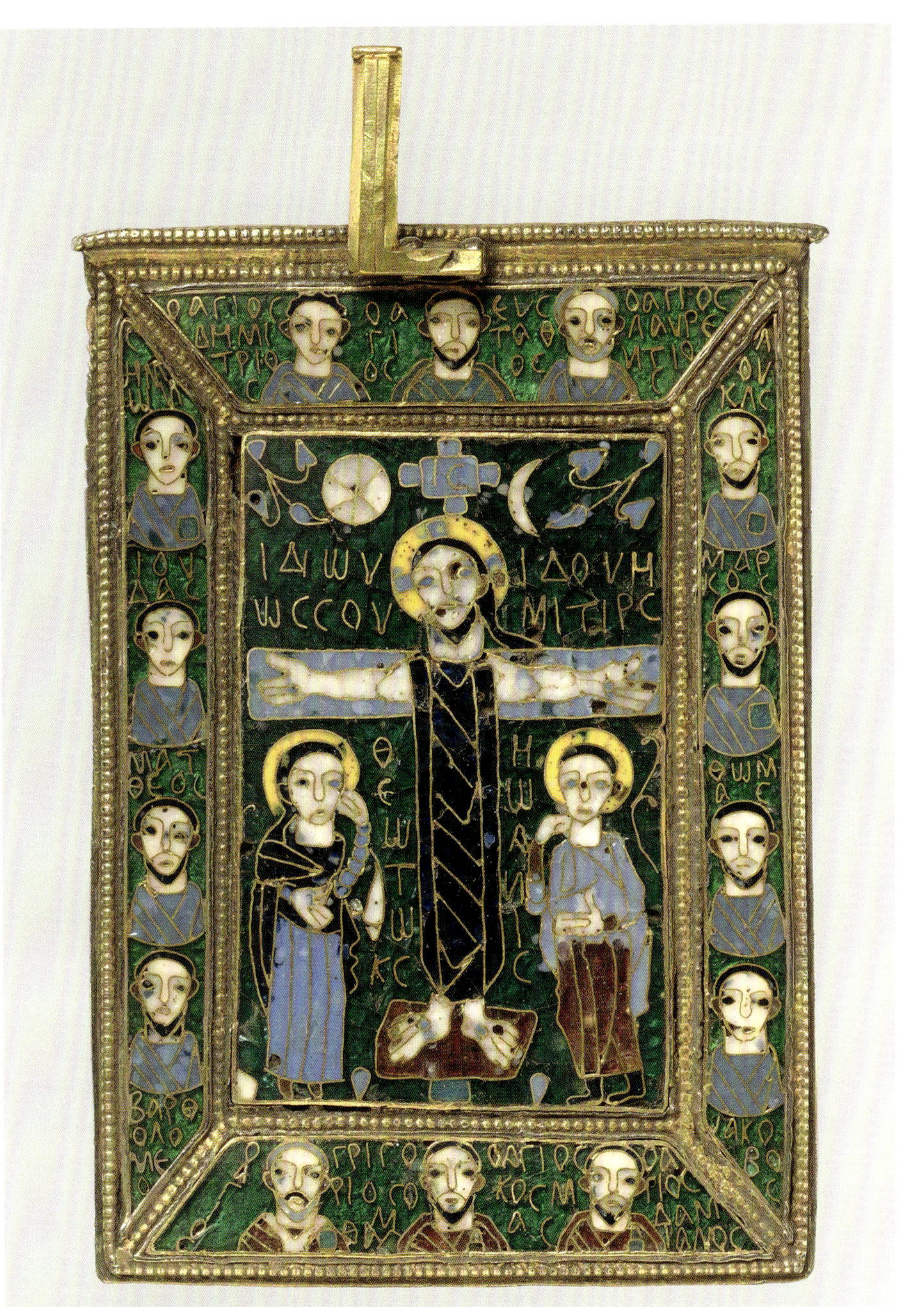

FIGURE 4.12. Cover of the Fieschi Morgan Staurotheke, ninth century, possibly Constantinople, Byzantine (made). Cloisonné enamel, gilt silver, gold, niello. Metropolitan Museum of Art, New York. Public domain.

planned and can afterward be filled with either precious stones or enamel. To fill the cells with enamel, Theophilus explains that you take single sheets of gold, fit them into the cells, and then remove them. These are the base-plates for the enamel. The gold plates must be fitted with a border made from a gold strip bent twice around.[88] In this compartment you then solder the various cells that will hold the enamel. According to Theophilus, your cloisons can have the shape of "circles, or scrolls, or flowers, or birds, or animals, or figures" (*sive circulos sive nodos sive flosculos sive aves sive bestias sive imagines*).[89] When you have thus created the cells, you are ready to fill them with molten glass. Earlier, in book 2 of his treatise, Theophilus had explained that to make your enamel colors, ancient mosaic stones, known as *tesserae*, were used:

> In the ancient buildings of pagans, various kinds of glass are found in the mosaic work—white, black, green, yellow, blue, red, and purple. They are not see-through but opaque like marble [*non est perspicax, sed densum in modum marmoris*] and are like little square stones. From these, enamels are made in gold, silver, and copper, of which we shall speak fully in their place. One also comes across various small vessels of the same colours, which the French, who are most skilled in this work, collect. The blue they melt in their kilns, adding to it a little clear and white glass, and make from it precious blue [*saphiri*] sheets of glass, which are very useful for windows. The purple [*purpura*] and the green [*viridi*] they also make use of in a similar way.[90]

These last three colors—blue, purple, and green—made translucent by the admixture of clear and colorless glass for use in stained glass windows, were also the most common translucent enamel colors.[91] After having tested your *tesserae* to see whether they all melt evenly, you heat them and throw them into a copper vessel containing water, and they "will immediately splinter into small pieces" (*et statim resiliet minutatim*).[92] The glass splinters are then further broken up by grinding them. You use a "round pestle until [the glass] is powdered, and you wash it, put it in a clean shell, and cover it with a linen cloth" (*rotundo malleo, donec subtile fiat, sicque lavabis et pones in concha munda, atque cooperies panno lineo*). Theophilus explains that the various colors of powdered glass must be moistened before you pick them up from their shells with a goose quill "cut as fine as if for writing but with a longer point and unsplit." You use a thin rod of copper to scrape the colors from the feather into their cells. The enamel is now ready to be fired, cooled, washed, and fired again until all cells are evenly filled. Theophilus devotes a separate recipe to the last step in making cloisonné enamel: grinding and polishing it. For this Theophilus describes a method similar to the one he provides for polishing precious stones.[93] The enamel is affixed to a piece of wax so that you can hold it while rubbing it on various surfaces to smooth and polish it until the colors become translucent and clear (*translucidi et clari*).[94] To describe the colors of his enamel after it has been polished, Theophilus uses *translucidus*. Significantly, this same term is used in only one other place in the *Schedula*; to describe a technique called *pictura translucida*.[95] To make this so-called translucent painting, linseed oil ground with pigments is applied to tin foil. It was from this technique that the art of glazing with oil paint on wooden panels was born. I will discuss the details of *pictura translucida* in chapter 5, but here let me direct attention to the optical similarities between applying oil paints to polished tin foil and fusing colored glass to polished metal. Both the technique of *pictura translucida* and that of cloisonné enamel were designed to approach the special optics of precious metalwork set with gems. It is therefore no coincidence that Theophilus describes the optical effect of both arts by *translucidus*, a term that, as we have seen, has a long history in descriptions of precious stones.

Precious Stories on Copper

For the goldsmith one advantage enamel has over precious stones is that it can be used to create intricate decorative and figurative patterns. Thus enamels were especially useful for incorporating histories and stories on metalwork. Champlevé enamel became popular for this purpose in the twelfth century in the Meuse Valley (now Belgium, France, and Germany) and Limoges (southern France). Different from cloisonné enamel, where the walls of the cells are built up with gold wire, champlevé enamel is made by gouging enamel beds into a piece of metal. The indented cells are then filled with enamel frit (in Theophilus's recipe the crushed mosaic stones) and fired.

A notable early example of the technique is on the Imperial Crown in Vienna, believed to have been made for the coronation of Otto I (Otto the Great) in 962 or for his son Otto II in 967 (fig. 4.13).[96]

The enamel plaques on the gold crown are champlevé enamel with a few cloisonné cells added. Various shades of translucent green, blue, and red enamel decorate some of the halos, the wings of the two seraphim that accompany Christ, and the robes of the various figures depicted on the four enamel plaques. Gold was certainly the most fitting material for a royal crown, but because the champlevé technique required such thick metal, alloys of copper were more commonly used; copper is malleable

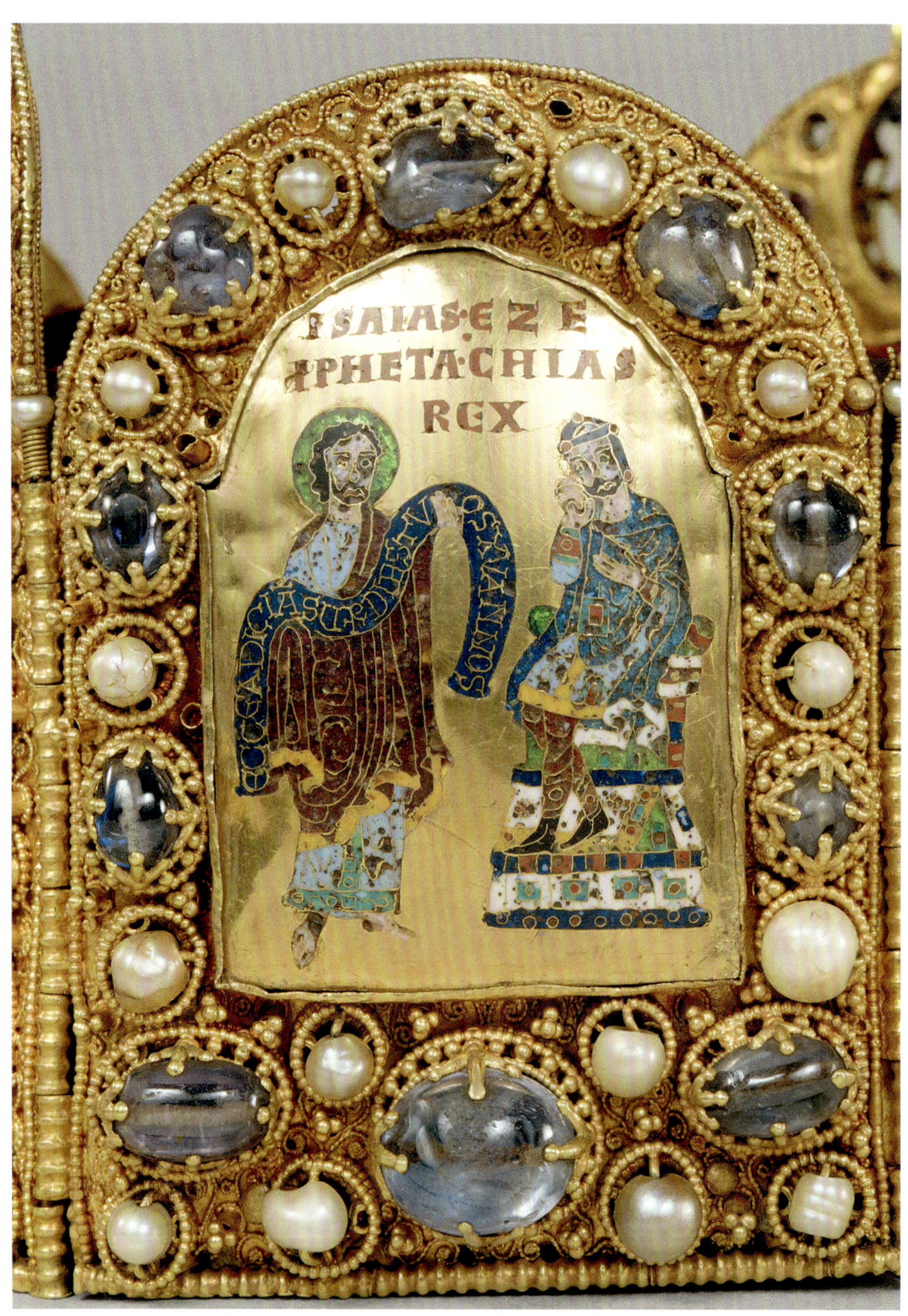

FIGURE 4.13. Detail of the Imperial Crown of the Holy Roman Empire, second half of the tenth century (or early eleventh). Gold, enamel, precious stones, pearls. Kunsthistorisches Museum, Vienna. © KHM-Museumsverband.

and, compared with gold, relatively cheap. The copper that remained exposed was gilded. Indeed, here too enameled copper was meant to look like solid gold.

When alloys of copper are heated, a discolored oxide layer forms between glass and metal, a disadvantage compared with gold.[97] Applied to such a dark and unreflective base, translucent enamel colors would thus have appeared dark and dull. Mosan craftsmen discovered a way around this problem by applying gold foil under certain motifs, so they could apply translucent enamel frit over it. This special technique can be found in a series of eleven copper plaques (ca. 1150–75) that today are dispersed across several museums around the world. The plaques were probably part of a larger ensemble, perhaps an altar, and contain several translucent enamel colors applied to a layer of gold foil. Technical research has shown that the copper likely was first covered with a base layer of glass to which the gold foil was applied. Afterward the translucent enamel color was fired onto this gold foil.[98]

A particularly beautiful example within the series is the plaque depicting Pentecost (fig. 4.14).

Here Saint Peter's halo is made of translucent deep red enamel over gold foil. It seems that the translucent enamel was reserved for the most important apostle, seated directly beneath God's pointing finger, because the others have opaque red and turquoise enameled halos instead. Saint Peter was indeed the first apostle who proclaimed the miracle of Pentecost—the Holy Spirit's descent upon Christ's disciples. The architecture of the scene contains some translucent enamel as well; the red floor the apostles' feet are resting on is translucent red enamel over gold foil, and there is translucent green enamel on gold foil in the architectural decoration encompassing the heads of all the apostles.[99] The other plaques belonging to this group also use translucent red and green enamel for important motifs such as Saint John the Baptist's halo, angels' wings, Christ's cross, the celestial regions, and the wing tips and eyes of the two mythical gryphons pulling Alexander the Great's chariot.

Perhaps the most important reason for the increasing popularity of enamel may be that, unlike precious stones, it allowed artisans to depict the stories that became such a crucial characteristic of medieval art. In these stories the translucent enamel colors, once designed to approximate the costliest gemstones, were now used to evoke the incorporeal glowing splendor of the Divine. The previous examples show that medieval artisans developed sophisticated ways to give the impression of a solid piece of metalwork using different materials. That so many medieval arts are characterized by

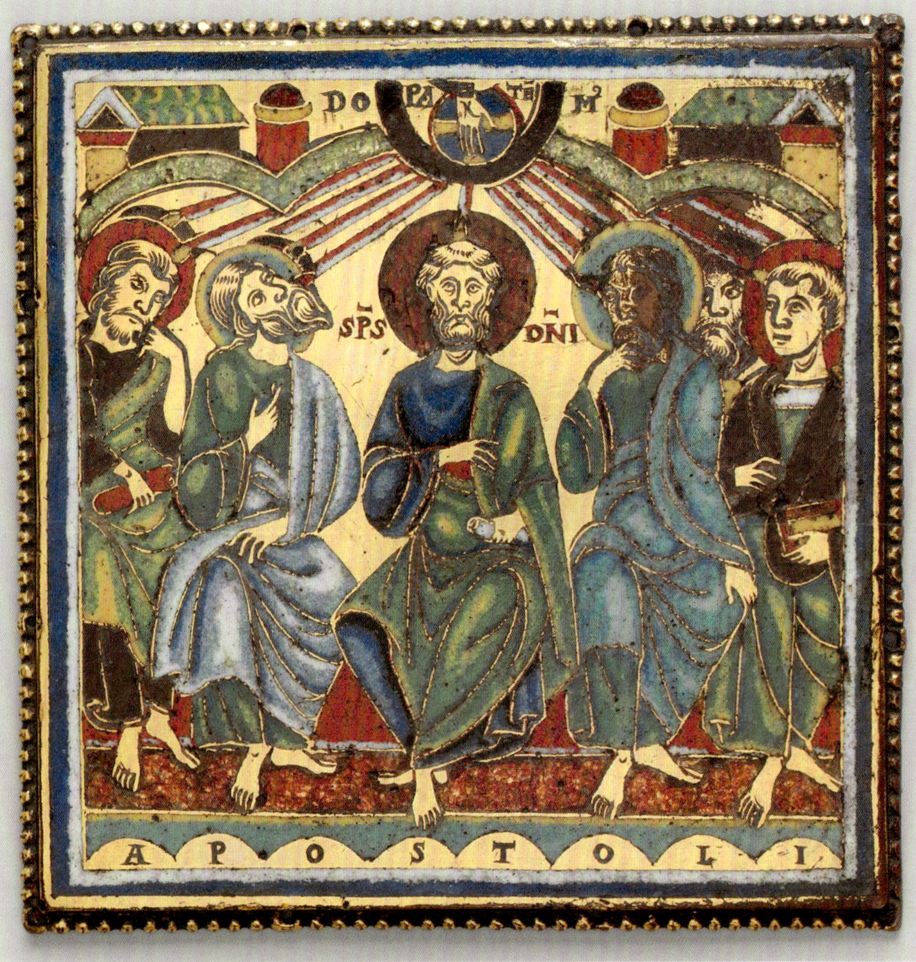

FIGURE 4.14. Plaque with Pentecost. Champlevé and translucent enamel on copper gilt. Cloisters Collection, Metropolitan Museum of Art, New York. Public domain.

this type of material mimesis suggests that such art objects were in demand. Depending on their intended use, the craftsmanship required to make these intricate optical substitutes must have been held in high esteem as well. The art of material mimesis was also appreciated for another reason; it allowed artisans to "vanquish" nature. We have seen that medieval natural philosophers believed the processes and materials produced by "art" were related to those produced by nature. This means that re-creating the splendor of gems and gold in different materials allowed artisans to creatively "increase" and “enlarge” the supply of those precious materials produced in the earth in small size and quantities, achieving a scale unseen in the natural world. A church window glazed with precious stones made from glass could evoke the divine light. And, as we will see in chapter 5, wood could be used to enlarge precious metalwork to create monumental altarpieces and frontals. In this context, and in imitation of many of the craft practices discussed in this chapter, painters explored the practical uses of oil’s ability to form a light-transmitting colored substance with certain pigments to imitate the appearance of enamel, gems, and other translucent splendors on the polished surfaces of tin, silver, and gold.

5

Making Glazes

Practices, Recipes, and Reconstructions

The earliest surviving medieval panel paintings and polychrome wooden sculptures were made from the eleventh century to the thirteenth, about the same time as the champlevé enamels just discussed. These art objects, however, have much more in common than their dates. Most champlevé enamels were made with pieces of copper gilded to look like solid gold, and most medieval panel paintings made from wood were covered with brightly polished gold leaf or silver or tin varnished to look like gold. The saturated glow so much admired in precious stones, translucent enamel, and stained glass was evoked on wood with translucent oil paint: the glaze. Such glazes were applied to polished metal foils so that, as with enamel and precious stones, light would be reflected through the paint.

This desire to imitate reflective and refractive splendor ensured a permanent place for the oil medium in painters' workshops. Even though translucent pigments and oil were likely first used to color crystals and other transparent minerals—as described in chapter 4—these materials eventually became the main ingredients of glazes, one of the most important resources of medieval painters. To fully understand the roots of the oil medium in the workshops of medieval panel painters, this chapter will therefore take a closer look at the materials they used to make glazes—the binding medium and the pigments; the techniques they used; and their reasons for applying this special translucent paint. This discussion also highlights a fundamental difference between varnishes and glazes. Unlike the varnish—a substance medieval artisans manipulated to make it so thick and lustrous that it could be applied only with the hands or with a sponge—glazes were prepared so they could be applied with a brush. Glazes thus were used to paint a variety of translucent details, including gems, or the intricate patterns and figurative details of enamel.

Thick and Clear as a Beautiful Crystal: Painter's Oils

A glaze can be defined as a smooth, translucent coat of light-transmitting paint of saturated color made with pigments that have a refractive index similar to that of their binding medium.[1] Drying oils are the only binding medium that allows for making a "true glaze," which allows light to be transmitted without interruption through binding medium and pigment. Oils are called "drying" when their components can form cross-linkages, initially by oxidation, the reaction with oxygen in the air. As a result of this polymerization process, which takes a very long time, the oil forms a solid film.[2] Even though the oil may be touch-dry in a few days, depending on the oil used and its preparation, complete drying may take many years.[3] The sixth-century Greek physician Aëtius of Amida explains how such a drying oil can be extracted from nuts either by "pounding" or "pressing" them or by bruising them and then throwing them into boiling water. Such accounts of oil extraction are rare, however, and recipe compilations recording the materials and techniques of the medieval painter typically do not explain how oily binders are extracted from raw material. The instructions for oil extraction in Theophilus's *Schedula diversarum artium* are thus rather exceptional in providing a glimpse into how oils used for painting might have been obtained in the twelfth century. In a recipe that explains how to stain doors red with linseed oil, Theophilus elaborates on a method for expelling oil from flax using a type of heat extraction:

> If you want to stain doors red, take linseed oil [*oleum lini*], which you make in this way:
>
> Take the seed of flax, and dry it in a pan over the fire without water. Then put it into a mortar and pound it with a pestle until it becomes a very fine powder. Replace it in the pan, pour in a little water, and heat it strongly. Afterward, fold it in a new cloth and place it in the press where olive oil [*oleum olivae*], walnut oil [*nucum*], or poppyseed oil [*papaveris*] is usually extracted, so that this oil may be extracted in the same way. Grind red lead or vermilion with this oil upon a stone without water and, with a paintbrush, apply it over the doors or panels that you want to stain red, and dry them in the sun. Then paint over them again and dry again. Finally

apply over it a coat of the sticky substance which is called varnish [*gluten quod uernition dicitur*], and which is made in this way.[4]

Theophilus's method for expelling oil from linseed involves heating and crushing the seeds, then reintroducing water before oil can be extracted.[5] The press he recommends is used for pressing a variety of other oils as well. If the seeds are heated before or during extraction, the oil is more easily expelled, yielding more product. Immediately after extraction the oil is still rather turbid (fig. 3.1) because freshly expressed oil contains a lot of impurities from the husks and other parts of the seeds. When the oil stands for some time, these impurities quickly settle and form a sediment at the bottom of the container, and then the oil can be poured off and or filtered (fig. 3.2). Reconstructions show that the oil extracted by Theophilus's method is slightly darker than oil obtained by cold extraction alone (fig. 5.1).

This is likely because the coloring matter from the husks of the seeds is darker after being heated.[6] Theophilus explains that oil paint can be applied

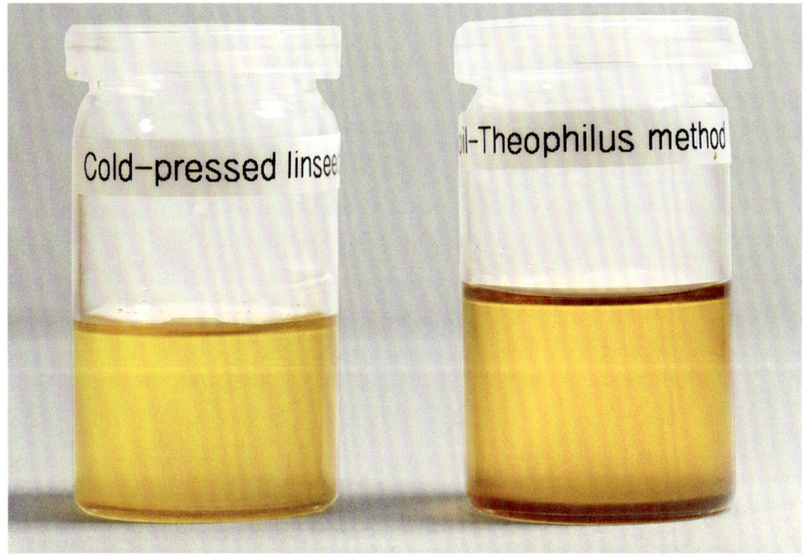

FIGURE 5.1. *Left:* linseed oil obtained by cold extraction. *Right:* linseed oil extracted by the Theophilus method. For this experiment I used a mechanical oil press. Expelling by pressure closely approximates the method that would have been used in the premodern period, when oil was typically expelled by placing it in a cloth between two wooden beams that were put under pressure. Note that the oil extracted by the Theophilus method (which uses heat) is slightly darker than the cold-pressed linseed oil pictured next to it. Photo by Marjolijn Bol.

only to objects that can be dried in the sun, and that each time you apply a color, you must wait until the first has dried before applying another over it: "One figures this is a particularly long and tedious process" (*imaginibus diuturnum et taediosum nimis est*).[7] This remark about oil's slow drying rate suggests that Theophilus's painter's oil received no further treatment after its extraction, meaning it was likely not "bodied," or cleared of its aqueous matter.[8]

Compared with regular cold-pressed drying oils, "bodied" or "prepolymerized" drying oils have different optical and handling properties that were of interest to premodern painters.[9] Bodied oils can be prepared in a variety of ways. Today they are typically manufactured by heating them in a vacuum, producing a "stand oil." In the past, drying oils were often prepolymerized by exposing them for a long time to oxygen and sunlight, producing "sun-thickened oil." Bodied oils were also prepared over a heat source, producing what we know today as "heat-bodied oil." Historical recipes suggest that a combination of both methods—for instance, exposing a heat-bodied oil to oxygen and sunlight—was likely used as well.

The presence of heat-bodied oil in a paint sample can be demonstrated by scientific analysis, but the painter's use of sun-thickened oil cannot be detected by instrumental means because this method of prepolymerization does not change the physical makeup of the oil. Recipes nevertheless suggest that it must have been common practice to expose both cold-pressed oils and heat-bodied oils to air and light. When untreated cold-pressed drying oils or heat-bodied oils are exposed to air and light for several days during a hot summer or, if the season is less sunny, for several weeks, the oils become not only thicker, but also lighter in color. After being exposed to sunlight for some time, cold-pressed linseed oil is bleached until it is almost colorless, whereas heat-bodied linseed oil—which turns relatively dark brown after being cooked—pales to a lighter brown (fig. 5.2).[10]

In addition to these optical changes, bodied oils have different handling properties than untreated oils; they dry faster, are relatively viscous, and have a slightly higher refractive index.[11]

One of the earliest recipes that explains how to make a bodied oil for "tempering" pigments (*quomodo aptatur ad distemperandum*) can be found in book 3 of the treatise of Eraclius.[12] The recipe in question does not specify what type of drying oil is used, but it teaches readers how to create a heat-bodied oil by cooking it with lime while continually "scumming" it—a procedure quite similar to the way varnishes were prepared, as we saw in chapters 2 and 3. The oil is bodied further by exposing it to the sun for at least a month:[13]

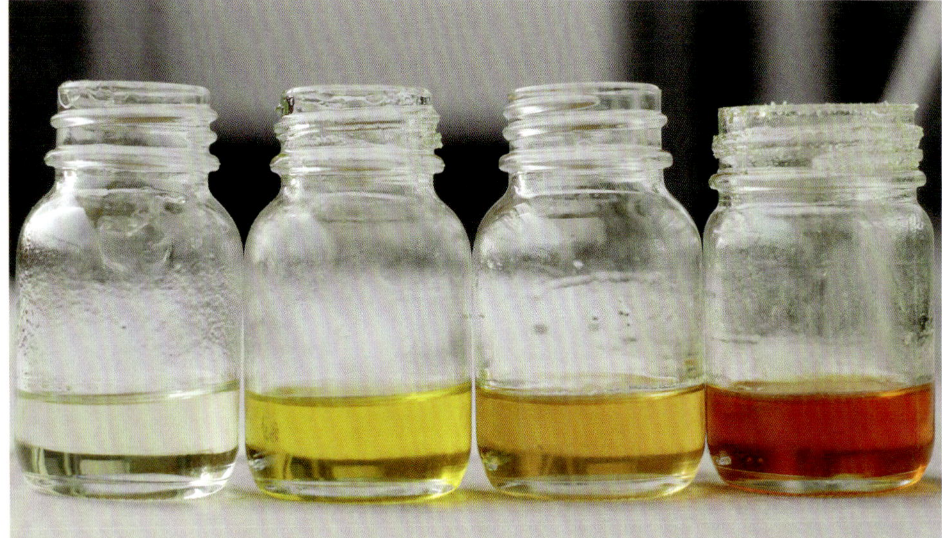

FIGURE 5.2. *Left to right:* cold-pressed linseed oil clarified in the sun; untreated cold-pressed linseed oil; heat-bodied linseed oil clarified in the sun; and heat-bodied linseed oil that received no further treatment. Photo by Marjolijn Bol.

> Put a moderate quantity of lime [*calcem*] into oil and heat it, continually scumming [*despumando*] it; add ceruse [*cerosium*] to it according to the quantity of oil, and put it in the sun for a month or more, stirring it frequently. And know that the longer it remains in the sun, the better it will be. Then strain and keep it, and distemper the colours with it.[14]

If *cerosium* is indeed to be translated as "ceruse,"[15] a white, lead-based pigment, then it would have helped increase the oil's drying rate together with its exposure to sun and air. The latter procedure also would have helped to clarify the oil. After exposing the oil to the sun for a time, during which it needs to be stirred to prevent a skin from forming on its surface, Eraclius recommends straining it, again a practice common in varnish making. These procedures would have produced a thick and clarified oil that would have dried relatively more quickly than an untreated cold-pressed drying oil.

Various other recipes similarly suggest that viscosity, clarity, and faster drying were the main concern of medieval panel painters when preparing oily binders. In the first decade of the fifteenth century, a recipe in the treatise of Alcherius explains how to mix oil with "quicklime" (*chaux vive*) and equal quantities of "ceruse" (*ceruse*). In this recipe the oil is not cooked with these ingredients but is put in the sun "without moving" (*ne le movez*) for a month or more because "the longer it remains the better it will be."[16] It is not clear whether this recipe describes a variation of Eraclius's method (not cooking the oil on a fire and not stirring it when exposed to sun and air) or whether the scribe made an error in copying it from a manuscript containing book 3 of *De coloribus*. Whatever the case, various other recipes dating to the beginning of the fifteenth century show that it was not uncommon to prepare painter's oils without heat.[17]

Cennino Cennini, for instance, describes in detail how to prepare two types of oil; the first is cooked on a fire, and the second is prepared by exposing it to the sun. In the first recipe, Cennini explains how to make a heat-bodied oil that is "good for a binder and also for mordants" (*buono per tenpera e anche per mor*).[18] To make this oil you need "somewhere between one and four pounds of linseed oil" (*olio di semenza*), which must be put in a new pot, preferably glazed. Cennini also explains how to construct a safe oven that perfectly encloses your pot of oil so that the fire cannot reach the oil and risk burning it and "even burning the house down." The oil should be heated on a "moderate fire" (*unfuocho tenperato*) because, so we learn, the more slowly the oil is brought to temperature "the better it will be." As with the many varnish recipes from the same period, this oil used for tempering colors must be "boiled down to a half" (*e fallo bollire per / mezo*). If you want to use it as a mordant, you add an "ounce of liquid varnish" (*oncia di vernicie liquida*). This last step in the preparation of Cennini's drying oil would have produced a substance that was more viscous, but especially stickier, typical for mordants. Mordants are strong adhesive compounds that were used, among other things, for fixing gold leaf and other materials to paintings. This is perhaps why Cennini includes a second recipe in which he explains that if you plan to use your oil as a mordant, it should always be cooked on a fire (*ma per mordenti vuole essere pur di fuocho cioe chotto abbi il tuo*), but that for painting (*più perfetto da ccholorire*) a better oil is produced by cooking it in the sun (*olio buono eperfetto cotto al sole*).[19] To make this second, more perfect painter's oil, you put linseed oil in a bronze or copper basin or in a tub and keep it in the sun: "And if you keep it there until it becomes half it is absolutely perfect for painting" (*tieni tanto che torni per mezo e perfettissimo da colorire*). This means that Cennini considered a relatively viscous oil, but of much lighter color than the heat-bodied oil he prepared for making mordants, the best binding medium for use with pigments.

A similar concern for producing a thick yet clear oil for painting can be found in another fifteenth-century recipe, this time from the Strasbourg manuscript. The recipe describes in remarkable detail how to prepare an oil that can be used to "temper all colours with oil much better and more masterfully than the other painters" (*Nun wil ich hie leren wie man alle varwen mit oli temperier sol bas und meisterlich denn ander moler*).[20] This painter's oil can be made from either linseed, hemp, or old nut oil, prepared so that they are "pure and clear and dry fast" (*luter und clor werde und dester gern bald troken werde*). Closely resembling the recipes for the preparation of varnish in the same treatise (see chapter 3), you cook your oil with "old white calcined bones" (*alt gebrent wis bein*) and the "same amount of pumice" (*als vil bimses*) while skimming off the scum. Finally, you dissolve zinc vitriol (*galicen stein*) in the hot oil so that it "becomes pure and clear" (*gar luter und ouch klar*). And, again as in making varnish, the second part of the recipe explains that the oil you have thus prepared must be filtered through a "clean linen cloth into a clean basin" (*durch ein rein lin tüchlin in ein rein bekin*). This oil is then put in the sun for "four days so the oil becomes thick and also as clear as beautiful crystal" (*dik und ouch luter als ein schöner cristall*). The Strasbourg recipe continues that this oil should be used to grind all your pigments to a consistency resembling porridge:

> [It] dries very fast [*trocknet gar bald*] and makes all the colours quite clear and bright [*macht alle varwe schön luter und ouch glantz*]. Not all the painters know about this oil. Because of its quality this oil is called *oleum preciosum*, as one *lot* costs at least one schilling. One should grind all colours with [this] oil and also temper all colours to the thickness of a porridge [*als ein halber bri*], neither too thick nor too fluid.[21]

Another recipe for painter's oil, contained in the Tegernsee manuscript, shows a particular concern with oil's drying characteristics. The recipe was recorded sometime during the mid-fifteenth century or the early sixteenth. It explains that oil paints take a long time to dry. To remedy this, you should dissolve alum on a fire until the water has evaporated and the foam remains. The alum should be added to oil paints "when you are grinding them on the slab" (*wenn du sÿ reibst aud dem stain*). For the same reason that alum was used in varnish recipes, the scribe hints that it is thought to influence oil's ability to dry better.[22]

A few things about the previous recipes for making painter's oil are particularly noteworthy. Except for Theophilus's recipe, all instructions for making painter's oil have been recorded in manuscripts dating from the fifteenth century to the early sixteenth. Even Eraclius's recipe for painter's oil—included in the third book of his twelfth-century treatise—was transmitted in a 1431 manuscript. These recipes recommend preparing the oil so that it becomes relatively pure and therefore would have had better drying properties than the untreated cold-pressed oil Theophilus included in his treatise. Just as in the varnish recipes I discussed in chapter 3, the oils are manipulated during heating by adding various substances thought to attract impurities—such as calcined bone dust, zinc vitriol, and alum—or by improving the oil's drying characteristics in other ways such as adding metal compounds. Finally, the oils are further manipulated by clarifying them by exposure to sunlight and by filtering them through a linen cloth. All the recipes for producing painter's oil share a concern for viscosity. The recommended preparation—cooking the oil and exposing it to oxygen and to the heat of the sun—makes the oil considerably more viscous than drying oils that receive no further treatment after extraction. Even though these recipes appear to reflect a mix of medieval practices and those developed between the fifteenth and sixteenth centuries, the manipulations of the oil medium they describe were to give the oils properties suitable for glazing. As we will see later in this chapter, it was indeed the oil's viscosity—thinner than a varnish but much thicker than a raw cold-pressed oil—that made it particularly apt for glazing.

The Colors without Body

Comparative analysis of premodern recipes shows that medieval artisans were intimately familiar with the special translucency of some pigments, especially when ground with oil, and used a rather particular vocabulary to describe the optics of their coloring materials. In this context the green pigment verdigris is discussed most frequently. Eraclius, for example, points out that verdigris should *always* be ground with oil when it is used on wood, and the fourteenth-century Icelandic recipe collection *Líkneskjusmið* states that while all colors can be ground with glair (the liquid that settles out from beaten egg whites left to stand), "green" *must* be ground with oil.[23]

In the past, variations of the appellation "verdigris" were used to describe a group of pigments made from the corrosive product of copper (and its alloys; see fig. 5.3). The basic method for making verdigris is always the same—exposing copper to organic acids, such as vinegar (fig. 4.7)—but historical recipes describe many variations of this procedure. As a result the chemical substances produced, variations of copper acetate compounds, also differ greatly.[24] Copper acetates can be divided into two groups: basic and neutral. Basic verdigris is formed by acetic acid vapor (vinegar), water vapor, and air acting on copper and its alloys, while neutral verdigris, which was made to purify the pigment scraped from the copper, is formed when

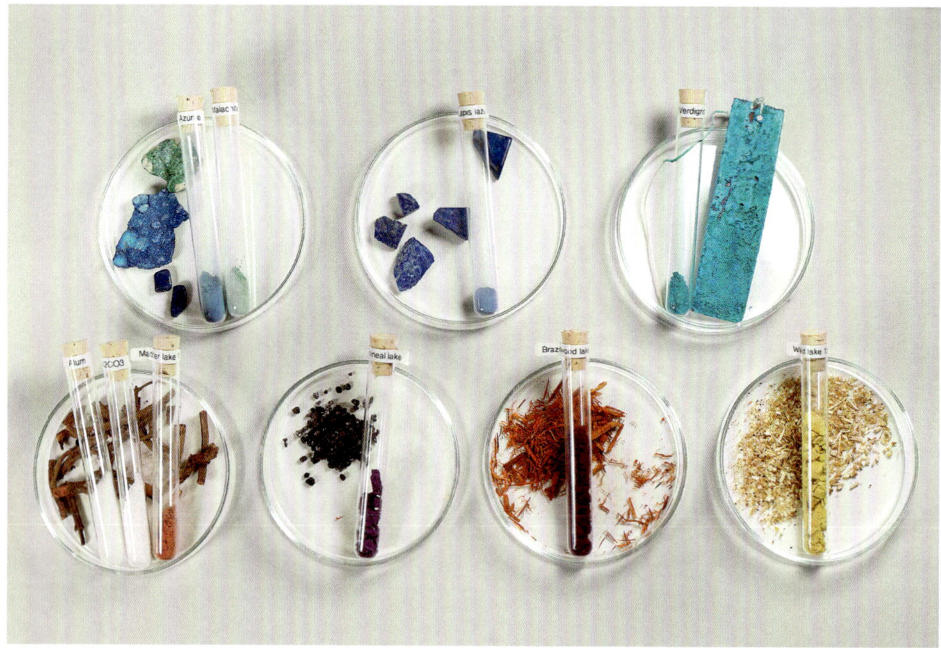

FIGURE 5.3. The various glazing pigments with their raw materials. *Top row, left to right:* azurite and malachite; ultramarine blue (lapis lazuli); and verdigris. *Bottom row, left to right:* madder lake; cochineal lake; brazilwood lake; and weld lake (note that depending on preparation the color of lake pigments may vary quite a bit). Photo by Marjolijn Bol.

basic acetates are dissolved in acetic acid (vinegar) or are ground with it (and afterward recrystallized from it).[25]

Until the discovery of synthetic alternatives, verdigris was the most common green pigment for easel painting. Its popularity had everything to do with its special optics. Writing in the fourteenth century, Alcherius gives a detailed description of the optical characteristics of verdigris, a pigment he categorizes as a color "without body," similar to the yellow dye saffron:

> To make a green clear in its nature, and without body [*in substantia clarum et non corpulentum*], that is, having no substance [*absque substantia seu corpori*], such, for example, as is the colour of saffron, i.e. of *crocus*, which does not cover up other colours so as to conceal them, on account of its thinness [*subtilitate*], clarity [*claritate*], and rarity [*raritate*], owing to which other colours appear through it, wherefore this colour as well as the said green is overpowered and shows little or not at all, nor can it be much seen over other colours.[26]

Alcherius thus considers the colors "without body," pigments that "have little substance" and that allow other colors to appear through them. In the course of the fifteenth and sixteenth centuries we often find a similar terminology used to describe the optical characteristics of translucent pigments and the natural dyes that were transformed into the "lake pigments" so they could be painted with. Before the introduction of synthetic dyes, the dyestuffs used to make pigments were exclusively organic: plant-based or insect-based.[27] Until the beginning of the seventeenth century, historical recipes and other sources on art technology typically mention red dyes extracted from female scale insects—most notably dyer's kermes, derived from *Kermes vermilio* Planchon; lac, produced by *Kerria lacca* Kerr and several other species; cochineal from Old World carmine scale insects of the genus *Porphyrophora* such as Polish cochineal, *Porphyrophora polonica* L., and *Porphyrophora hamelii* Brandt; and, since the sixteenth century, also from the Mexican cochineal insect *Dactylopius coccus* Costa. Since the Middle Ages red dyes were also obtained from various plants, including the dyestuff brazilwood from sappan wood, *Biancaea sappan* (L.) Tod., and, since the sixteenth century, from New World species such as *Paubrasilia echinata* (Lam.) Gagnon, H. C. Lima & G. P. Lewis (fig. 5.3).[28] The lake pigments derived from the various species of soluble redwood are not very lightfast and were therefore more frequently used in book illumination than in panel painting, as is evinced by both art objects and historical recipes.

Other important sources for lake pigments were textile shearings and clippings, notably discarded wool and silk. Recipes and scientific examination show that such textiles were recycled to obtain the dyes for making red lakes, especially kermes and madder.[29] This last dyestuff is derived from the roots of *Rubia tinctorum* L., cultivated across Europe (fig. 5.3).

The various yellow dyestuffs available to medieval painters are all characterized by poor lightfastness. This is likely why there are not many recipes that explain how to prepare them for painting with.[30] As we have seen, yellow dyestuffs did play a prominent role in the recipes for yellow varnishes. In these examples the dyes were dissolved in their binding medium. But this is not the way natural colorants were used in paint.

Before they could be painted with, dyestuffs had to be stabilized by precipitating them onto a colorless particle—the substrate.[31] The chemical bond between dyestuff and substrate transforms the soluble dye into an insoluble pigment known as a lake. The reagent used to form this substrate was generally potassium aluminum sulfate (known as alum or potash alum), which was made to react with an alkali such as lye prepared from wood ash (fig. 5.3). The dyestuff was first extracted from its raw material in the alkaline solution, then potassium aluminum sulfate was added to produce a form of hydrated alumina that can precipitate with the dyestuff to form the lake pigment (fig. 5.3).[32] Of the substrates that could be used for this purpose, hydrated alumina is perfect for producing translucent glazes, since it has a refractive index of 1.5. This means that natural dyes precipitated to this material are fully light transmitting when ground with oil (compare fig. I.8).

A recipe in LeBègue's fifteenth-century treatise for making "a rose colour from Brazil wood" (*ad faciendum colorem ligni Brexilii rosaceum*) shows a perfect understanding of how the translucency of the dye depends on the substrate used to transform it into a pigment.[33] Having first explained how to make a pigment out of the red dye, the recipe adds that if gypsum is not used but only hydrated alumina is added, the color produced can be used only for "shading," not as an opaque color with "body" (*corpus*):

> For it has no body or substance [*quia corpus seu substantiam non habet*]; and when the *bracha* [white lead] or gypsum is added, then it can be used as a body colour as well as for shading, because the gypsum or *bracha*, which have body, give their body to the colour [*qui corpus habent incorporant colorem ipsum*].[34]

Other recipes similarly use *corpus* to describe the relative covering power of pigments. The sixteenth-century Marciana manuscript even categorizes pigments and dyes based on having body, meaning opaque, or "being without body," that is, translucent:

> The colors are of two kinds, one which consists of those which have no body [*che non hanno corpo*], and which do not conceal the colours laid under them, but only tinge them, as saffron [*zafferano*] for instance; the other consists of those which have body [*che hanno corpo*] and which cover every other colour over which they are laid [*e' quali quoprono ogni altro colore essi truovono sotto*]. . . . The lake has no body [*la laccha non ha corpo*], therefore take cinnabar [*cinabro*], and according as you wish the colour to be more or less dark, take more or less lake.[35]

This terminology for describing the optical characteristics of colors appears to have been firmly established by the end of the sixteenth century. Besides the Marciana manuscript, it can be found in numerous other recipe collections. Colors are described in this way in BnF. MS Fr. 640, by Vasari in his *Vite*, and in the many recipes recorded at the beginning of the seventeenth century by Theodore Turquet de Mayerne (London, British Library, Sloane MS 2052), to name but a few examples. Significantly, painters' vocabulary for describing the optics of their colors appears to have been disseminated to other fields of knowledge as well. In his *Dell'historia naturale* (Naples, 1599), the apothecary and naturalist Ferrante Imperato (ca. 1550 to ca. 1631) explains how the colors used in painting can be divided into three categories: pigments from minerals, from earths, or from plants. He then continues that painters further distinguish between colors that "have more or less body" (*l'havere più o meno corpo*). The painters say that colours have body when they cover (*che soporo*) and are without transparency (*senza trasparenza*); and of those that do not have body, they say they "have transparency" (*c'hanno trasparenza*).[36]

A recipe from MS Fr. 640 gives a perfect description of the optical characteristics of a glaze paint made with pigments translucent in oil. We learn that "one usually glazes with colors lacking body, such as lake and verdigris." To use other pigments with body to create a light-transmitting paint, the author of MS Fr. 640 advises mixing in "plenty of calcined and ground crystal, which also has no body and reduces their thickness."[37] Elsewhere in the same manuscript, the author observes that *crystallin*, after it has been made into a very fine powder by heating and quenching in cold water, looks like crushed lead white (with a refractive index of 1.94–2.09, lead white is one of the most opaque—that is, not light-transmitting—pigments in oil), but in reality "it has no body."[38] He explains this point in more detail in another recipe: "Crystallin having been crushed in water appears to have body, but in oil it does not have any" (*Le cristallin estant broye a eau semble estre avoyr corps mays a huile il nen ha point*).[39]

In MS Fr. 640 *crystallin* refers to perfectly colorless and transparent glass, such as that made in Venice at the time. This type of glass has a refractive index of about 1.5, so when crushed *crystallin* is mixed with water their different indexes of refraction would result in a white substance, whereas when ground with a drying oil their shared refractive index would indeed have caused the crushed *crystallin* to become close to invisible in the medium it was immersed in (see also fig. I.8).[40] When *crystallin* is mixed with pigments that would otherwise have been opaque in oil, as the recipe recommends, it gives some translucency to the paint because light can be transmitted through the transparent glass particles.

Beyond the pigments discussed thus far—verdigris for green and the various dyestuffs that could be used to produce red and yellow lake pigments—painters also used a precious blue pigment extracted from lapis lazuli (lazurite) to make glaze paints. Before the fifteenth century, this pigment—also known as ultramarine blue—was more often used by painters south of the Alps because shorter trade routes with the East made lapis lazuli more available (fig. 5.3).[41] In the North ultramarine was not entirely uncommon, but scientific analysis shows that another expensive blue pigment, made from the copper carbonate mineral known as azurite (found in copper ore deposits), was used more frequently (fig. 5.3).[42] Azurite's refractive index (1.73 to 1.83), however, is not as close to oil's as is that of ultramarine, which has a refractive index of 1.5.[43]

Azurite and ultramarine were valued for their intense blue color, but when these pigments are ground with oil, they turn relatively dark and lose much of their brilliance, especially when glazed over other colors. Medieval painters therefore might have tried to preserve their brilliant blue by, for example, using an aqueous binding medium for ultramarine or, as happened more often with both ultramarine and azurite, by mixing it with lead white. The fifteenth-century Montpellier *Liber diversarum arcium*, for instance, explains that while all types of colors may be ground with oil, azurite darkens, "whence white is added until it is recalled to pristine."[44]

Besides its dark appearance, azurite is a challenging glazing pigment for another reason. To preserve its color, azurite must retain a certain particle size. When ground too fine it turns pale, almost gray.[45] Paints made from azurite therefore tend to be rather gritty (fig. 5.4). And particles protruding from the painted surface cause the visible light to scatter, reducing potential color saturation and translucency.

With verdigris, the various lake pigments, ultramarine, and azurite as the only colors that approach the refractive index of the oil medium, painters were limited to green, red, and blue for making glazes (fig. 5.4). Glaze paints were typically made from one pigment only, because mixing colors would reduce the paint's translucency and saturation.[46] To make purple, azurite or ultramarine was sometimes mixed with a red lake (fig. 5.5).[47] More commonly, however, glazes were used to "mix" colors in another manner, and this same method was also used to influence the tone and saturation of underlying paint. Paintings show that, to create purple, red lakes were glazed over azurite blue (fig. 5.5). And by the same method, red, blue, and green glazes were used to "deepen" a brighter color—to influence its "value." For instance, when red lake is glazed over a bright, opaque red pigment such as vermilion, its value intensifies to a darker red (fig. 5.5), while a copper green glaze intensifies in value when it covers a green or yellow underlayer (or polished gold leaf). The Montpellier manuscript contains a recipe that describes these two techniques:

> It is noted. Lac may be placed over all colours [*Lacca omnibus coloribus potest superponi*], apart from green and glaucus. If it is put over azure or light blue it makes a muted violet [*superponatur azuro vel blaueto fit violatum obscurum*]; if over just red, a greater red [*si supra rubeo solo magis rubeum*].[48]

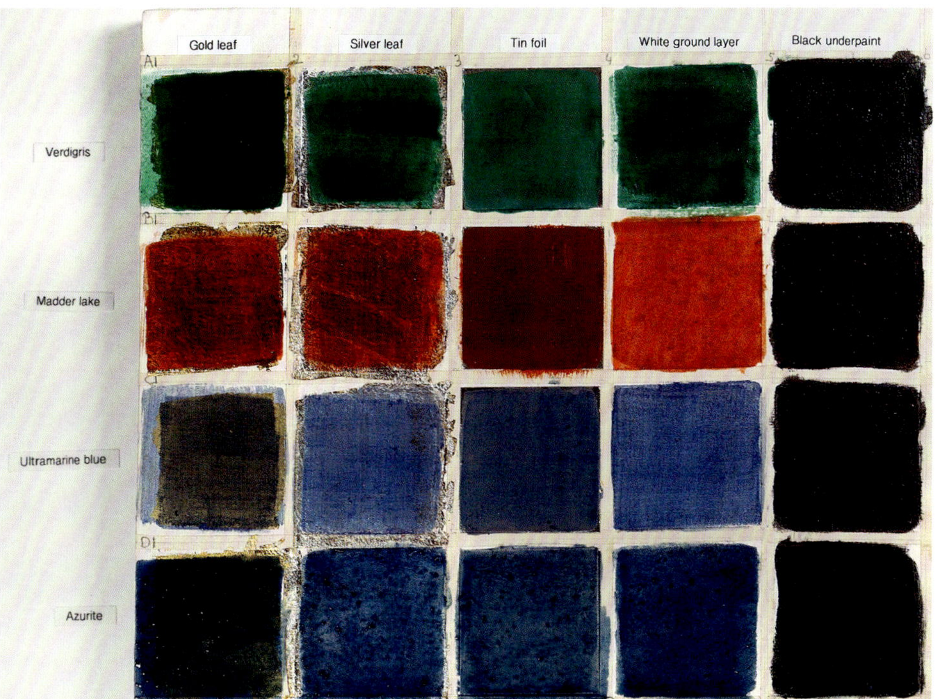

FIGURE 5.4. Oak panel with glazes applied over different underlayers. Note that gold leaf enhances the color of verdigris and red lake paint but that the blue glaze looks much better on a cool underlayer of tin, silver, or white. Recipes and objects show that green and red were indeed more often glazed over gold, while blue was typically combined with silver (or tin). These same metals were also used in jewelry, where translucent green and red gems are set on gold foils whereas translucent blue gems are more often set on silver foils. Photo by Marjolijn Bol.

Recipes like these are rare, however, and we typically encounter descriptions of the glazing technique in another group of recipes that explain how to decorate gold, silver, and tin by painting on them with oil colors, which I will discuss below.

Pictura Translucida: Imitating Gems and Enamel

Instructions for glazing on metal can be found in the same group of treatises discussed in the chapters on varnishing, the earliest of which is in Theophilus's *Schedula*. He describes the technique—which he calls *pictura translucida*—as follows:

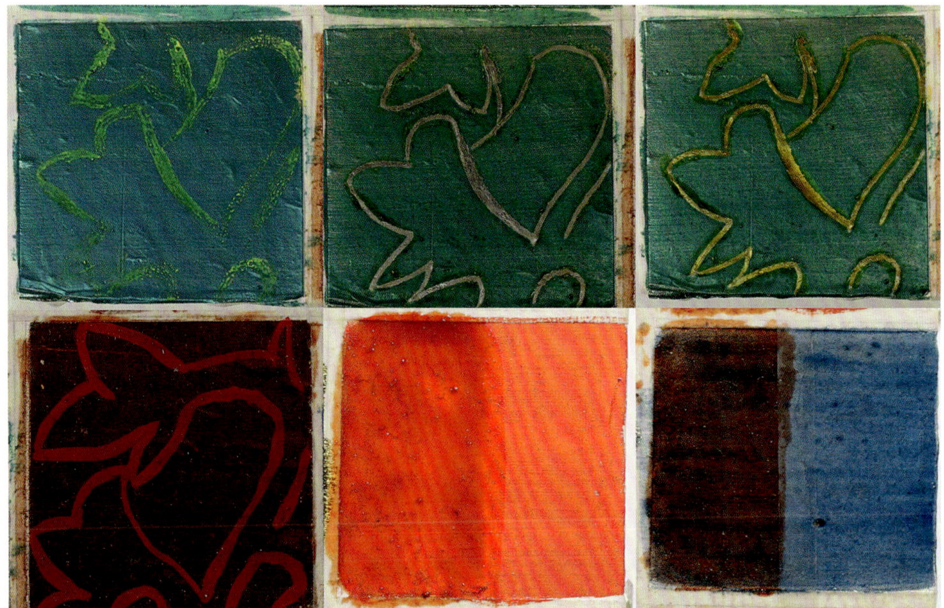

FIGURE 5.5. *Top row, left to right:* green glaze over yellow honey pattern on tin foil; honey pattern removed with tip of brush; and pattern varnished with linseed oil-sandarac-saffron varnish (notice also the change in color from the fresh application of copper green on the left to the varnished pattern on the right). *Bottom row, left to right:* madder lake glaze over red pattern on tin foil; madder lake glaze over "opaque red" underlayer; and madder lake glaze over azurite. Photo by Marjolijn Bol.

Translucent Painting [*Pictura Translucida*]

A painting which is called translucent [*quae dicitur translucida*], is also made on wood, and by some it is described as *aureola*. You should make it in this way. Take some tin foil, not coated with varnish [*glutine*] or colored with saffron [*croco*] but just plain and carefully polished [*sed ita simplicem et diligenter politam*], and with it cover the place you want to paint in this fashion. Then very carefully [*diligentissime*] grind the pigments that are to be applied with linseed oil [*oleo lini*] and when they are very thin draw them on with a brush [*ac ualde tenues trahe eos cum pincello*] and so let them dry.[49]

In this recipe Theophilus clearly distinguishes painting on tin foil with oil colors from coloring such foils yellow by means of varnish. For this last technique, which, as we have seen, can be found in recipe books as early as the ninth century, Theophilus had earlier supplied three recipes.[50] On closer inspection, there are indeed a few noteworthy differences between the practice of varnishing metal foils or leaves (fig. 2.3) and *pictura translucida*. Unlike the yellow varnish—which is applied in the sun using your hands—Theophilus explains that *pictura translucida* ought to be applied with a brush. And rather than using a mixture of oil and resin with a yellow dye added—as in the art of varnishing metal to give it the appearance of gold—a "translucent painting" is made by diligently grinding pigments with linseed oil alone. Because of their handling properties, the varnishes used to imitate gold could function only as surface coatings, since their viscosity required them to be applied warm and spread with one's hands. When linseed oil is not made into varnish it is less viscous, and this lets the painter handle his paints with a brush and thus allows him to make decorative and figurative designs.

Indeed, *pictura translucida* and the yellow varnish served quite different purposes in panel painting. Thick yellow oil-resin varnishes were used to create a material mimesis of an opaque material—gold—by evoking its color with a translucent yellow substance and using a polished metal underlayer to imitate its gleaming splendor. *Pictura translucida*, on the other hand, was used to create a material mimesis of translucent, colored materials such as precious stones, designs in colored glass, and the patterns and stories depicted by the art of enameling.[51] In these instances the polished metal underlayer brings out the translucency of the glaze the way a translucent precious stone would light up when set on polished silver or gold (fig. 5.4).

Let us take the example of the thirteenth-century Westminster Retable (fig. 2.6), which was lavishly decorated with all kinds of mimetic materials (fig. 5.6). The microarchitecture of the retable's frame was embellished with several hundred imitations of cloisonné and champlevé enamel, numerous factitious precious stones, and imitations of stained glass windows. Most of the imitation gems have been lost, but the colors of the glass stones that remain in the retable's frame suggest it once shone with translucent green, blue, red (flashed glass, which is a thin layer of translucent red glass fused to a base of colorless glass), and colorless glass stones. These glass gems were pressed onto putty (chalk and lead white mixed with a drying oil) covered with silver leaf.[52] In some cases the silver underneath the imitation gems was glazed with a red lake pigment ground with a drying oil, perhaps as a mordant to stick the glass gem on or to give the red glass an even deeper, more saturated glow.[53]

As with the glass gems, the retable's imitation enamel is a perfect material mimesis of the optical characteristics of real enamel (fig. 5.7).[54]

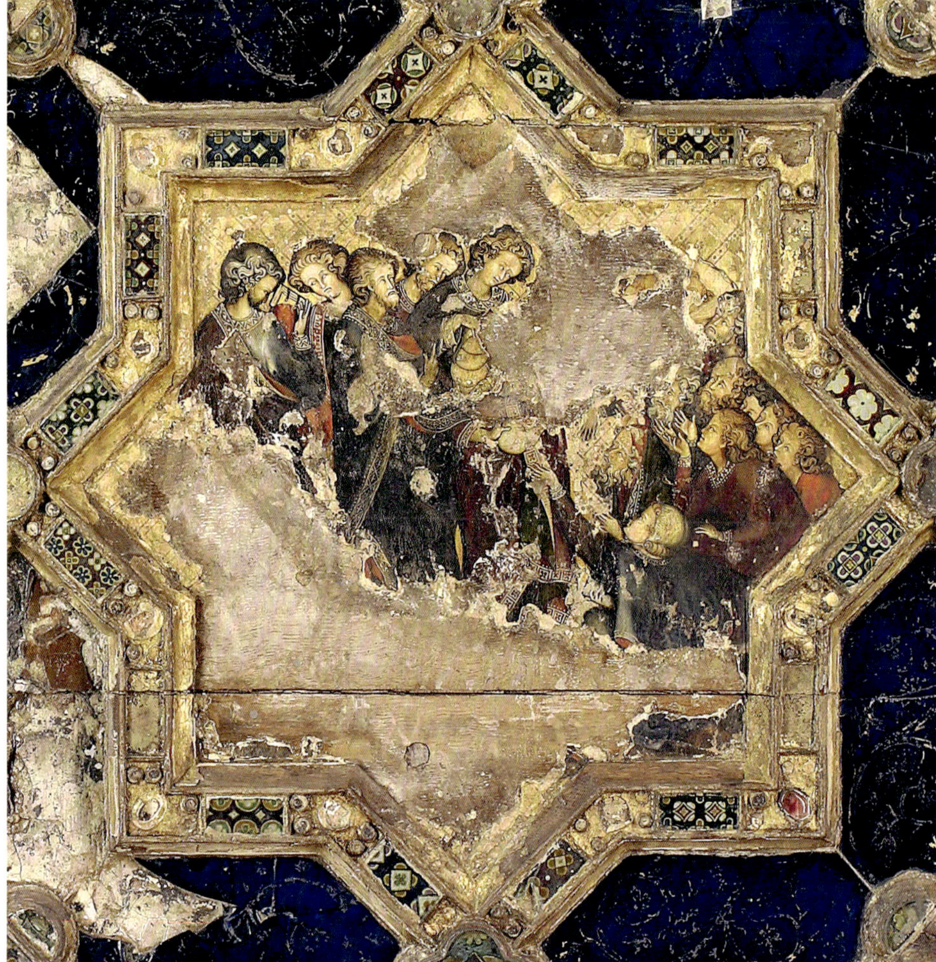

FIGURE 5.6. Detail of the Westminster Retable (see fig. 2.6) showing one of the main scenes framed by imitation enamel. © Dean and Chapter of Westminster.

First, gold leaf was applied to a thick layer of white putty covered with a thin layer of reddish brown mordant (carbon black, red lead, and lead white ground with linseed oil and perhaps a little resin). The gilded areas were then painted with oil colors: glazes made with ultramarine blue, copper green, and red lake, but also with more or less opaque pigments such as yellow ochre, vermilion, red lead, black and white lead (and the mixtures of these pigments). With these paints, decorative patterns suggest the separate cells of the cloisonné enamel, and as with real enamel their colors vary

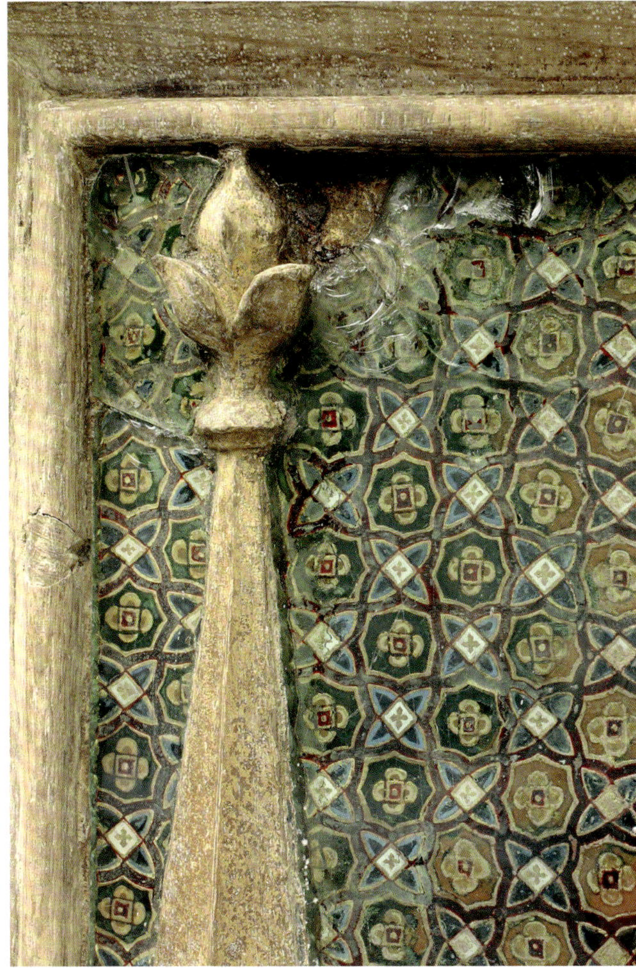

FIGURE 5.7. Detail of the imitation enamel that can be found on the Westminster Retable. © Dean and Chapter of Westminster.

between translucent red, green, and blue and more opaque paints. Finally, the painted gold leaf was sealed off with a piece of transparent, colorless glass. The glass not only protected the paint but also ensured a richly saturated effect on the painted enamel underneath. In this way the glass helps imitate the smooth, polished surface of real enamel and functions optically in the way varnishes were used to bring out the colors of a painting.[55] In fact it was quite likely the other way around; during its early history, the three thick layers of oil-resin varnish applied to panel paintings helped

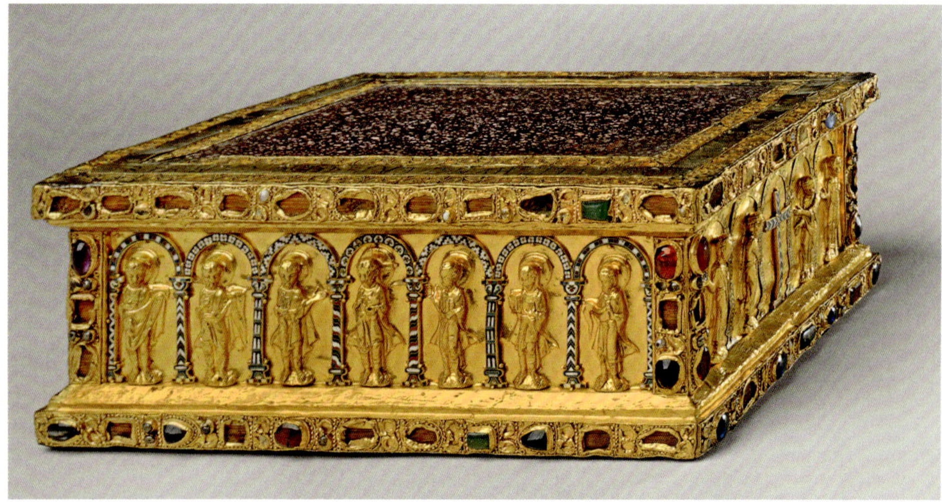

FIGURE 5.8. Portable Altar of Countess Gertrude, shortly after 1038. Gold, cloisonné enamel, porphyry, gems, niello, wood core. Cleveland Museum of Art.

give them a shine like the polished surface of enamel and other reflective materials.

Architectural frames like the one embellishing the Westminster Retable are common in medieval painting and closely resemble the microarchitecture decorating reliquary shrines and other types of metalwork. The saints in the gold relief on the Portable Altar of Countess Gertrude (shortly after 1038), for example, are surrounded by a similar, albeit much smaller, microarchitecture made from cloisonné enamel framed by a decorative border, which had originally been set with ninety-two mostly glass imitations of gems, almost half of them now lost (fig. 5.8).

The lid of this portable altar, one of the many of this type to survive, is a slab of porphyry framed in gold by a Latin dedication: "Gertrude offers to Christ, to live joyfully in him, this stone that glistens with gemstones and gold" (*GERTRUDIS XPO. FELIX UT VIVAT IN IPSO OBTULIT HUNC LAPIDEM GEMMIS AUROQ NITENTEM*).[56] The back of the Westminster Retable—though it is so degraded that it is difficult to recognize today—was likewise decorated with porphyry; not with the real stone, however, but with a two-dimensional painted imitation.[57]

Glazes over metal can be found on numerous surviving medieval panel paintings and sculptures, in the North as well as south of the Alps. For instance, besides being embellished with varnished tin foil, the Spanish altar frontals mentioned in chapter 2 were typically decorated with glazed foils. A beautiful example is on a frontal dating to the end of the twelfth century or beginning of the thirteenth. The central scenes of the frontal are again made from three-dimensional stucco reliefs that evoke the appearance of repoussé metalwork (fig. 5.9).

The glazed metal surfaces behind the figures are applied to cloth mounted on wood. When we compare this work with a copper-gilt repoussé book cover made in Limoges (late twelfth century or early thirteenth; fig. 5.10) and a silver-gilt repoussé book cover embellished with precious stones made in England about 1065 (fig. 5.11), it becomes clear that not only do all three objects share their visual vocabulary, but the Spanish frontal and the enameled book cover create optical substitutes for the material appearance of repoussé metalwork done with more precious materials.

The fifteenth-century Montpellier *Liber diversarum arcium* contains a

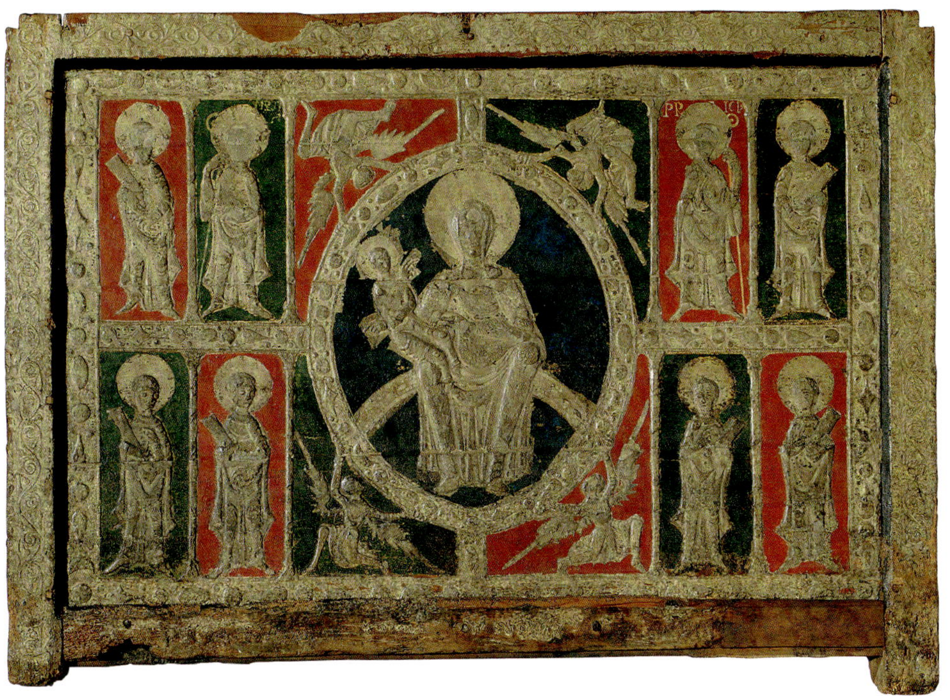

FIGURE 5.9. Altar frontal from Alós d'Isil, first quarter of the thirteenth century. Stucco reliefs and remains of varnished metal plate on cloth-covered wood. Museu nacional d'art de Catalunya, Barcelona. Website of the Museu nacional d'art de Catalunya of Barcelona, www.museunacional.cat. CC BY-NC-SA 3.0.0.

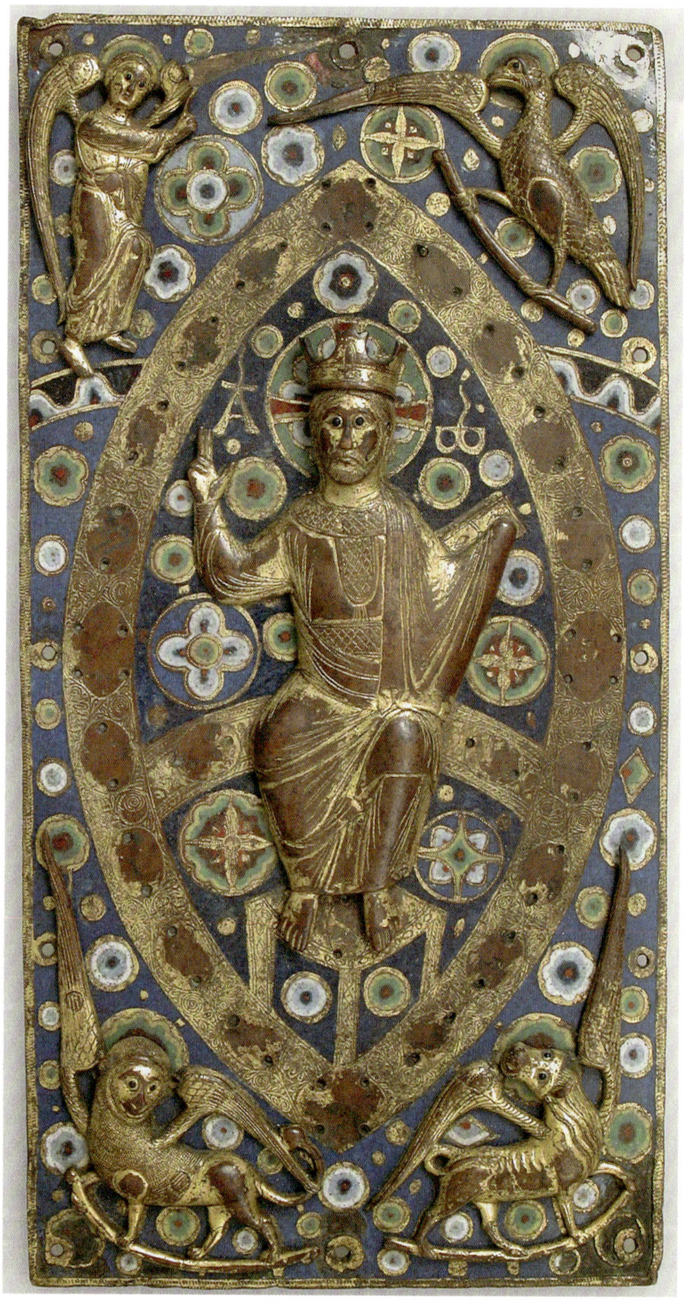

FIGURE 5.10. *Christ in Majesty*, made in Limoges, France, ca. 1185–1210. Champlevé enamel on gilded copper. Metropolitan Museum of Art, New York. Public domain.

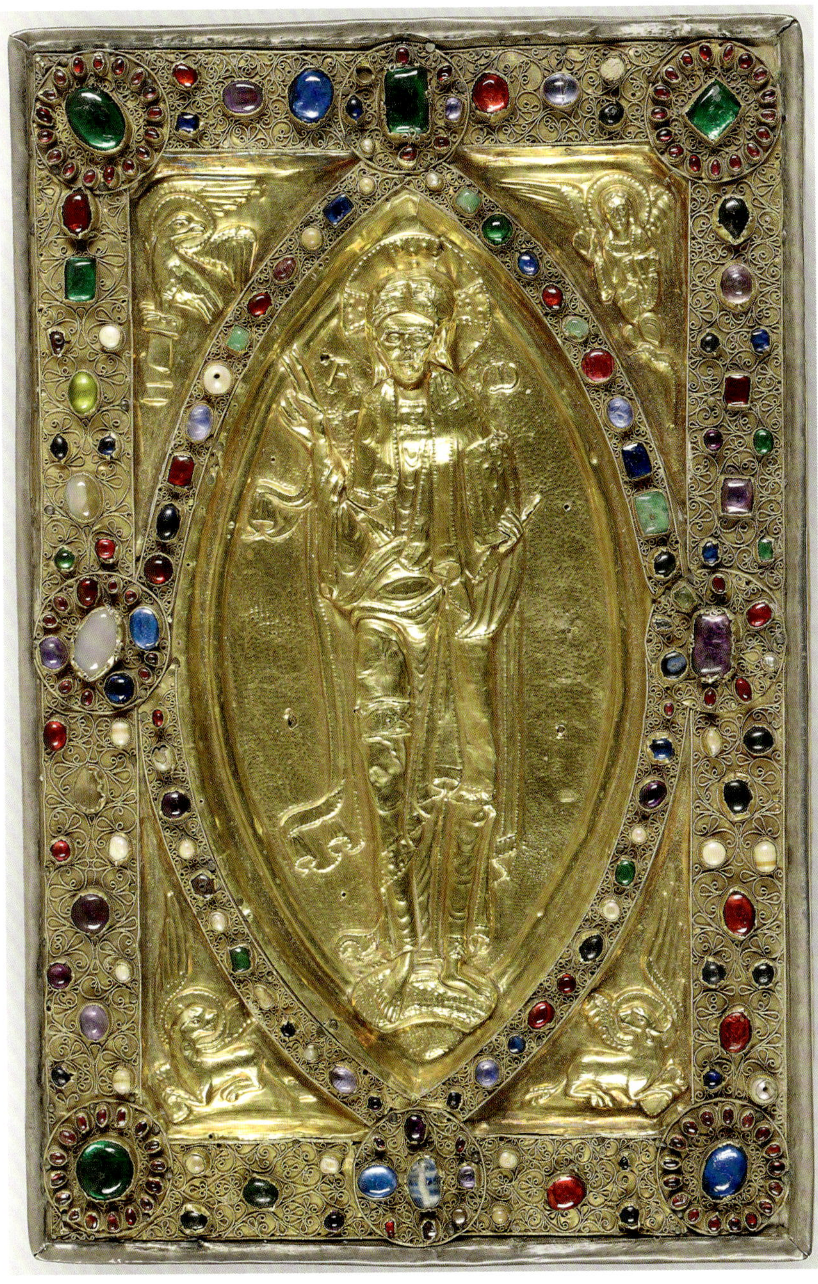

FIGURE 5.11. *Christ in Majesty* (cover of the Gospels in Latin), England (Canterbury), ca. 1060, for Judith of Flanders. Silver-gilt repoussé metalwork set with precious stones (many of them nineteenth-century replacements). MS M. 709, Morgan Library and Museum. Purchased by J. P. Morgan (1867–1943) in 1926. © The Morgan Library and Museum, New York.

version of Theophilus's *pictura translucida* recipe that gives a few more details about the nature of the techniques used to glaze on metal:

> Also is made a certain painting, which is called "translucent" or "brilliant" [*translucida sive auriolla*], which is made in this way on polished material [*super politum*]: and thereupon cover whatever you want to paint with saffron [*crocum*], then carefully apply the colours with linseed oil [*oleo lini*], very thinly [*valde tenues*]; and draw with a sharp brush, then make smooth with a broad [brush] made from ass's hair; for this in fact the most appropriate color is green [*viridis*].[58]

The Montpellier recipe for making *pictura translucida* contains several significant changes from that of Theophilus. For instance, we learn to use two types of brushes; a pointed/sharp brush (*pincelo acuto*) to paint with pigments ground with linseed oil and a broad brush made from ass's hair (*amplo facto de comis asini*) to "smooth" (*aplano*) with "green." The pointed brush was used to apply a pattern while the broad brush made from very soft hair was used to apply a layer of translucent green paint over this pattern (fig. 5.5). Such a uniform glaze is more easily created with an oil that is thicker than a regular cold-pressed oil, because a relatively thick oil makes it easier for the painter to apply a layer free from brushstrokes. Smoothness, as I noted in the introduction, improves the translucency and color saturation of the glaze. It is therefore quite likely that the recipes for making a thick painter's oil reflect a painterly tradition strongly tied to the art of glazing and, more generally, to creating a smooth and lustrous painting.

There is another interesting difference between the Montpellier recipe for making *pictura translucida* and Theophilus's recipe. Contrary to Theophilus, who points out that the tin should not be colored yellow, the Montpellier manuscript instructs readers to color the metal with saffron before painting on it. The simplest explanation for this difference may be that the scribe who recorded the Montpellier recipe made an error, changing the word *locum*, "place," where Theophilus says the tin should be applied, into *crocum*, "saffron."[59] But it is also possible that the Montpellier recipe records a different practice where patterns were drawn on imitation gold rather than on real gold such as we have seen in the Westminster Retable.

This last technique can in fact be found on some of the earliest surviving medieval panel paintings. Among the thirty-one Norwegian altar frontals studied by Unn Plahter and her coauthors, three of the earliest, made for churches in Kaupanger, Hauge, and Heddal (ca. 1250), contain figurative scenes painted almost entirely over silver leaf.[60] In this way these paintings on silvered wood imitated the contemporary champlevé enamel discussed in chapter 4. In the Kaupanger frontal, for instance, the ground is covered with silver leaf that was coated with a yellow varnish to give it the appearance of gold. The figures and ornaments in the pictorial plane have been outlined with black paint. In some places the yellow varnish was covered with green and red glazes, the dominant colors of the frontal. Copper green, for instance, is glazed over varnished silver to paint gems, garments, and cushions, while red lake glazes were mainly used in the background of the painting.[61]

The Heddal frontal is an even more exquisite example of how these early Norwegian paintings play with the optics of metalwork (fig. 5.12). In this work almost the entire pictorial plane is covered with silver leaf varnished to look like gold. Analysis of a sample of this yellow varnish showed it contained heat-bodied linseed oil and pine resin.[62] On top of this varnish the figurative scenes—draperies and backgrounds—were delicately glazed with ultramarine blue, copper green, and red lake (identified as lac dye extracted from sticklac, a resinous secretion deposited on certain trees by a female scale insect, probably *Kerria lacca* Kerr.) As with the Westminster Retable, the empty holes in the frames of the Hauge, Heddal, and Kaupanger

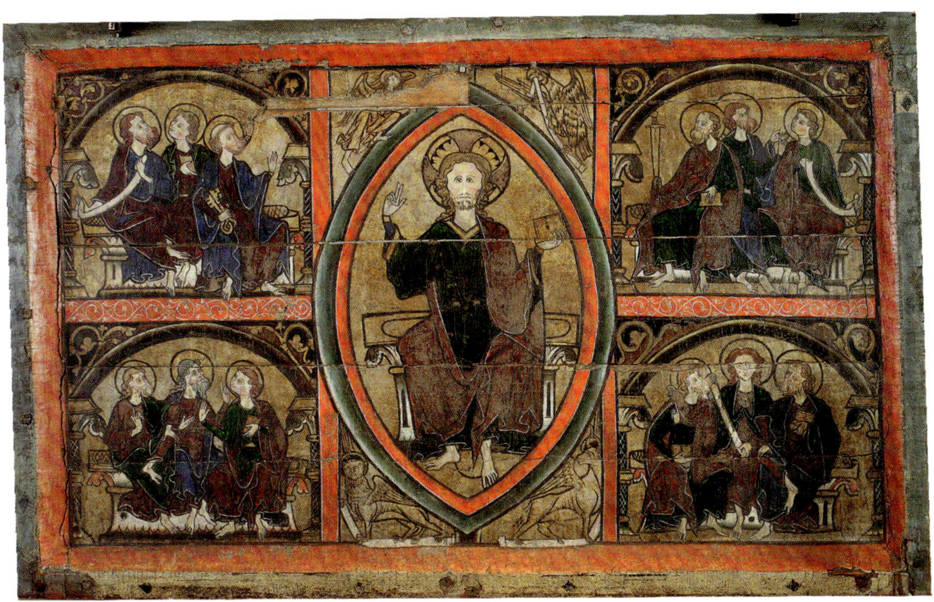

FIGURE 5.12. The frontal from Heddal, ca. 1250. Kulturhistorisk Museum, Oslo. CC BY-SA 4.0.

frontals remind us of the three-dimensional imitations of precious stones that would once have been mounted there. Most of the other Norwegian altar frontals examined were not painted entirely on metal foil or leaf, however. Instead, they contain locally applied metal foils and leaves, often silver, that were decorated with glazes. Just as with the translucent details in the champlevé enamels studied in chapter 4, the painters of the Norwegian frontals selected backgrounds, architectural features, gemstones, halos, and garments for local application of these glaze paints.[63] Precious stones were usually glazed within an outline of black directly over burnished silver leaf, or over an intermediate layer of yellow varnish.[64]

Besides wooden panels, artisans also embellished transparent glass panes with glaze colors to imitate enamel, a method quite similar, in fact, to the technique used to make the imitation enamels in the Westminster Retable. This technique, known today as "reverse glass painting," or *verre eglomisé*, was especially popular in Italy during the thirteenth and fourteenth centuries. It was used to make small devotional objects, especially reliquary shrines, and to embellish paintings and their frames.[65] The technique can be found, for instance, on a small panel (originally part of a larger ensemble) from the Victoria and Albert Museum, London. The transparent gilded glass pane is decorated with Christ on the cross and several saints all engraved in gold and backed with colored paints (fig 5.13).

Cennino Cennini gives a detailed explanation of how to make such gilded and colored glass. He explains that gold foil first has to be applied, using egg white, to "white glass, with no green cast, very clean and free from bubbles" (*divetro biancho che non verdeggi ben netto senza*).[66] As soon as the gold is dry, the glass is laid on a panel covered with a black cloth. A design is created by scraping into the gold, "drawing" on it with a needle. The black cloth behind the glass would have made it easier to see the pattern being created. After the design is finished, Cennini advises "laying in a ground" with ultramarine blue ground with oil. The drawing is then "backed up" with colors ground in oil, notably with black, and with the glazing pigments ultramarine blue, verdigris, and lac lake. We learn that black is the most striking of the four pigments because, like the black cloth, "it models the figures for you better than any other color" (*le fighure meglio che nessuno altro cholore*). We have seen that ultramarine was often considered too dark when ground with oil, but in *verre eglomisé* this property was desirable; as Cennini explains, against the dark ground of ultramarine and black ground with oil, the translucent figurative design would stand out more.

In addition to imitating enamel, recipes suggest that glass was also embellished with glaze colors to imitate stained glass for use in buildings or

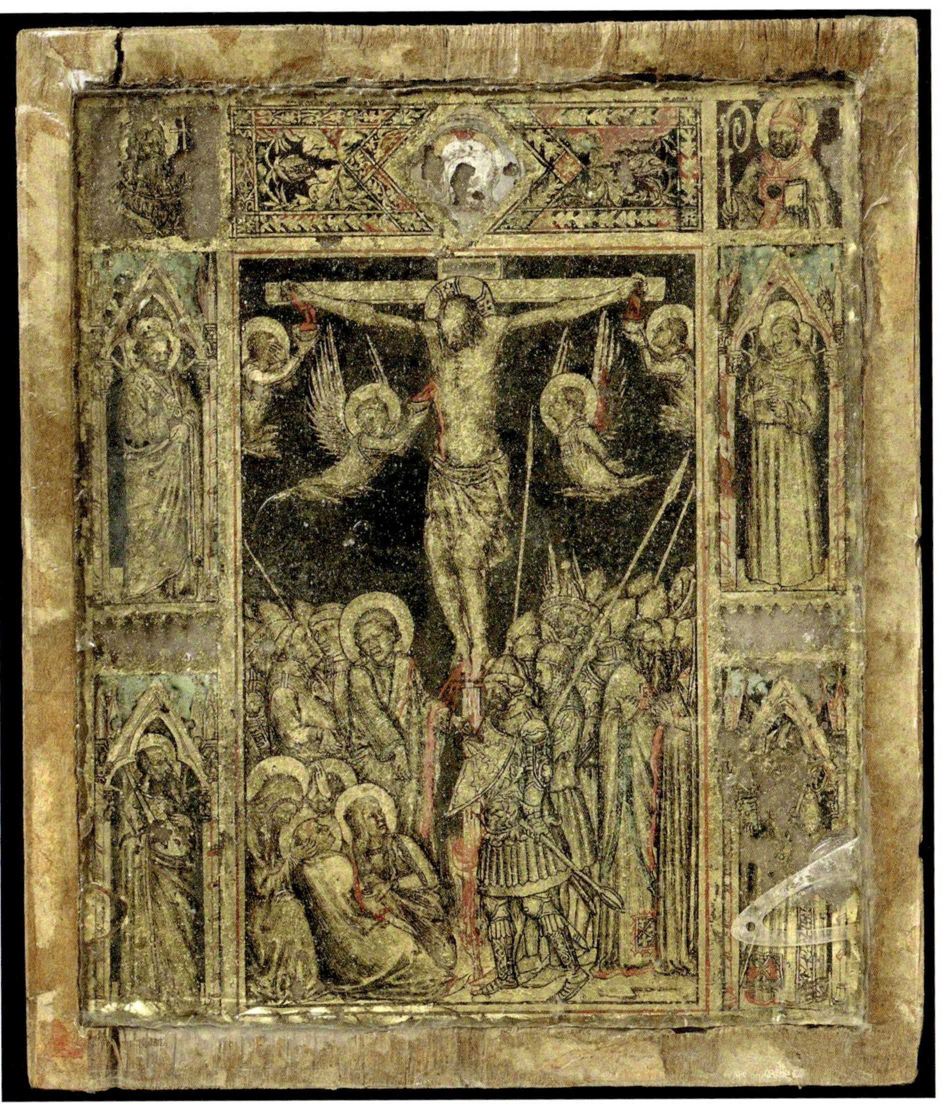

FIGURE 5.13. *Crucifixion*, Umbria (probably, made), ca. 1370–80. Glass backed with engraved gold foil and paint. Victoria and Albert Museum, London. © Victoria and Albert Museum, London.

for smaller objects such as glass tableware. The Marciana manuscript, for example, contains a recipe describing this practice: "If you want to paint your glass with colors that, although beautiful, do not penetrate into the glass" (*Se vuoi colorire d'altri colori e quali saranno begli ma non penetrano nel vetro, ma stanno in superficie*).[67] We learn that the glass needs to be painted with colors that "have no body" (*che non hanno corpo*), meaning, of course, the pigments suitable for glazing. Similarly to Cennini's recipe for making a reverse glass painting, the Marciana manuscript recommends using *verde-rame* ("verdigris"); *laccha fine* ("fine lake"); *azurro fine* ("fine azure"); and for the blacks *nero di noccioli di pesca arsi* ("peachstone black," that is, charred remains of a variety of fruit stones and nutshells) and *carbone* ("charcoal black"). The glass should first be given a coat of nut oil or linseed oil and allowed to dry in the shade. The pigments to be used are ground with the same oil used to coat the glass, then the glass can be decorated with this paint. The Marciana recipe explains that even though the color does not penetrate the glass as it would with real stained glass windows, the oil-painted glass will "remain beautiful for a long time" (*durera assai tempo bello*), and, if you want, it can also be varnished.

Pictura Translucida: Imitating the Glow of Silk

In addition to its first recipe for *pictura translucida* discussed above, the Montpellier manuscript includes a second method—a unique recipe that cannot be traced to other recipe collections—for making a translucent painting by using a "colour that does not dry" (*de colore qui non siccatur*):

> Also is made another colour, which does not dry, in this way. Temper any colour you like with honey; that "colour," indeed is especially used in translucent pictures [*pictura translucida*], namely when you want to make images of silver or tin green, and yet the silver itself to remain [*um enim volueris ymaginem argenteam vel stagneam facere viridem / et de ipsomet argento relinquere*], detail the details with that same colour, then fill in the whole image with green, complete the work and varnish [*vernicato*] it; then with a pointed brush as an instrument, remove and clean away the colour that is put there; and if you want, with yellow varnish [*doratura*] the same lines of silver or tin may be gilded [*inaurare*].[68]

Rather than using pigments ground with linseed oil, the pattern for this translucent painting is made by applying a pigment tempered with honey. Over this honey pattern, a green oil glaze is applied and then varnished (fig. 5.5). By means of a *sgraffito* technique, the honey color is removed with a pointed brush (fig. 5.5). This creates a metallic pattern against a green background. According to the recipe, this pattern can also be filled in with the yellow varnish, here referred to as *doratura*, a term we have encountered in other recipes for making yellow varnishes (fig. 5.5).[69] This last step would have been particularly important when silver was used, because silver would have tarnished if left uncovered. The Montpellier manuscript does not explain what kind of motifs were painted by the technique described here, but medieval paintings suggest this recipe may reflect a method for rendering the color, patterns, and glow of silk-brocaded textiles.

Figured silks were worn by the richest ecclesiastical and political rulers, decorated the architecture of churches and palaces, or were used on liturgical objects and reliquary shrines. These textiles were made by brocading, a special weft technique that allows the weaver to add a raised pattern to the fabric. In their most precious form such brocades were patterned with gold or silver thread. The materials used and the time required to make figured silks put their price on par with that of gold.

Silk fiber is produced by the mulberry caterpillar (*Bombyx mori* L.). It is not only incredibly strong but also very smooth. That smoothness allows the fabric both to effectively reflect the light—causing silk's luster—and to effectively refract the light—giving it a saturated color when colored with certain dyes. Silk also has another interesting optical property. In cross section a silk fiber is translucent and shaped like an irregular triangle that refracts light like a prism, producing the characteristic shimmer of woven silk fabric.[70] In fact the optics of the saturated, lustrous colors of silk fibers closely resemble the glow of precious stones and the materials used to imitate them—glass and glazes.

The optical resemblance between silk and oil glazes is likely why glaze paint came to play a prominent role in imitating silk brocades on medieval paintings. Medieval panel painters and polychromers used a great variety of techniques to imitate these precious textiles, especially gold silk brocades: from applying glazed metal ornaments to the paint surface to various gilding techniques, often using intricate punched designs and colorful translucent glazes. A much-used technique for imitating brocade was applying tin foil reliefs. These were often used on garments in the background of paintings and, as we have seen, to sculpt entire figures (fig. 2.4).[71] The technique can be summarized as follows: tin was pressed into a mold with an incised pattern—usually in wood or metal—then it was strengthened with a filler on one side. This filler was typically a varnishlike substance. The tin relief was either gilded with gold leaf, covered with yellow varnish to make it look like gold, or left bare to imitate silver. The flat zones were usually painted with translucent or opaque pigments to imitate the colors of the silk textile, while the raised parts of the relief remained uncolored, evoking the glitter of the gold threads.[72]

The *Liber illuministarum* from the monastery in Tegernsee explains how to make such an applied relief brocade and tells how to decorate it.[73] After you finish making the tin relief, you can cover it with gold leaf or paint on it immediately. We learn that oil is ground with cinnabar (*zinober*) to make red; with simple blue (*ring plab*, likely a medium-quality azurite) or azurite (*lasur*) to make blue; and with verdigris (*spangrün*) to make green.[74] Verdigris, the author of the manuscript explains elsewhere, is always ground with oil when it is used to paint on silver.[75]

Cennini's *Il libro dell'arte* also contains various detailed instructions for using oil paint to imitate gold- and silver-brocaded fabrics. One recipe explains how to paint cloths of gold and silver in various colors. To imitate a green gold-brocaded fabric Cennini recommends a glaze of verdigris ground with oil over a pattern previously applied on a gold-leaf ground: "Likewise, gild the ground, draw the design that you want on it. Lay in the grounds with verdigris [*verde rame*] in oil [*adol(i)o*], shading some of the folds twice. Then apply it everywhere, evenly all over the grounds and over the designs."[76]

Cennini also explains what pigments to use for imitating brocades with silver. To paint a red drapery, you first apply vermilion—a bright red opaque pigment—ground with pure egg yolk. The motif or ground in vermilion is then glazed with one or two layers of "good quality lacca [red lake] in oil" (*laccha fine o d olio*).[77] For painting a drapery in "ultramarine blue" (*dazurro oltre marino*), Cennini recommends using a ground layer of silver leaf.[78] But in this case oil is not the medium of choice. Unlike the *verre eglomisé* painting, where the dark blue of ultramarine in oil was desirable, Cennini now recommends grinding the precious pigment with glue so it does not lose its blue brilliance.

Since brocaded textiles are easily recognized in paintings because of their patterns, they have been extensively studied in terms of style, iconography, dissemination, and painting technique.[79] Plain and unadorned fabrics have received far less study than imitation brocades, both as textiles and on paintings, probably for two reasons: not many plain textiles have been preserved, and without a pattern it is difficult to compare these textiles—if they do survive—with their painted counterparts. In the collections of the Kunsthistorisches Museum in Vienna, a twelfth-century royal gown made from blue silk provides a rare insight into what such a silk textile of uniform color might have looked like (fig. 5.14).

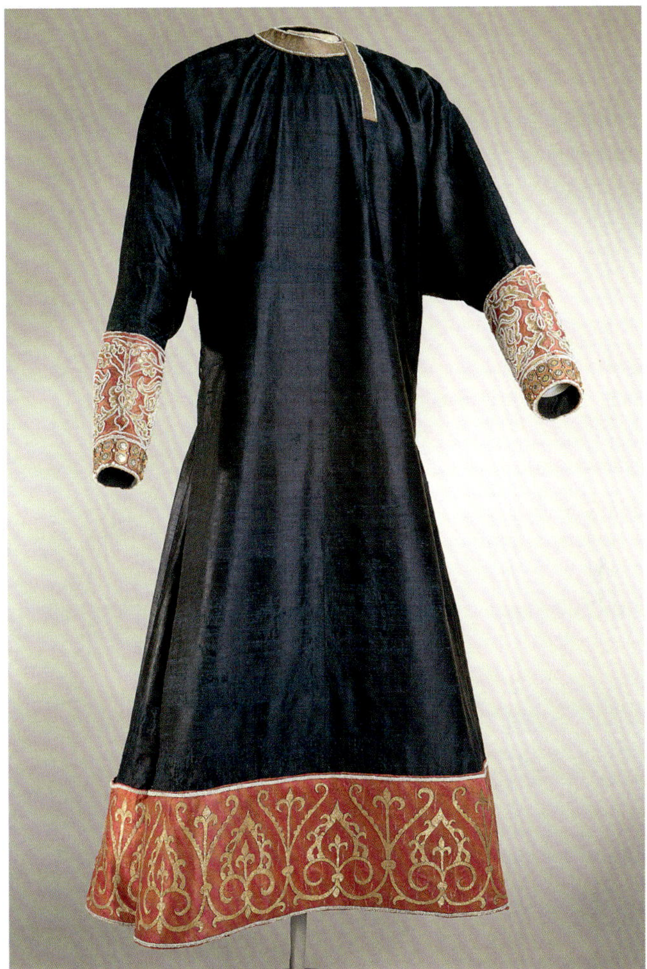

FIGURE 5.14. The Blue Tunicella (Dalmatic), Palermo, Royal Court Workshop (made), first quarter of the twelfth century. Samite, gold thread, gold piping, cloisonné enamel, gold filigree, pearls. Kunsthistorisches Museum, Vienna. © KHM-Museumsverband.

As the light plays over the textile, the silk appears light blue with an almost white luster at the highest relief of the folds, while in the deepest recesses it looks a deep and saturated blue, almost black.

In medieval paintings, uniformly colored textiles can be found glazed over metal foil or painted over a white ground. In the latter case fabrics are always depicted according to a specific "formula," also described in medieval recipes. Recipes use the Latin terms *matizare* or *illuminare* ("to lighten") and *incidere* or *umbrare* ("to shade" or "to outline") to explain how to depict the difference between the light and the dark parts of a fabric by applying light and dark colors over a base color.[80] In various recipes for modeling drapery on wood and on the wall, for instance, Theophilus explains how to use this method, which in the *Schedula* is called *illuminare* and *umbrare*. In a recipe telling how to depict draperies on colored glass, he explains that it is done by the same method painters use:

> If you are diligent in this work you can make shadows [*umbras*] and lights [*lumina*] on draperies, as in a coloured painting [*pictura colorum*], in this manner. When you have made your strokes of the aforesaid colour on the draperies, spread the colour with the paintbrush so that the glass is clear [*perspicax*] in that part where you normally make lights [*lumina*] in a painting. And let the brush strokes be thick in one place, light in another and then lighter, and distinguished with such care that they give the appearance of three shades of colour being applied. . . . It should have the appearance of a painting composed of a variety of colours.[81]

Here Theophilus writes not about painting on glass with colors—a much later invention—but about applying a dark pigment to colored glass. In a late twelfth-century stained glass window, originally made for the Collegiate Church of Saint-Étienne in Troyes, this technique is used as Theophilus describes. The lightest color (*lumina*) is imitated by the color of the glass, while the shadows (*umbras*) of the folds of the draperies are imitated by locally making the glass opaque to varying degrees (fig. 5.15).[82]

This medieval formula for rendering the form, color, and relief of draperies and other materials can be found in nearly all figural arts, from glass painting and enameling to painting on wood and in manuscripts (compare, for instance, figs. 5.15, 5.12, and 4.14). In the literature this formula is typically discussed as a system for rendering differences between light and shade, but not for depicting specific fabrics. And certainly, to our modern eye, the medieval paint system of "lightening" and "shading" might not make it immediately apparent whether a garment was made from wool, linen, or silk. Here I will argue that by varying the contrast between the colors used to paint the lighter and the darker parts of the textiles, medieval artisans (not just painters) did in fact attempt to express particular materials. When artisans depict draperies by showing the stark contrasts between the darkest and the lightest parts, they imitate the similar sharp contrasts we perceive in silk garments. In using this method to imitate silk

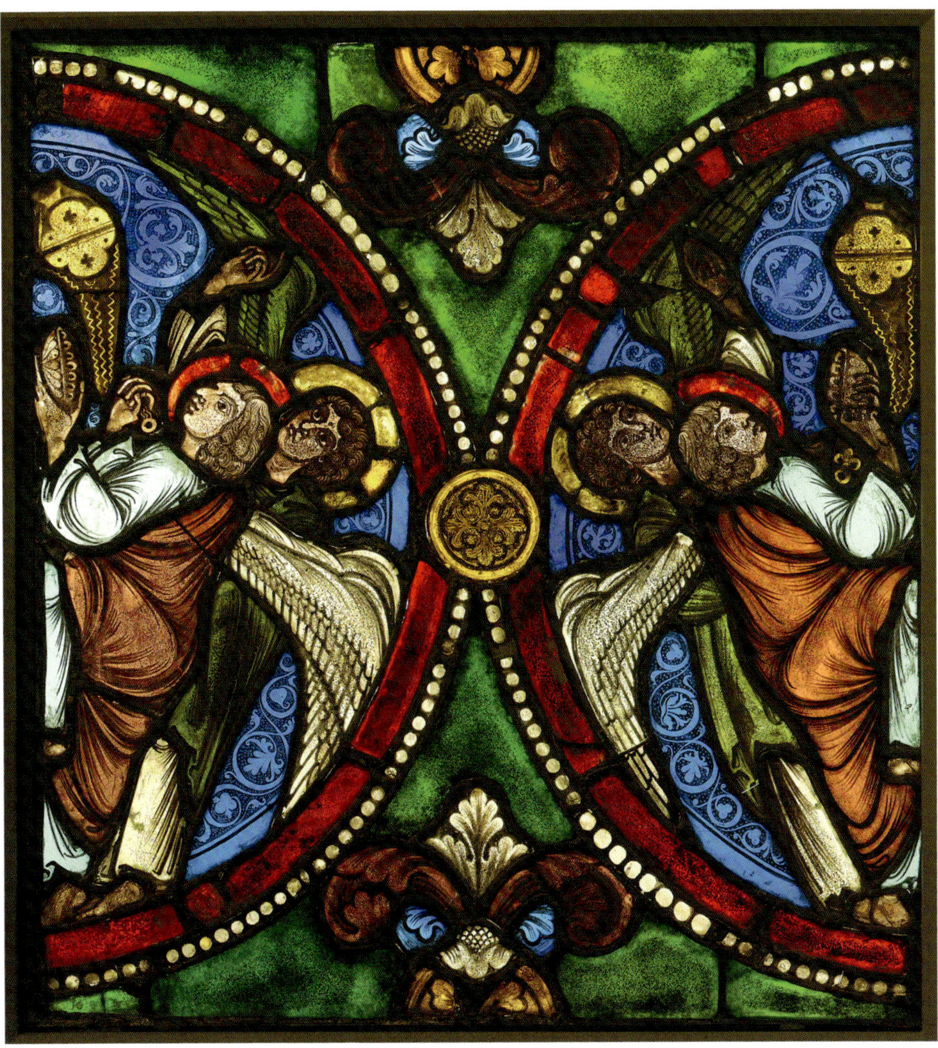

FIGURE 5.15. *Angels Swinging Censers,* ca. 1170, made in Troyes for Saint-Étienne. Pot-metal glass, vitreous paint, and lead. Metropolitan Museum of Art, New York. Public domain.

textiles, glazes played a crucial role. Medieval painters used oil glazes to imitate the saturated, glowing colors in the deepest folds of green, red, or blue silks. This paint system comes very close to the mimesis of the visible world as it developed in the fifteenth century, which will be the subject of chapter 6. In the previous examples, however, the technique was still part of an art that was defined by material mimesis. Glazes were used to create a painted optical substitute for silk that was not brocaded with metal thread.

Terms for Glazing

The strong link between the art of glazing, the imitation of textiles, and the decoration of metal foils and leaves is also demonstrated by studying early vernacular terms used to describe the technique. One of the earliest such references is in the 1434 contract of the Ghent citizen Willem de Busoen with the painter Saladin de Stoevere.[83] The contract describes an altarpiece depicting the birth and death of the Virgin on the interior of its wings and stipulates that all panels be painted with "good and fine oil paints, as is custom" (*goeder fijnder olijverwen alsoet behoert*). The Virgin's gold brocade mantel (*guldin lakin*) has to be painted on one side with fine azure (*finen aijsuere*). The inside of her mantle is to be glazed (*gheglatsiert*) with sinoper (*sinopere*). In the past *sinopere* referred to a red pigment made from a dark red lake.[84] This is further supported by the technique the contract recommends for applying the pigment: *glatsieren*. The term *glatsieren* is likely a compound from *glatt*, "smooth," and *sieren*, "to decorate." In this meaning the term *glatsieren* is directly related to the history of glazing in medieval practices of material mimesis, being a smooth decorative layer of paint used to imitate gems, stained glass, and enamel and, as here, the luster of silk textiles. In the 1470 ordinances of the Antwerp Guild of Saint Luke, a variation of *glatsieren* appears in a similar context, this time as a glaze over metal. In its eleventh rule, the ordinances explain that the painter or sculptor should not combine silver with polished gold leaf. Silver leaf should be "varnished or glazed with paints" (*veryst oft gelaetseert syn van verwern*).[85] French guild regulations use a related term to describe the glazing technique: *glacer*. This use appears to be slightly later than the Dutch ordinances and contracts that use the term *glatseren*.[86] One of the earliest instances can be found in the 1480 Tournai ordinances of the painters and glassmakers guild. Like the Antwerp guild of Saint Luke, the Tournai ordinances include a rule stipulating that part gold (gold mixed with an alloy) is allowed only when it is covered with a colored glaze (*glacié de coulleur*).[87]

Textual evidence shows that by the sixteenth century a variation of *glatsieren* was also used in the German vernacular to describe glazes on polished metal. As such it can be found in a 1518 contract with the Basel painter Hans Herbst (ca. 1470–1552), which orders the production of an altarpiece for the church of Mary Magdalen. The contract stipulates that precious materials be used, including gold leaf worked in a variety of ways. The altar's frame, draperies, and *Gesprenge* (a decorative wooden pinnacle) are to be gilded.[88] The landscape too is to be covered with polished metal, which the contract specifies can be either gold or silver leaf. But if silver leaf is used, the "landscape" should be *glasiert*. As with the ordinances of the Guild of Saint Luke, the translucent glaze paint would have been mandatory to prevent the silver from tarnishing on exposure to air and humidity. The 1518 contract also stipulates that the *paviment* (tiled floor depicted on the altarpiece) is to be silvered and glazed with red, blue, or green.[89]

About the same time, German-language sources also use *lasieren/laseren* to describe the art of glazing. It is found, for example, in the 1516 Strasbourg regulations for registering as a master painter in the city.[90] To register, the painter had to produce three masterpieces, one consisting of a carved sculpture with a drapery, such as the Virgin Mary or an angel. The sculpture was to be decorated (*zierung*) by gilding (*vergulden*) and glazing (*laßeren*) it. Ernst Ploss has argued that *laßeren* might have derived from *laziuren*, which originally meant to paint with an expensive blue such as azurite or ultramarine.[91] But when we take into account that *glatsieren* was already in use about a century earlier specifically to denote the art of glazing, the German *lasieren* might have been derived from this term as well. Whatever the case, by the seventeenth century *glats[i]eren* had established itself in the Dutch vernacular to describe translucent paint layers. In his 1604 *Schilder-boeck*, Karel van Mander even includes a definition preceding the artists' biographies in a poem meant to instruct young painters: *Grondt der Edel vry Schilderconst*. Van Mander explains that to make beautiful draperies (*om schoon te traperen*) you should use the glazing technique (*hem behelpen met glasseren*), which, when skillfully done (*leght bequamelijck aen u met glatseren*), helps to give a glowing translucency to velvet and beautiful satin (*Behelpt alst past tot een gloedich doorschijnen, Om maken Fluweelen en schoon Sattijnen*).[92] Van Mander uses another variation of this term in the life of the painter, engraver, and glass painter Hendrick Goltzius (1558–1617). We learn that Goltzius, when glazing (*glaceren*) a draped cloth with ultramarine, conveniently used a technique of "pushing" (*stooten*) his brush that was very common with the glass painters (*Glaes-schrijvers*).[93]

The previous sources all bring up glazing in relation to two of its most important functions, decorating metal leaf and foils and glazing draperies. Van Mander's passage about Goltzius even suggests a relation between applying glazes and the art of glassmaking. South of the Alps, accounts of the glazing technique are similarly connected to decorating foils, imitating textiles, and imitating gemstones, but here a different terminology developed to describe the technique. Since the late sixteenth century, Italian sources use the appellation *velare*, "to veil," to describe thin layers of paint that reveal what lies below. Two of the earliest occurrences are in Armenini's treatise and in the work of the painter and art theorist Giovan Paolo Lomazzo, *Trattato dell'arte de la pittura, scoltura, et architettura* (1584). Lomazzo includes an extensive discussion of the technique and purpose of glazing in a chapter titled "Transparent Colors and How to Use Them" (*Dei colori trasparenti, & come si adoprano*).[94] Lomazzo details how to use colors "most devoid of body" (*più privi di corpo*), such as red lakes (*lacca*), verdigris (*verderame*), and *verdetto*. About this last pigment he explains elsewhere in his treatise that it is also called *santo*, but is yellow (*che si chiama santo mà tira al giallo*).[95] *Verdetto* was likely a lake pigment extracted from buckthorn berries (various species of *Rhamnus*) that could produce either a yellow (unripe berries) or a green lake (ripe berries).[96]

Ground with oil (*ad oglio*), these *colori trasparenti* are used to "depict all clear, see-through [*trasparenti chiari*] bodies, such as *carbonchi, Rubini* and the like." Rubies, for example, are first sketched (*abbozzate*) with colors that are devoid of "clearness of transparency and vivacity" (*senza il lucido della trasparéza e sua vivacità*), after which they receive a layer of "a pure, clear and beautiful lake" (*la lacca pura, netta, & bella*). This way the painted gems appear to be covered by "a glass of shiny fire, as are the real, & natural" precious stones: "And this you cannot do in fresco, where the lights and the shadows of transparency can only be given with drawing. In the same manner verdigris [*verderame*] and *verdetto* [yellow lake pigment from Rhamnus spp.] enliven the tempering of the emeralds [*smeraldi*] and similar transparent materials."[97]

Lomazzo continues to explain how these pigments are also used to depict the "lusters and shinings" (*il lustro, et la vivacità*) of satins or silks. He nevertheless laments that the painters of his time are using glazes on just about any drapery that does not require the luster of silk (*vivacità di seta*). As a result, "pure cloths" such as wool or linen—"which do not want to be veiled in transparent colors to give them shine" (*che non si voglia avelare di colori trasparenti per dargli il lucido*)—can no longer be depicted. According to Lomazzo this has corrupted contemporary painting so much that using "transparent colors to give it shine" (*i colori trasparenti per dargli il lucido*) is so esteemed that you cannot see a painting without it, no matter how good it is.[98]

Armenini, whose treatise was published only three years after Lomazzo's work, stressed the importance of "veilings" to harmonize a painting in its final stages of being finished. But he also writes about a practice in which draperies are first sketched and then uniformly glazed, a technique quite similar to the one Cennini had also described for painting brocades. Just like Lomazzo, Armenini does not approve of this practice and stresses that "good artists scorn veiling and take offense at seeing draperies of only one color."[99] He nevertheless still explains how the technique works: To make a green drapery, it is first painted lightly with verdigris, *smalto* (blue pigment made from ground glass, deriving its color from cobalt), and *giallo santo* (yellow lake from ripe buckthorn berries, discussed above) mixed with "common varnish."[100] After this is dry, the drapery is "veiled" uniformly with verdigris, which is again mixed with common varnish and applied with a "brush of miniver" (soft squirrel hair brush).

Armenini's and Lomazzo's *velatura* eventually established itself in the Italian language to describe any type of light-transmitting paint. As such it was not just used to describe the glazing technique but also described the technique of "scumbling," which uses pigments opaque in oil, that is not light transmitting, to create a translucent paint layer.[101]

In this last meaning we find it, for instance, in Filippo Baldinucci's (1624–97) *Vocabolario toscano* (1681):

> Veil. To cover with a veil [*Velare. Coprir con velo*]. Lat. *Velare.* Almost all our artists veil well by applying pigment with little color and a lot of tempera (or as is commonly said, watery or thin) onto canvas or panel in a way that this is not lost from view but remains to a degree quite unclear, and nicely obscured, almost so that from above itself [there is] a thin veil [*in modo che questo non si perda di veduta, ma rimanga alquanto mortificato, e piacevolmente oscurato, quasi che avesse sopra di sè un sottilissimo velo*].[102]

Baldinucci explains that *velare* received its name from the paint layer's optical resemblance to veil-like textiles, *velare* literally meaning to cover with a veil. Since in this case the *velare* was applied with a lot of binding medium and little pigment, the veil-like paint Baldinucci describes is a scumble.[103] With this technique, the painter could create a translucent paint with pigments that are opaque in oil. Lead white, for instance, was often used for this purpose to paint thin, translucent head veils or loincloths. In scumbling, the paint is translucent because light is transmitted *alongside* the pigment particles and not, as with a glaze, *through* pigment particles *and* binding medium.[104]

The recipes and treatises discussed in this chapter show not only that the art of glazing was perfectly understood by painters working as early as the twelfth century, but that this technique lay at the heart of their practice as an art of material mimesis. A relatively viscous oil medium was combined with pigments translucent in oil to evoke the glowing radiance of precious stones, enamel, and stained glass windows and the saturated color and glow of silk. In fact, the popularity of the oil medium in medieval painting may have developed in part because only oil can so perfectly capture the translucent splendor that painters and their patrons desired. The refractive index of other binding media available at the time was too low to create the saturated, glowing paint possible with oil. In chapter 6 we will see that, when panel painting changed from the mimesis of materials to the mimesis of the visible world, the glaze was no longer the most important technique for painting translucent materials. Fundamental to the new paint system introduced by Jan van Eyck and his contemporaries was the way the play of light on colored translucent and colorless transparent materials was now depicted by pigments fully opaque in oil.

6

The Eyckian Turning Point

Glazing and the Imitation of the Visible World

Who will he be who'll ever imitate
The ruby's shining bright transparent color,
Or counterfeit the beauty of its splendor?

[*Chi serrà quel che possi el chiar colore*
lucido e trasparente de un rubino
contrafar mai, o el suo vago splendore?]

GIOVANNI SANTI, 1482–87[1]

Art around 1400: Specular Precedents

In 1398 Philip the Bold (1342–1404), Duke of Burgundy, commissioned a carved altar shrine for the monastery he had founded as Chartreuse de Champmol, outside Dijon, France. The wings that closed this altarpiece were painted by Melchior Broederlam (born ca. 1350, died after 1409). On feast days these painted doors opened to reveal an elaborate carved relief by the sculptor Jacques de Baerze (born before 1384, died after 1399) that, in imitation of metalwork, was also gilded and glazed by Broederlam (figs. 6.1 and 6.2).[2]

The wings of the Crucifixion Altarpiece are often discussed as a textbook example of pre-Eyckian panel painting. Just a few decades before the earliest-dated paintings by Van Eyck, Broederlam's wings show how painters' approach to space and light was slowly changing. The central scenes

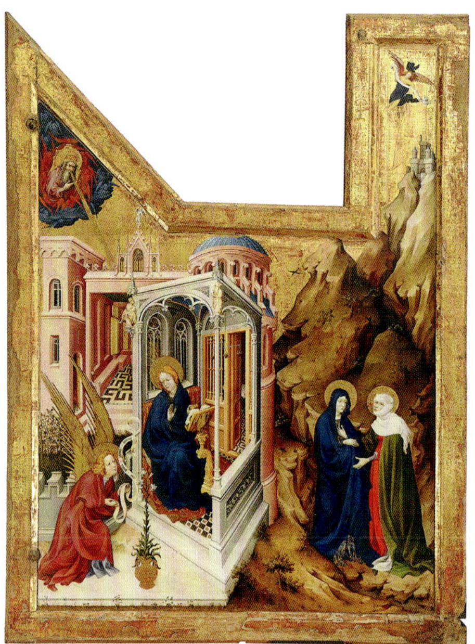
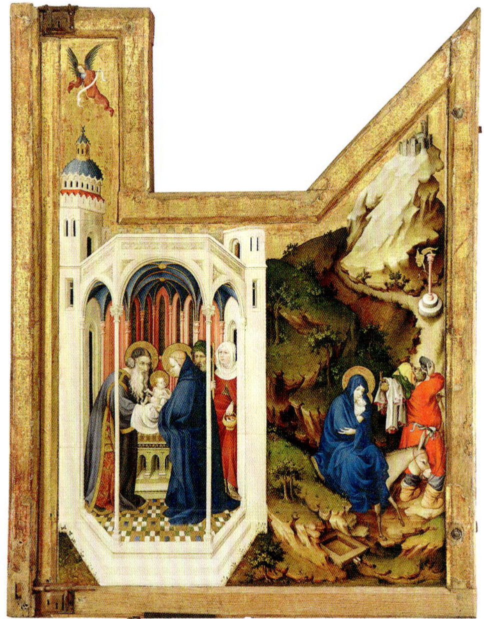

FIGURE 6.1. Jacques de Baerze and Melchior Broederlam, Retable de Champmol (open), from the chapel of the Chartreuse de Champmol, Dijon, France, 1393–99 (made). Musée des Beaux-Arts, Dijon. © Musée des Beaux-Arts de Dijon / François Jay.

are situated in elaborate landscapes, and light-dark contrasts create a sense of three-dimensionality in the figures and in the architecture. But despite Broederlam's innovative approach toward light and space, he deals with refraction and reflection in the same manner as the painters who came before him. Broederlam imitates reflection with materials that reflect light, and he imitates refraction with paint that transmits light. Metalwork, for instance, is imitated with gold leaf, as can be seen in the gold leaf eagle holding the Virgin's lectern, a gilded vase in the foreground, and small details such as the clasp that closes the Virgin's book. And the flame of an oil lamp is imitated by means of a translucent red glaze over polished gold leaf.[3] Something similar can be said of a small triptych named after the duke who owned it about 1900, which is now kept in the Museum Boijmans Van Beuningen, Rotterdam (figs. 6.3 and 6.4).

The Norfolk Triptych (ca. 1415–20) was likely used for personal devotions in the household. In its closed state, its shutters display painted scenes from the Bible set in landscapes similar to those of the Crucifixion Altarpiece. But unlike Broederlam's paintings, the Norfolk Triptych's landscapes

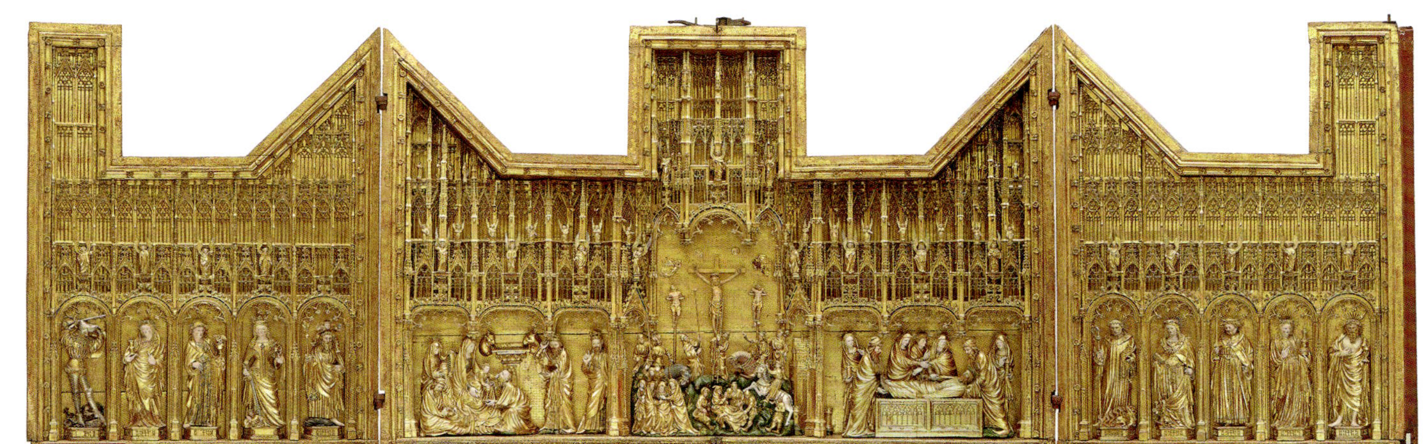

FIGURE 6.2. Jacques de Baerze and Melchior Broederlam, Retable de Champmol (closed), from the chapel of the Chartreuse de Champmol, Dijon, France, 1393–99 (made). Musée des Beaux-Arts, Dijon. © Musée des Beaux-Arts de Dijon / François Jay.

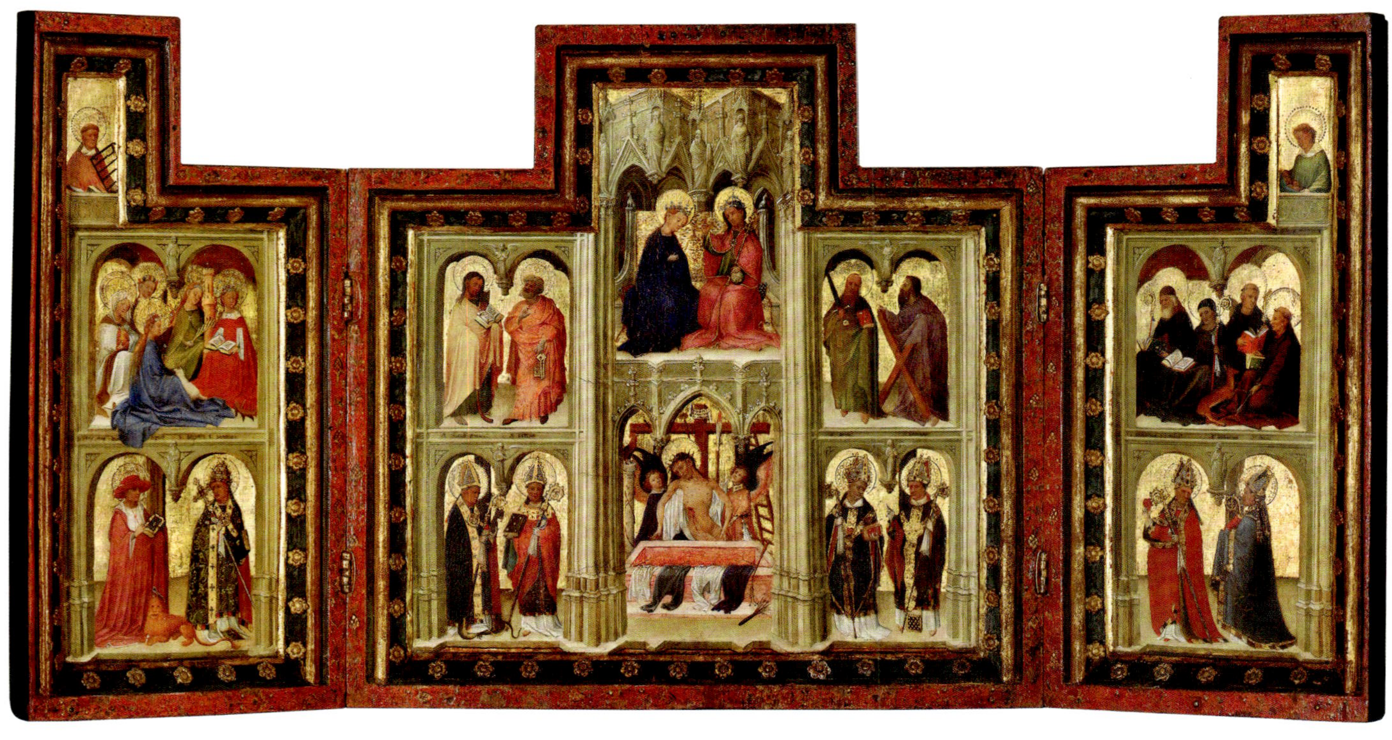

FIGURE 6.3. Norfolk Triptych (open), ca. 1415–20. Museum Boijmans Van Beuningen, Rotterdam. Photo by Studio Tromp.

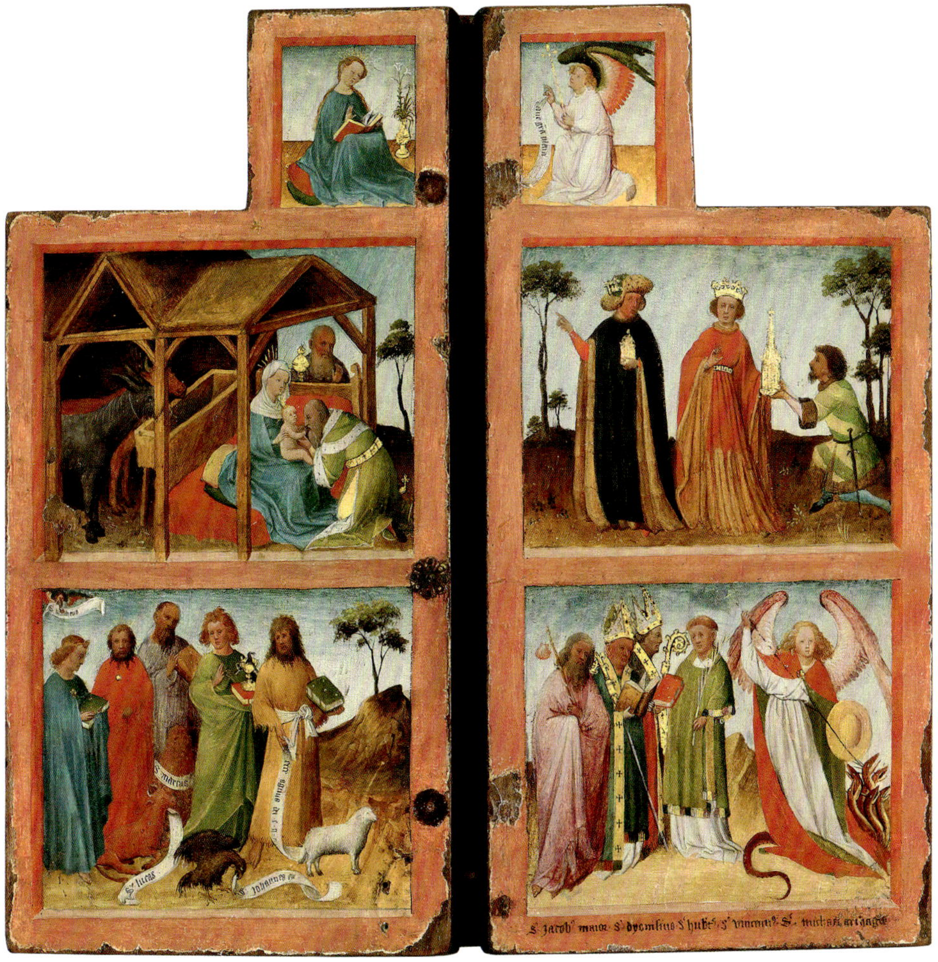

FIGURE 6.4. Norfolk Triptych (closed), ca. 1415–20. Museum Boijmans Van Beuningen, Rotterdam. Photo by Studio Tromp.

are not set against gold leaf backgrounds; they are in fact among the earliest panel paintings to have blue skies. On the inside, however, the Norfolk Triptych imitates metalwork in a manner similar to the carved relief of the Champmol Crucifixion Altarpiece. It is decorated with paintings of saints and scenes from the life of Christ that are placed within a painted Gothic architectural frame against a gold leaf background. Their robes are embellished with gold leaf borders, while—now much degraded—glazes over gold leaf are used to imitate precious stones (fig. 6.5).

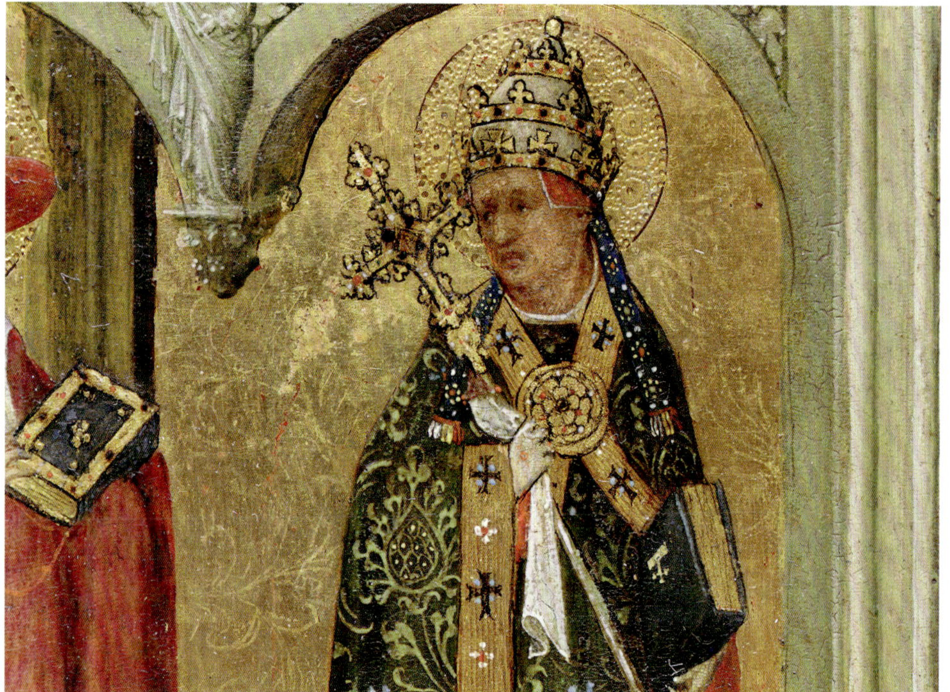

FIGURE 6.5. Detail of the Norfolk triptych with discolored glazes on the decorations on the mantle, the miter, the cross, and the book on the left. Photo by Marjolijn Bol.

With the Limbourg brothers as its most famous representatives, the art of book illumination shows similar changes in the conception of light and space at the beginning of the fifteenth century. The three Dutch brothers from the city of Nijmegen—Herman, Jean, and Paul (all d. 1416)—emigrated to France, where they worked for the Valois princes and came to be exclusively employed by Jean de France (1340–1416), duc de Berry, one of the greatest patrons of the arts at the time. Under his auspices, the Limbourg brothers created one of their most magnificent works, the *Belles heures*, from 1405 to 1408/9. This illuminated manuscript (Cloisters Collection, New York) was intended for private devotions and, as was customary for a book of hours, contains readings from the Gospels and prayers to the Virgin. It is filled with the Limbourg brothers' experiments with light, space, and surface textures, such as depicting Saint Eustace knee-deep in a small river that winds into a faraway landscape (fol. 164v) or showing Saint Louis, king of France, approaching Damietta aboard a ship with billowing sails (fol. 173r; fig. 6.6).

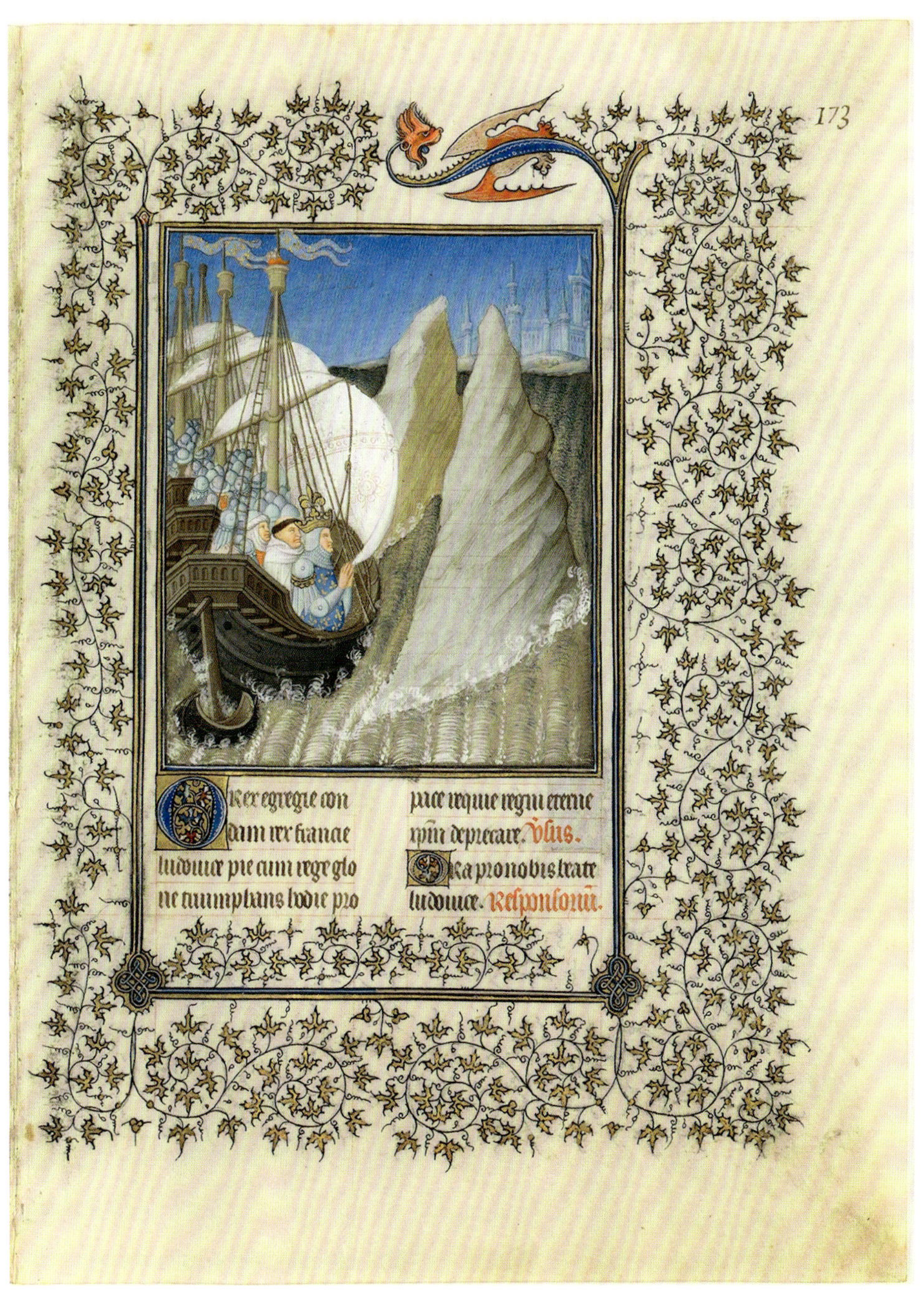

FIGURE 6.6. Limbourg brothers, *Saint Louis approaching Damietta*, from *The Belles Heures* of Jean de France, duc de Berry, 1405 to 1408/9, fol. 173r. Metropolitan Museum of Art, New York. Public domain.

But like panel paintings made during the first two decades of the fifteenth century, the Limbourg brothers' approach to reflected and refracted light is one of material mimesis. The waves in the miniature depicting Saint Louis, for example, have touches of gold and silver, and Saint Louis's gold crown is depicted with gold leaf as well.[4] In fact, throughout the miniatures most metallic surfaces are still imitated with real metal while gemstones, although on gold-leaf crowns and other precious objects their settings are sometimes outlined, are conspicuously absent throughout the *Belles heures* miniatures.[5]

Weaving Gems

There was one art flourishing around 1400 that does show a different approach toward depicting translucent materials, and precious stones in particular. During the fourteenth century tapestry had become an important art form for illustrating biblical stories and depicting historical events. Because the materials of tapestry are by definition opaque, weavers faced a different challenge than painters did in trying to render the optical qualities of certain materials. They could, of course, add reflections with gold thread, but similar material imitations of translucent materials such as precious stones—which the medieval painters imitated with translucent paint over reflective foils—cannot be made with silk or wool. Silken vestments were therefore often embroidered in relief to provide settings for precious stones or glass imitations (fig. 5.14), but it appears that figurative tapestries were not typically embellished in this way. Instead, precious stones were woven. Two important early examples with a plethora of such woven gems are the tapestry series known as the *Nine Heroes*, kept in the Metropolitan Museum of Art, and the two tapestries depicting the legend of Saint Piat and Saint Eleutherius, still kept in their original location, the Tournai Cathedral.

The Nine Heroes Tapestries were made about 1400 to 1410 in the southern Netherlands and originally comprised three tapestries depicting nine or ten heroes or "worthies," including King Arthur, Joshua, David, Hector of Troy, and Julius Caesar. The two Tournai tapestries are especially interesting because they can be precisely dated based on documentary evidence. The tapestry series was commissioned in 1402 (finished in 1404) by Toussaint Prier, a canon at Tournai Cathedral.[6] The garments and jewelry of the figures on both tapestry series are embellished with an abundance of precious stones. The weavers used a rather particular visual system to depict these gems, always juxtaposing a light and a dark color of the same hue (fig. 6.7).

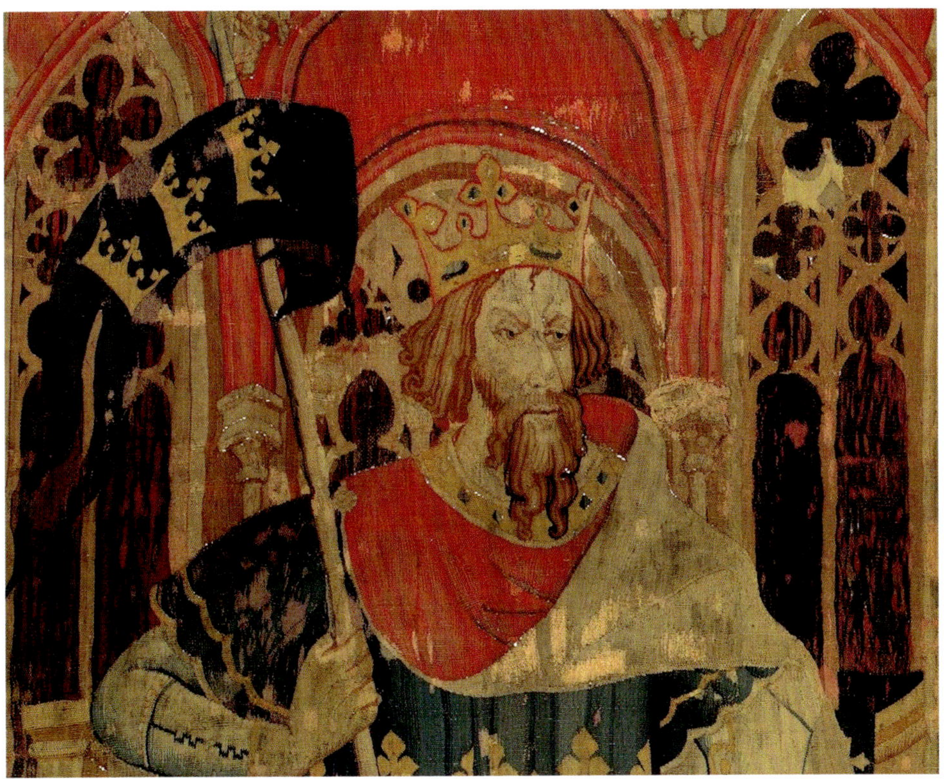

FIGURE 6.7. Detail of *King Arthur* (from the Nine Heroes Tapestries), South Netherlandish, 1400–1410. Wool warp, wool weft. Metropolitan Museum of Art, New York. Public domain.

This method for weaving gems might have derived from a much older method found in some of the earliest surviving medieval mosaics. Both the fifth-century mosaic in the Santa Maria Maggiore in Rome and the sixth-century mosaic in the San Vitale in Ravenna (fig. 6.8), for instance, depict gemstones with the same visual system used by the fifteenth-century tapestry weavers, juxtaposing light and dark tesserae of the same hue. Seven hundred years later, the top of the mosaicked cupola of the narthex of the Basilica di San Marco in Venice is decorated with depictions of precious stones, still rendered by the same method.

It is not entirely clear where this visual system for portraying gems in mosaics came from. One possibility is that a simplified scheme survived from ancient methods used to depict the action of light on precious stones and other translucent materials, such as can be seen on the Fayum mummy

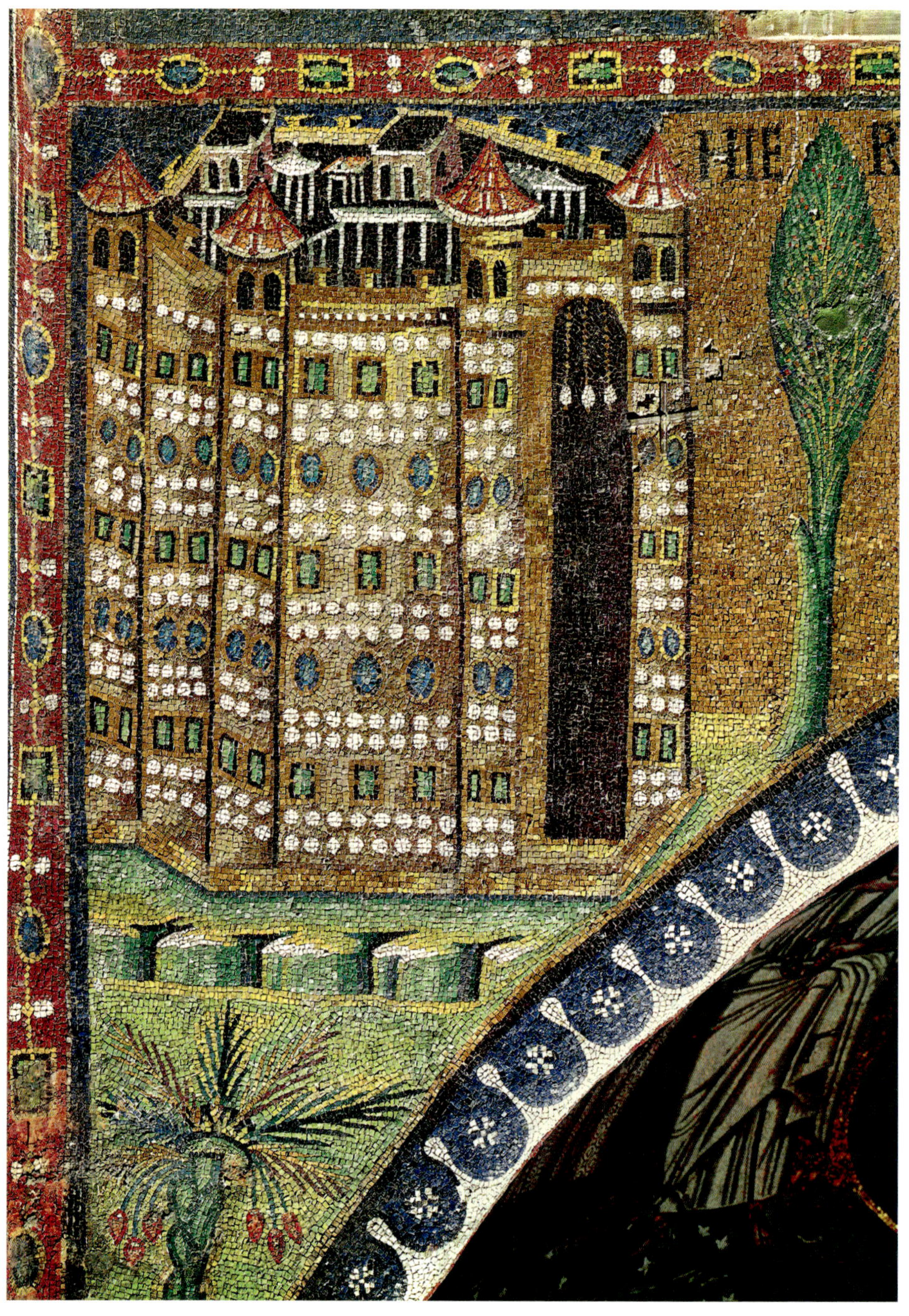

FIGURE 6.8. *Jerusalem*, detail of Justinian mosaic, mid-sixth century. Church of San Vitale, Ravenna (Italy). © 2023 Photo Scala, Florence.

portrait I discussed in the introduction (fig. I.2). Ancient painters used a similar juxtaposition of light and dark colors of the same hue to depict the translucency of precious stones, and in the second mummy portrait I discussed (I.1) it was also used to paint the woman's translucent brown eyes. This visual system for depicting translucency is not complete, however, without the specular reflection, the bright white fleck of light shimmering on a smoothly polished gem or in someone's eyes. We do find this bright little light in both mummy portraits, but it had been lost since antiquity. The gems depicted on early Christian mosaics and on fourteenth-century tapestries, juxtaposing only light and dark colors, therefore appear to be schematic, an echo of a paint system once used in ancient art. In the art of painting too, not a single work before Van Eyck that can be dated with certainty—not on panel, in books, or on the wall—depicts reflected light without a reflection's actually being there. Instead, the medieval painters explored translucent oil glazes and varnishes over polished gold, silver, or tin to render the splendor of gems and metalwork, closely tied to their ambition to produce a work of material mimesis. They almost certainly would have noticed specular reflections on all kinds of materials in the world around them, but since their work was meant to literally refract and reflect light—through translucent glazes and from polished metal—it was not necessary to depict these actions of light by means of the paint system that uses specular reflections.

As I argue in this chapter, it was Van Eyck's rediscovery of the power of this little white light that enabled early Netherlandish painters, for the first time since antiquity, to depict an array of reflective and light-transmitting materials without using real gold or translucent paint. More than anything else, it was this rediscovery of painted light that changed the painters' ambition from the mimesis of materials to the mimesis of the visible world. In this new paint system varnishes and glazes were still important, but since they were no longer required to imitate gold, enamel, and gems, they assumed a new role.

Jan van Eyck and the Specular Reflection

As I explained in the introduction, specular reflections occur when light reflects off the surface of a (microscopically smooth) material at the same angle as do the incident light rays (fig. I.3). These small white glimmers help the human eye interpret a surface as "glossy." For the earliest-dated work that contains painted specular reflections, we have to turn to Jan van Eyck's Ghent Altarpiece (figs. 6.9 and 6.10). This large polyptych, constructed to open and close like a triptych, is still in its original location, Saint Bavo's

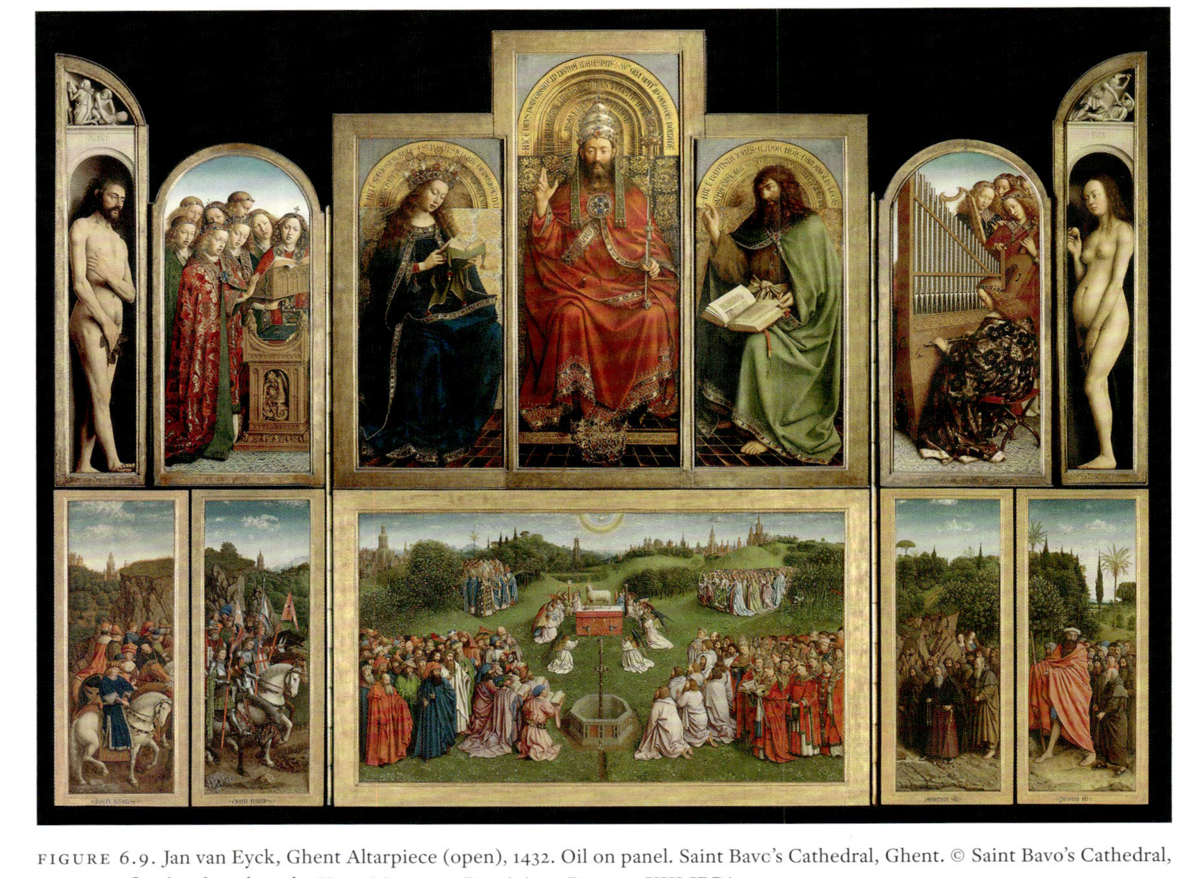

FIGURE 6.9. Jan van Eyck, Ghent Altarpiece (open), 1432. Oil on panel. Saint Bavo's Cathedral, Ghent. © Saint Bavo's Cathedral, www.artinflanders.be, photo by Hugo Maertens, Dominique Provost, KIK-IRPA.

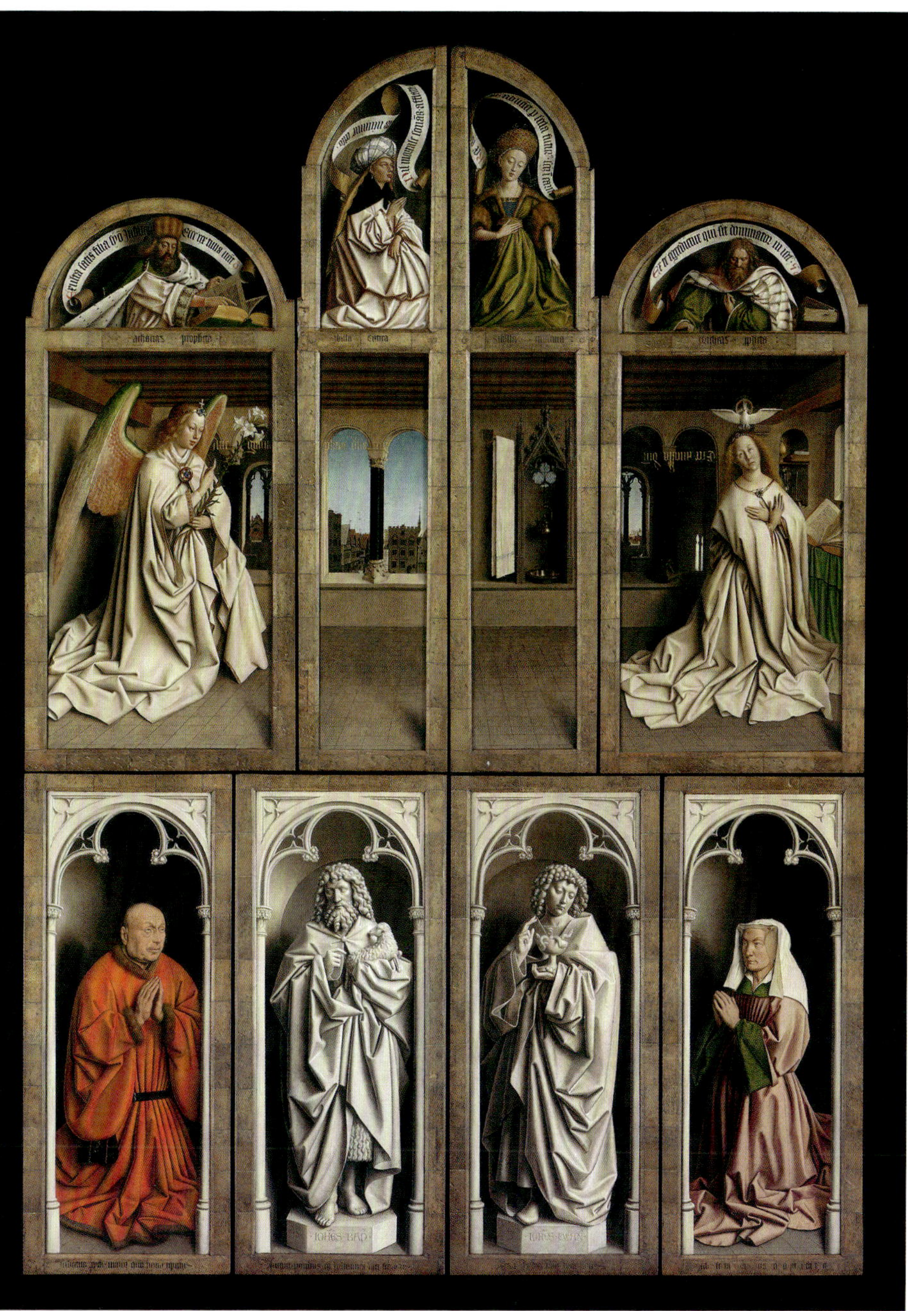

FIGURE 6.10. Jan van Eyck, Ghent Altarpiece (closed), 1432. Oil on panel. Saint Bavo's Cathedral, Ghent. © Saint Bavo's Cathedral, www.artinflanders.be, photo by Hugo Maertens, Dominique Provost, KIK-IRPA.

Cathedral in the city of Ghent. The quatrain on its frame, recovered during an 1823 restoration, gives a date: "The painter Hubert van Eyck, greater than whom no one could be found, began [this work]; Jan, second in art, completed the heavy task at the wish of Jodocus Vijd, who by this verse places under your protection what was completed on the sixth of May."[7]

When the letters painted in red in the last verse are combined, they give the year 1432. The Ghent Altarpiece consists of twelve interior panels, depicting heavenly redemption in the panels of the upper register and the Adoration of the Lamb in the lower panels. The closed altarpiece also comprises twelve paintings. Here Van Eyck painted its donors, Jodocus Vijd and Elisabeth Borluut, beside Saint John the Baptist and Saint John the Evangelist in the lower register. In the upper register we find the Annunciation, crowned by four lunettes depicting prophets and sibyls. Most of the figures on the Ghent Altarpiece wear splendorous jewelry and rich silk gowns set with precious stones. In the upper register of the opened altarpiece, Van Eyck embellished the miter of God the Father with an abundance of pearls and a variety of precious stones in translucent red, green, and blue (fig. 6.11). The paint system he used to evoke the reflective surfaces and glowing translucency of these gems is entirely different from that used by the painters working in the centuries before.

It is instructive to analyze the way Van Eyck paints gems using Willem Beurs's 1692 *De groote waereld in 't kleen geschildert* (*The Big World Painted Small*). In this treatise, rather uniquely, Beurs explains in detail how to paint a variety of motifs with oil colors. He teaches readers to paint butterflies and other insects, various kinds of fruit, lions, horses, humans alive and dead, metals, light and fire at night, and much more. Beurs also explains how to paint precious stones (*kostelijke gesteenten*), and although it was written more than two hundred years after the Ghent Altarpiece was finished, the method he describes is remarkably similar to the one Van Eyck used and can easily be reconstructed (fig. 6.12). Using the example of the ruby, Beurs provides a recipe that, he argues, can be used to paint all types of colored translucent precious stones:

> Rubies [*robijnen*] are painted like red wines: but they, there where the day falls [*daar den dag valt*], have to be heightened [*gehoogt*] with white. . . . There are other precious stones [*edele gesteenten*], that we do not explain separately because from what was said they can now be made with ease.[8]

According to Beurs, rubies are painted like red wines, but they also need to be heightened (*hogen* or *hoogsels*) with a white specular reflection (*daar*

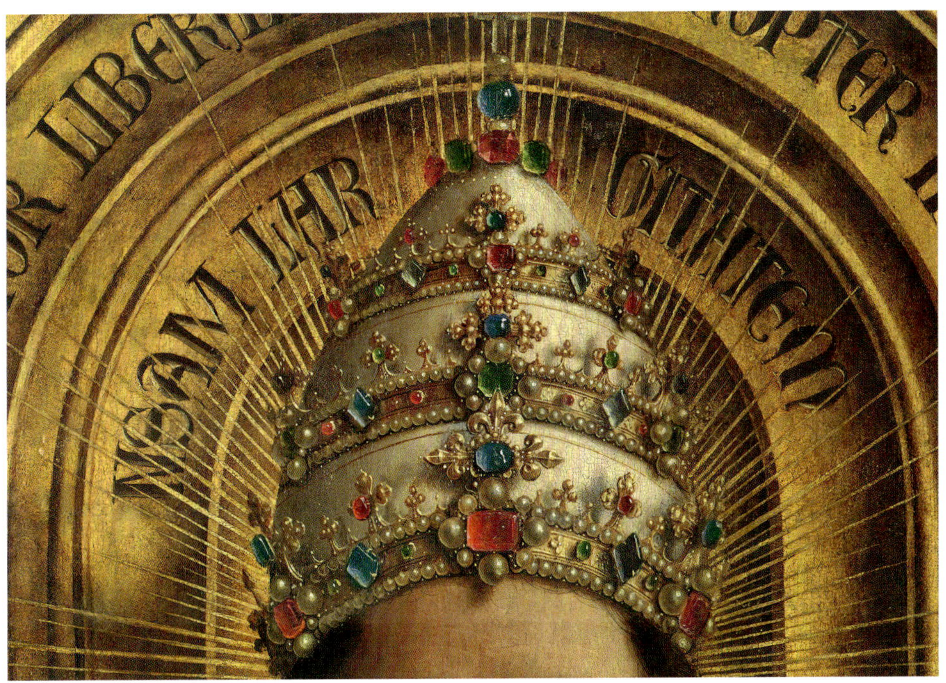

FIGURE 6.11. Jan van Eyck, gems on the miter of God the Father. Detail of the Ghent Altarpiece (before restoration). © Closer to Van Eyck, http://closertovaneyck.kikirpa.be.

den dag valt, met wit gehoogt worden—the modern Dutch term for the specular reflection is *hooglicht*).

In telling how to paint a glass filled with red wine, Beurs explains that the painter uses *lak* (a red lake pigment) and black for the darkest color of the wine (*tot de diepsels*); a red lake glaze depicts the intermediate color (*tusschenkoleur*), and for the internal reflection (*weersteutinge*) that occurs in a colored translucent liquid, a mix of red lake and vermilion (*vermilioen*) is used "so that you come as close to life as is possible" (*datge het leeven zoo na by komt, als 't mogelijk is*).[9]

These exact elements can be found in Van Eyck's precious stones on the Ghent Altarpiece. The darkest parts, or *diepsels*, of the red gems on the miter of God the Father are painted with a dark red glaze that is modulated to become lighter toward the intermediate color (*tusschenkoleur*) of the stone (compare figs. 6.12 and 6.13). Van Eyck painted the *weersteutinge*—the internal reflection that helps our eye recognize the translucency of the gem—with a bright red paint that is likely vermilion or a mixture of red lake and

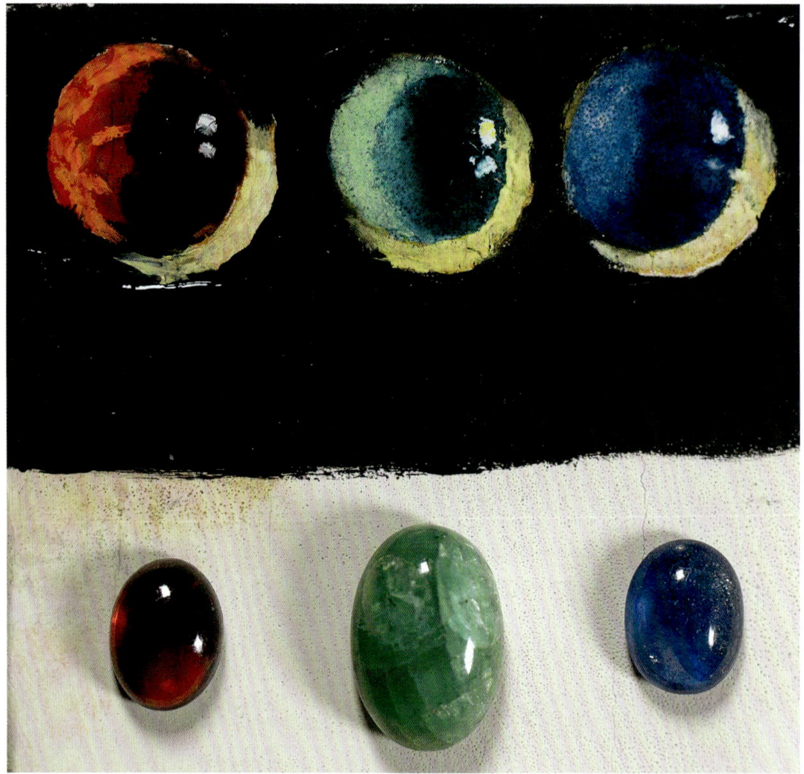

FIGURE 6.12. Reconstruction of painting a gem according to Willem Beurs. *From left to right:* Ruby, emerald, and sapphire (shown with a cabochon ruby, emerald, and sapphire). Photo by Marjolijn Bol.

vermilion, as Beurs suggested. Where the gem would have sparkled with specular reflections because of its microscopically smooth surface—the *hoogsels* in Beurs's recipe—Van Eyck applied small specks of white paint. In this system for painting light-transmitting precious stones, translucent paint is used to depict the saturated color of the ruby. But the stone's actual translucency is suggested with a paint that is completely opaque—in the case of the ruby, vermilion. When we study the paint system of the blue and green gems on the miter of God the Father, it becomes clear that Van Eyck used the same visual system to depict translucent gems of other colors, as Beurs indeed explained is possible (fig. 6.11). This system for depicting translucent gems can be found on all paintings attributed to Van Eyck. What is more, although before the Ghent Altarpiece not a single gem has been painted

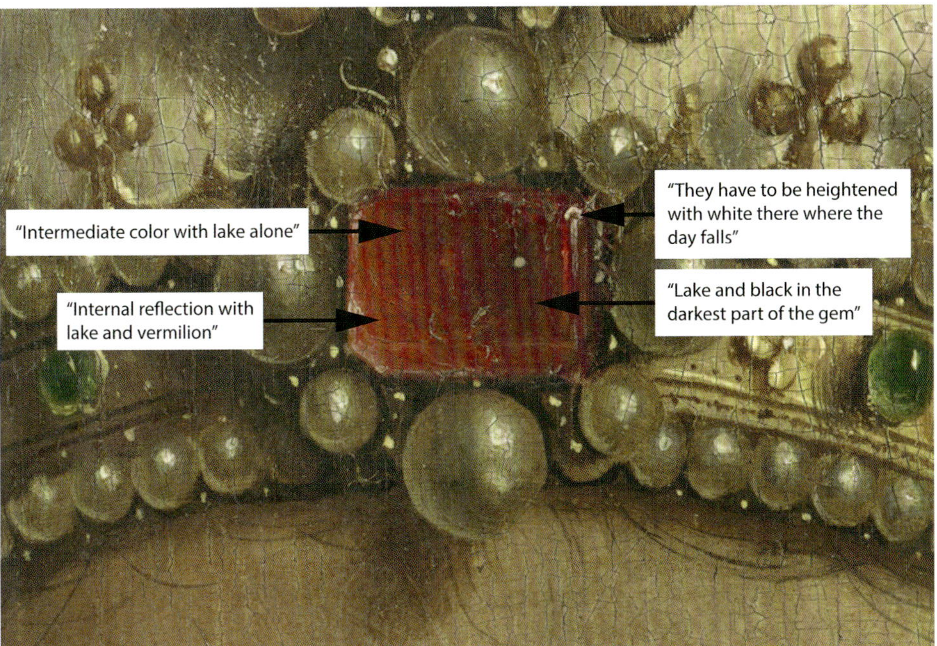

FIGURE 6.13. Jan van Eyck, Red gem on miter of God the Father, Detail of the Ghent Altarpiece (before restoration). © Closer to Van Eyck, http://closertovaneyck.kikirpa.be.

by this method, soon afterward almost every other painter knows how to apply the same visual system to depict precious stones and other translucent materials. Still, just because it was easily understood by other painters and could be codified in a recipe like Beurs's, Van Eyck's method for painting gems should not be considered a mere "trick." Quite the opposite: the new paint system completely transformed ideas about the nature of the picture until the advent of modernism, and not just for painting but for many other art forms as well.

Hybrid Gems

The introduction of the specular reflection can without doubt be called a revolution in the art of image making. Yet the older system of glazing gems and other translucent materials over polished metal foils and leaves was not soon forgotten. Both paint systems would be used side by side for centuries to come, and in a few cases they were combined to create "hybrid" forms of the old and the new approaches to mimesis. Examples of such hybrid gems can be found in the works attributed to the Master of Flémalle, to Rogier van der Weyden, and to some of their followers. Rogier van der Weyden's fame and work as a painter can be firmly established through archival and documentary evidence. We know that Van der Weyden was active in the workshop of Robert Campin (active in Tournai from before 1406 until his death in 1444), who took in four apprentices between 1426 and 1428: Rogelet de le Pasture (Rogier van der Weyden), Haquin de Blandain, Jaquelotte Daret, and Willemet.[10] But even though both Van der Weyden's and Campin's careers as painters can be reconstructed in some detail from written sources, not a single surviving painting can be attributed to either of them with absolute certainty. Thus there is no firm consensus about the oeuvre of the Master of Flémalle and Rogier van der Weyden, or about the identity of the other masters leading the workshops that produced the paintings we now group based on visual, technical, and in a few cases (but always later) documentary evidence. Considering this, we should take into account that it is at the very least possible that some of the Master of Flémalle–Van der Weyden group of paintings might have been made quite close to, or even before, the time Van Eyck started working on the Ghent Altarpiece. This we must keep in mind before we credit Van Eyck with yet another invention that might not have been entirely his own.

An early painting attributed to the Master of Flémalle, *Christ and the Virgin* (Philadelphia Museum of Art), is decorated with precious stones painted by a combination of the old and new systems for depicting translucent materials (fig. 6.14).

Some of the gems are glazed over the polished gold leaf halos of Christ and his mother. Red and blue precious stones, their settings outlined with black, are painted with a modulated glaze over gold leaf halos so convincingly that they indeed appear to protrude from the two-dimensional background. The painter of this work modulated his glaze to be lighter on one side of the gem and darker on the other. In this way he cleverly regulated the amount of light reflected from the gold leaf and thus, by juxtaposition of the darker and lighter colors of the glaze, approached the effect of internal reflection (which Van Eyck depicted with opaque paint instead). Significantly, each gem is also finished with one or two specular reflections to evoke the appearance of a glossy surface. The painter shows that he also knew how to use the new technique of painting gems as Beurs described it; he used it to paint the red precious stone on the Virgin's ring and the crystal on Christ's brooch. This suggests that, though he was familiar with both methods, this painter still preferred the older, and in a way more lustrous, technique for a divine motif such as a halo.[11]

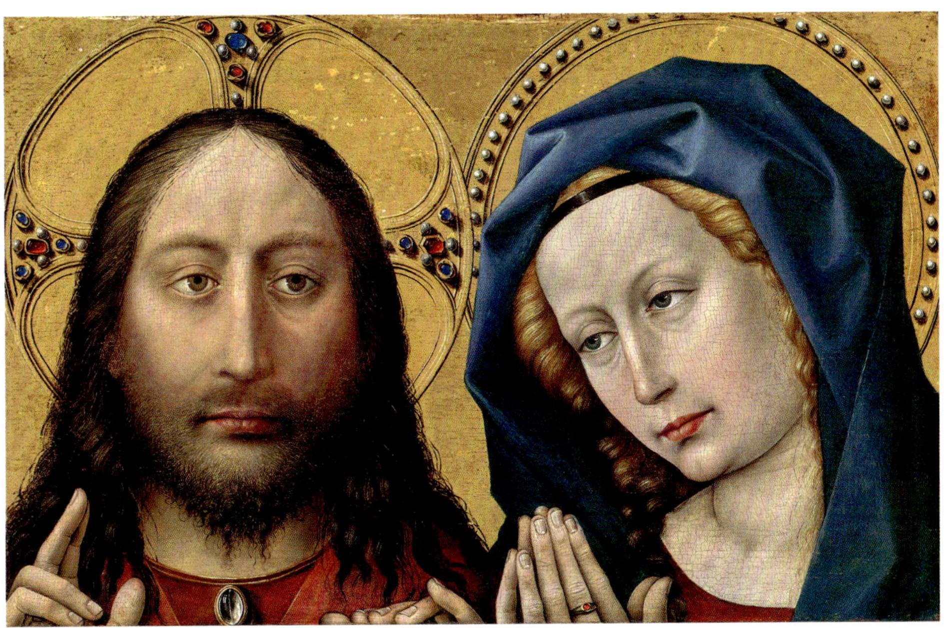

FIGURE 6.14. The Master of Flémalle (attr.), *Christ and the Virgin*, ca. 1427–32. Philadelphia Museum of Art. www.philamuseum.org.

Jacques Daret, a pupil or assistant of Robert Campin (known from his records from 1418 onward), uses comparable "hybrid" techniques to depict precious stones. Daret's Altarpiece of the Virgin, painted for the abbot of the monastery of Saint Vaast in Arras, should be singled out here because documentary evidence allows for its precise dating between 1434 and 1435.[12] This makes these four panels, apart from Jan van Eyck's oeuvre, the earliest Netherlandish paintings with firmly established dates: they were finished only two years after the Ghent Altarpiece. On the panel *Presentation in the Temple* (fig. 6.15), Daret uses gold leaf to imitate the ornaments embellishing the altar, the halos, and the borders adorning the cloaks of some of the figures. Gems are glazed over this gold leaf but, as with the Philadelphia *Christ and the Virgin*, Daret also renders their internal luster, specular reflection, and even shadows. In a panel belonging to the same altarpiece—*Adoration of the Magi*—Daret shows that he also knew how to use the new paint system to depict reflective and refractive materials without gold leaf.

To paint gems on the hem of the midwife Salome's sleeve, for example, Daret uses the same visual system that, as we have seen, Van Eyck and the

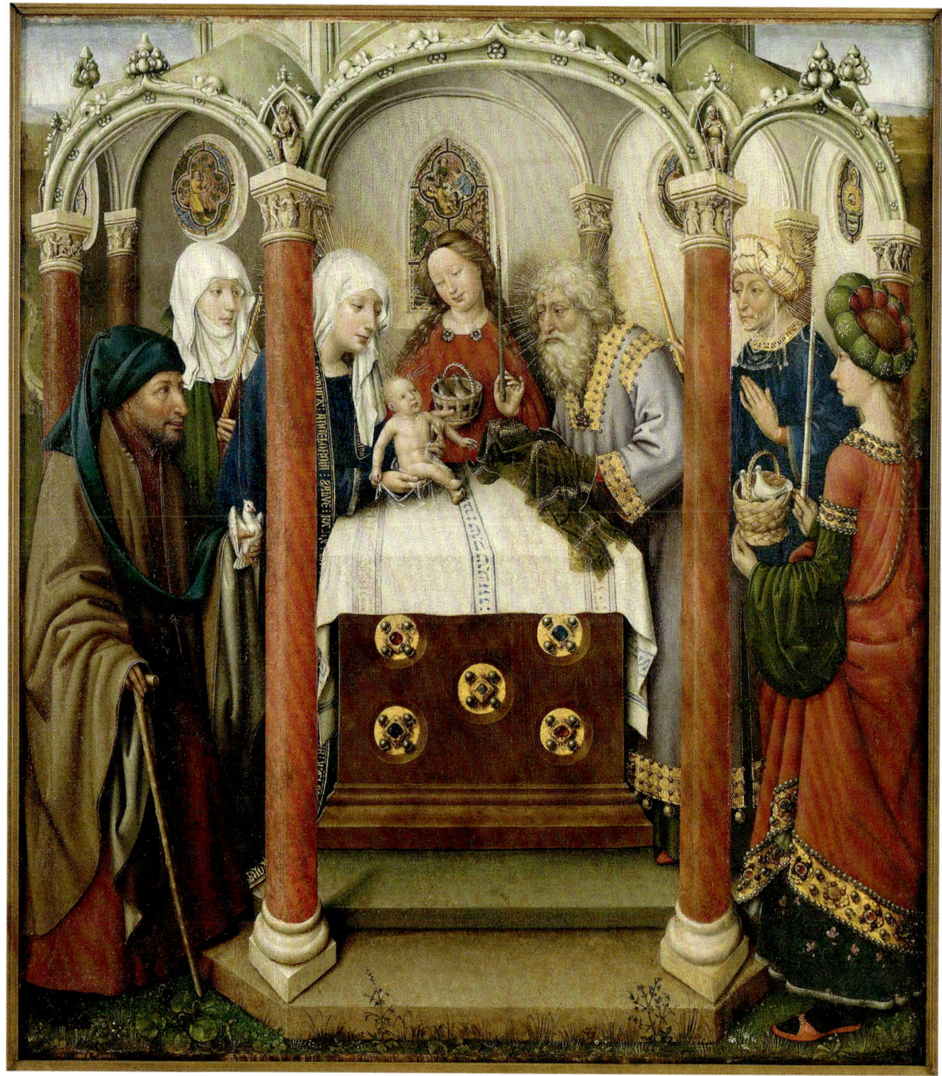

FIGURE 6.15. Jacques Daret, *Presentation in the Temple*, between 1434 and 1435. Petit Palais, Paris.

Master of Flémalle also used to paint precious stones without an underlayer of gold leaf.

Besides gems, textiles too were sometimes still painted by a combination of the old and new paint systems. A beautiful example can be found in a work attributed to Rogier van der Weyden, the *Portrait of Philippe I de*

Croÿ. Behind the sitter, Van der Weyden painted a green cloth that looks as if it was hung just after being unfolded, the creases in the fabric still visible. The painter depicts the cloth with a copper green glaze over a layer of polished silver leaf. According to Metzger and Steyaert, who investigated this portrait in the context of its restoration, the shapes of the folds have been incised with marks in the silver leaf, perhaps made with a metal point.[13] This is indeed a rather sophisticated method for adding three-dimensionality to a technique that had been used since early medieval times to create a material imitation of luscious silks and velvets.[14] Yet this method of engraving metal leaf below glazes is not new. As we saw in chapter 4, it was used centuries earlier to enhance translucent gemstones, enamel, and glazes (see, for instance, figs. 4.10 and 4.11).

A similar technique can be found in the Ghent Altarpiece. Jan van Eyck painted green (copper green), red (madder lake), and blue (azurite) tiles glazed over silver and gold leaf beneath the feet of the Virgin, Saint John the Baptist, and God the Father in the upper tier of the retable (fig. 6.9).[15] Like the silver leaf underneath Rogier's green cloth, the metal leaves under the tiles have been incised. In this case, however, it appears that Van Eyck did not intend to suggest their form "from below" the glaze paint, as with Rogier's cloth, but likely used it as a type of underdrawing. To demarcate their shape, Van Eyck left a line of gold leaf exposed between the tiles. To suggest the tiles' three-dimensionality, he painted a white line alongside the gold leaf left in reserve. In some of the corners of the tiles, he even painted the internal luster of the glazed tiles by using a lighter, more opaque color. In part Jan van Eyck, just as Rogier van der Weyden, inherited this method from earlier imitative techniques. Medieval artisans often imitated floor tiles with glazes over gold, silver, or tin. This is perhaps not surprising when we consider that actual medieval floor tiles were typically glazed with colored molten glass over a clay base. Yet there might have been a more religious reason motivating Van Eyck to use *pictura translucida* in the upper tier of his altarpiece as well. Medieval paintings and enamels show that it was common to make a translucent pavement below the feet of God the Father and accompanying saints. In the champlevé enamel depicting Pentecost that was discussed in chapter 4 (fig. 4.14), for example, a translucent red enamel frit was applied on gold foil below the feet of God the Father and the apostles.[16] That this tradition must have been well established and long-lasting becomes clear when we briefly return to the 1518 contract with the painter Hans Herbst discussed in chapter 5, which stipulated that the floor tiles depicted on the commissioned altarpiece were to be silvered and glazed over with red, blue, or green. In the Ghent Altarpiece, the lustrous floor under the feet of God the Father, Saint John, and the Virgin was therefore likely painted with the previously described hybrid form of *pictura translucida* because it was considered appropriate for celestial beings. This is further supported in that precisely in the upper tier of the polyptych, Van Eyck also used gold leaf in the background of the paintings.

In addition to the upper tier of the polyptych, Van Eyck also used a form of material mimesis on the polychrome frame of the Ghent Altarpiece. Recent conservation treatment has revealed that the polyptych's frame was originally embellished with imitation stone glazed over silver leaf, which had been obscured by later overpainting.[17] Painted imitations of stone and other materials can be found on several frames by Van Eyck and on the backs of some of his paintings as well, with some frames resembling weathered stone or marble, or with the backs imitating porphyry. This was not something new; slabs of porphyry not only were commonly used in portable reliquary shrines but were also imitated on the backs of paintings in earlier centuries. The back of the Westminster Retable, as we have seen, was embellished with a painted imitation of porphyry, and the twelfth-century French sculpture discussed in chapter 2 used porphyry and marble imitations for the seat of the Virgin's throne (fig. 2.5). Van Eyck's contemporaries therefore likely would have easily recognized and understood his references to the art of material mimesis on the backs of his paintings, since they were firmly established in centuries-old traditions. But by bringing these stone materials to the frames of his paintings—and in the case of Van Eyck's *Virgin and Child with Canon van der Paele* (1434–36, Groeningemuseum, Bruges) even copper incised with an inscription—Van Eyck also plays with the medieval tradition of applying gold leaf and gem imitations to frames for the same purpose. Indeed, it appears that when wooden panels became windows, characterized by the mimesis of the visible world, the frames and backs of paintings remained an active playground for the art of material mimesis.

Painting Transparent Materials: Tears and Bottles

The new approach to rendering with paint the optics of reflective and light-transmitting materials also allowed painters to depict transparent materials such as crystal, colorless glass, water, tears, and, in Daret's case, even icicles hanging from the stable in his nativity scene (fig. 6.16).

In the medieval system of optical substitutes, it was impossible to paint something see-through and colorless—such materials could be imitated only by applying a transparent layer and therefore would have been invisible. Indeed, the specular reflection is a prerequisite for depicting something

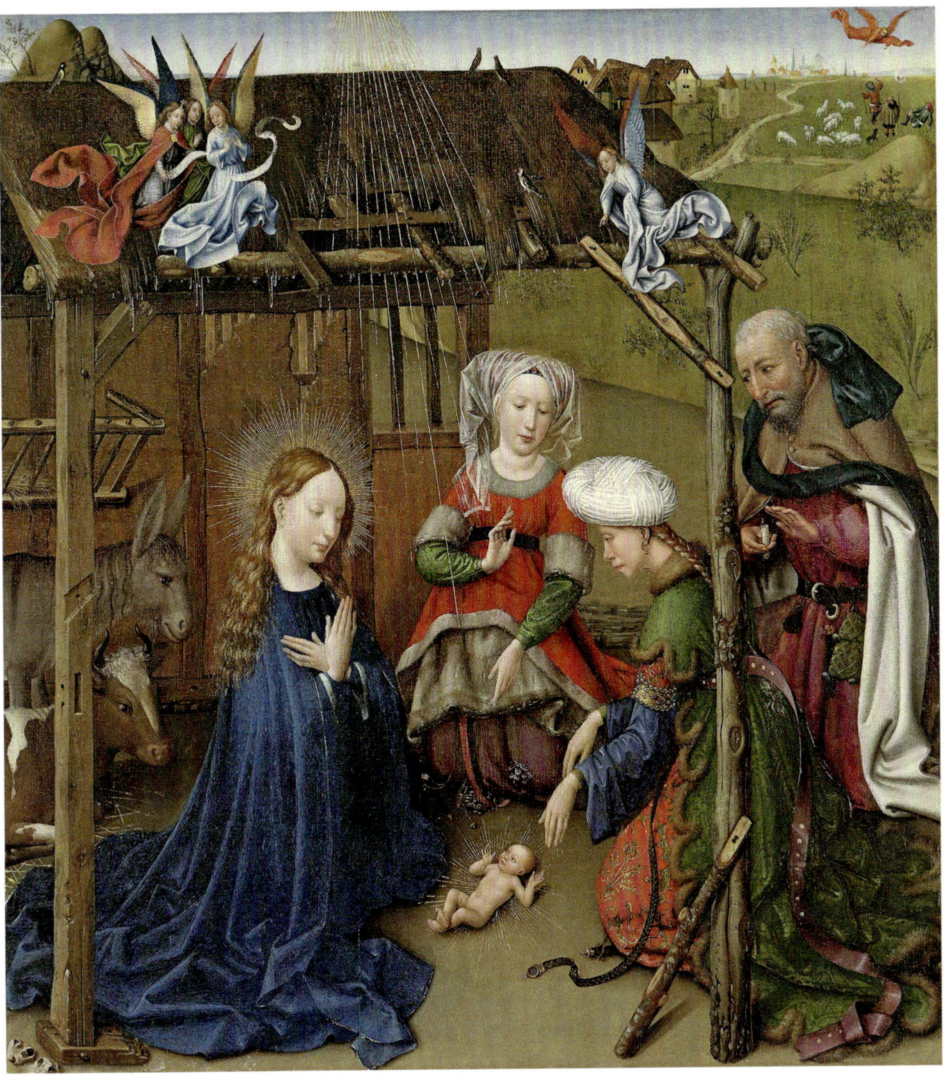

FIGURE 6.16. Jacques Daret, *Nativity*, between 1434 and 1435. Thyssen-Bornemisza Museum, Madrid. © Fundación Colección Thyssen-Bornemisza, Madrid.

transparent and colorless. One of the frescoes painted by Giotto di Bondone in the Cappella degli Scrovegni, Padua, is a beautiful example of how medieval painters struggled with this problem. The Scrovegni frescoes established Giotto's fame as the first master of human emotion since antiquity; something he was already heralded for by the late fourteenth-century chronicler

Filippo Villani: "For pictures formed by his brush follow nature's outlines so closely that they seem to the observer to live and breathe and even to perform certain movements and gestures so realistically that they appear to speak, weep, rejoice and do other things."[18] Besides the movements and gestures Villani speaks of here, Giotto also made some of his figures weep by painting tears (fig. 6.17).

In the fresco *The Massacre of the Innocents*, tears stream down the faces of the desperate mothers witnessing the slaughter of their children ordered by King Herod the Great of Judea (Matthew 2:16–18). Giotto's tears literally gush from the mothers' eyes in streams of a rather murky grayish scumble. Without the specular reflection, this was indeed the only way Giotto could paint this transparent liquid of human emotion.

For the earliest imitation of the colorless transparency of tears by painted specular reflections, we have to return to Rogier van der Weyden. As Villani had praised Giotto for his ability to make people weep, Bartholomaeus

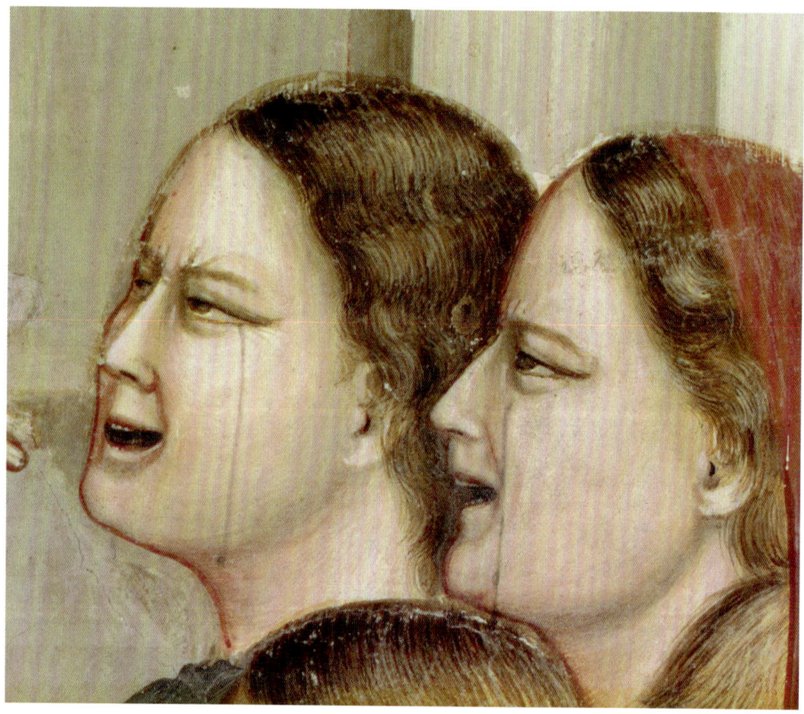

FIGURE 6.17. Giotto, *The Massacre of the Innocents* (detail), ca. 1305 (completed). Fresco. Cappella degli Scrovegni, Padua. © 2023 Photo Scala, Florence.

Facius praised Van der Weyden for the same achievement about a century later. Writing about the center panel of a triptych, *Christ's Descent from the Cross*, Facius remarks that "Mary his Mother, Mary Magdalen and Joseph, their grief and tears so represented, you would not think them other than real" (*Maria mater, Maria Magdalena, Josephus ita expresso dolore ac lacrimis ut a ueris discrepare non existimes*).[19] The work Facius wrote about is lost, but its theme was similar to the *Descent from the Cross* (before 1443), today kept in the Museo del Prado, Madrid.[20] This painting, unsigned and undated, has been considered one of Van der Weyden's most important works since at least the second half of the sixteenth century. The figures on the Prado *Descent* express their grief, mourning the death of Christ, who is held by Nicodemus and Joseph of Arimathea. Teardrops glisten on several faces. The painter's rendering of these tears is a world apart from the way Giotto approached the motif. With his recipe for painting "water drops," Beurs again gives us the basic formula for our analysis of Van der Weyden's paint system:

> To say something about water drops, those are painted as follows: one scumbles [*schommeltze*] them thin with black on the day [*op den dag*] over the painted body where the drop will be depicted, then where the internal reflection is [*in de reflexie of weersteutinge*] you apply some brighter color [*helderder koleur*] than the body below; now on the darker color toward the day, they should receive only a white specular reflection, and finally there has to be a shadow behind the internal reflection that is similar to the painted body that it looks like [*moet 'er agter de reflexie een schaaduwe zijn, overeekomende met het lichchaam, daarze op leit*].[21]

A closer look at the tears of Van der Weyden's Virgin shows that they have been painted exactly as Beurs prescribes (fig. 6.18).

To imitate the internal reflection (*weersteutinge*) in the transparent teardrops, Rogier outlined them with a flesh color slightly lighter than the Virgin's face. On the darker part of the tear "toward the day," which in Rogier's case is depicted by the flesh color of the Virgin's face, a bright specular reflection has been applied. And, finally, behind the bright color that imitates the internal reflection, Rogier depicted the tear's shadow, which indeed, as Beurs prescribed, follows the outline of the tear. Similar tears and water drops can be found on many other works attributed to Rogier van der Weyden and his followers. It is interesting that, even though in these works not all tears are rendered as delicately as the ones on the Prado *Descent*, they are all painted by the same visual system for depicting transparent

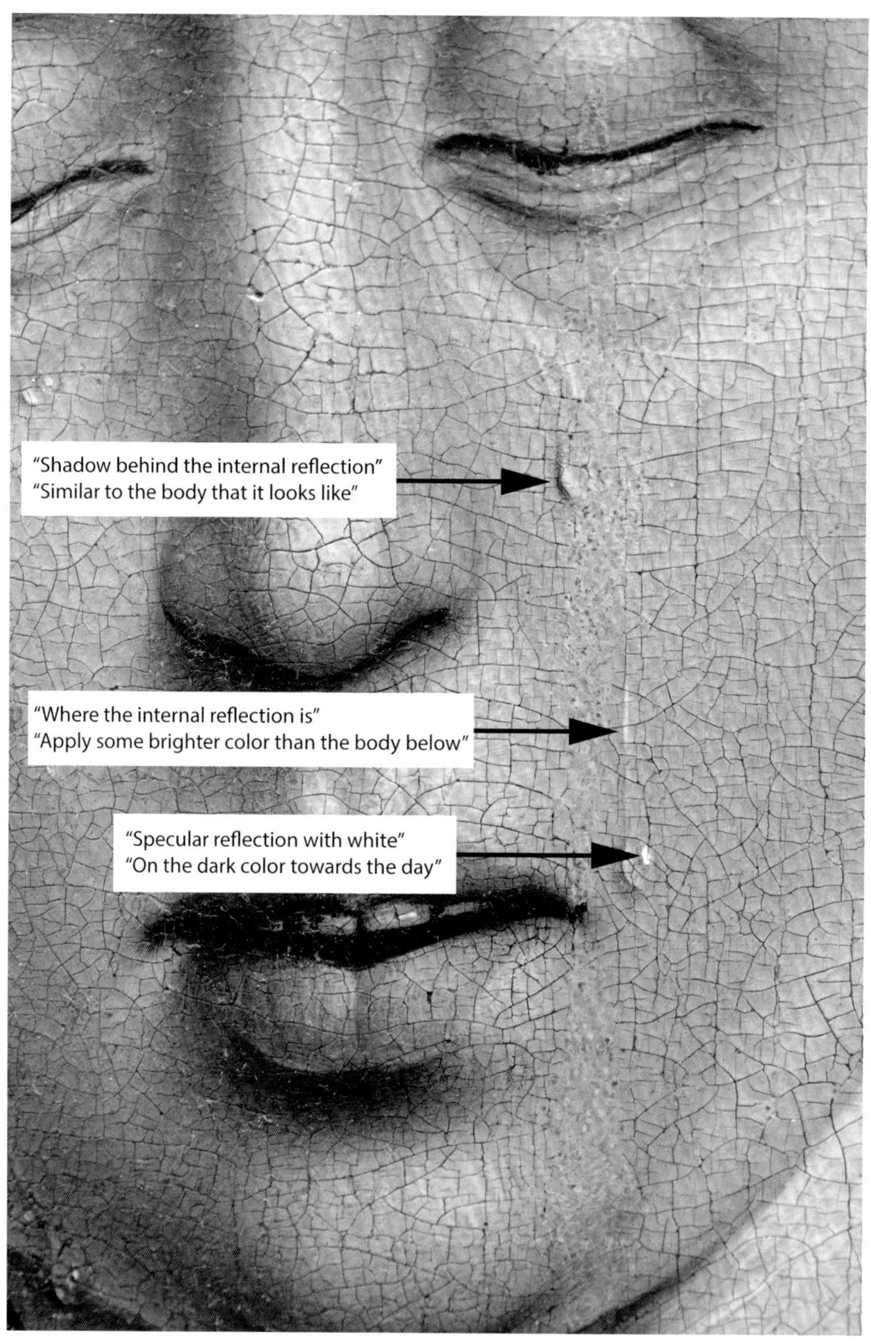

FIGURE 6.18. Rogier van der Weyden, Detail of *Christ's Descent from the Cross*, before 1443. Oil on panel. Museo del Prado, Madrid. Photo by Marjolijn Bol.

materials. Let us take as examples the tears of the weeping figures on the Mauritshuis *Lamentation* and the water droplets on the Berlin Saint John Altarpiece (figs. 6.19 and 6.20).

In both paintings, each tear and droplet has been depicted by internal reflections, specular reflections, and shadows in the same place as the tears of the Prado *Descent*. But there is one difference: the tears and water drops on the Mauritshuis *Lamentation* and the Berlin Saint John Altarpiece have been painted "on the day" with the thin black scumble recommended by Beurs, not the subtle flesh color used in the Prado *Descent*. Because of this darker outline, the result looks a little more schematic than the tears depicted on the Prado *Descent*.

Van Eyck is not known for depicting tears, but to portray the various optical characteristics of water and other transparent materials, he used the same paint system. In the *Fountain of Life* on the center panel of the Ghent

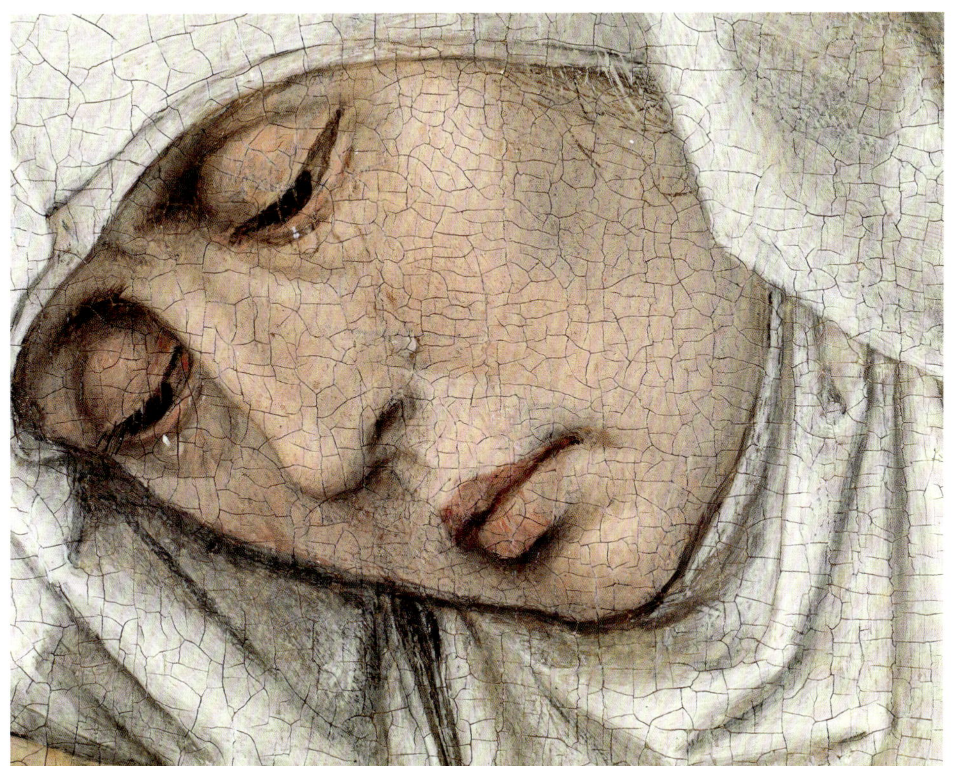

FIGURE 6.19. Rogier van der Weyden (attr.), *Lamentation* (detail). Oil on panel. Mauritshuis, The Hague. Photo by Marjolijn Bol.

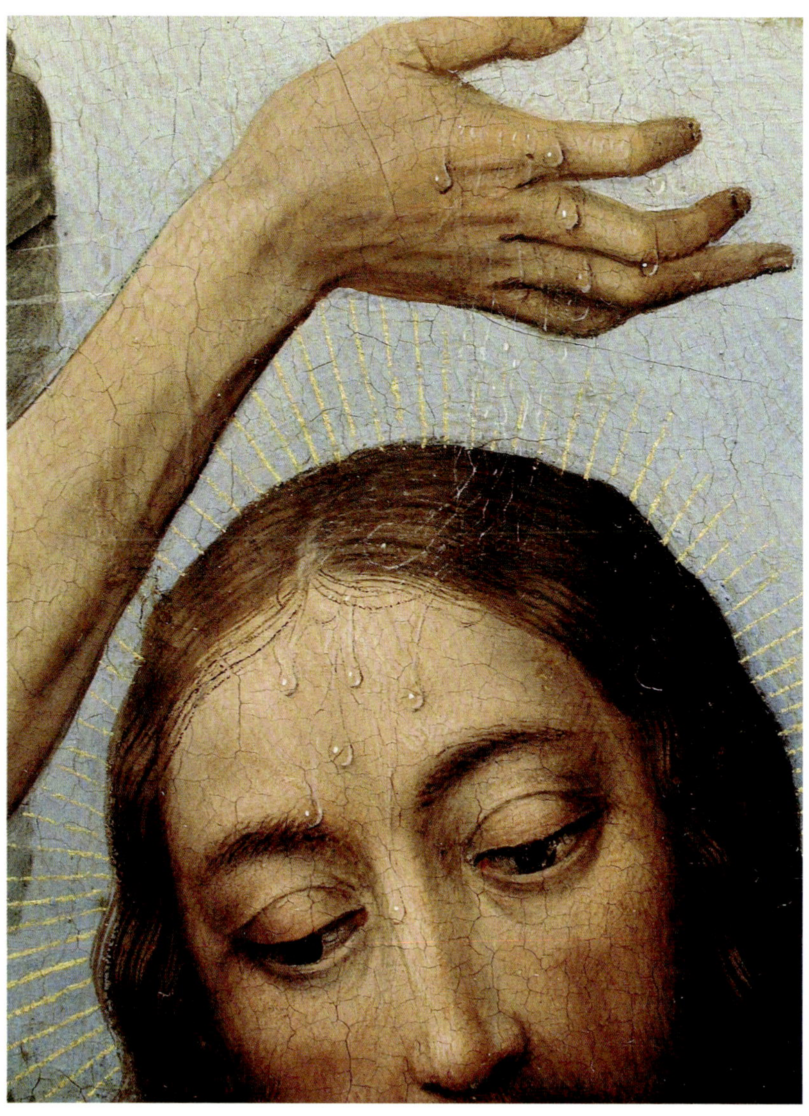

FIGURE 6.20. Rogier van der Weyden (attr.), *Baptism of Christ* (detail of the Saint John Altarpiece), ca. 1455. Oil on panel. Gemäldegalerie der Staatlichen Museen, Berlin. Photo by Impact of Oil.

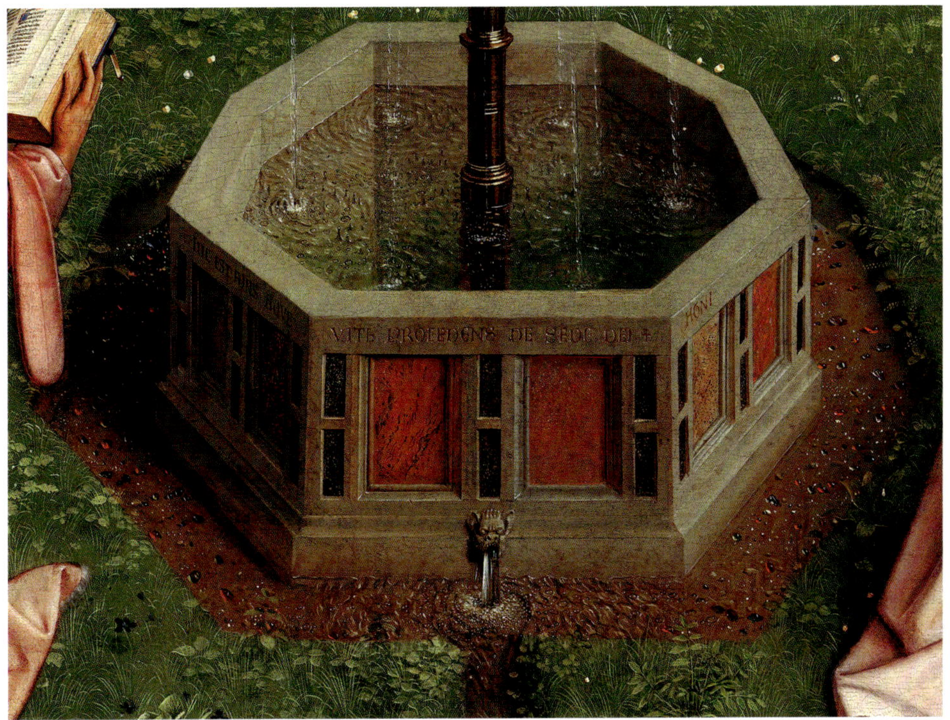

FIGURE 6.21. Jan van Eyck, *Fountain of Life*, Inscription*: Hic est fons aque vite procedens de sede Dei + Agni* (detail of the *Adoration of the Lamb*, Ghent Altarpiece, after restoration). © Closer to Van Eyck, http://closertovaneyck.kikirpa.be.

Altarpiece, for instance, Van Eyck juxtaposed white and black to depict the different behaviors of the water in the fountain (fig. 6.21).

From the mouth of a small creature sculpted at the base of the fountain, water streams into the canal surrounding it. The gushing water forms wavelets that are painted by setting curved brushstrokes of a brownish color against white brushstrokes. Water from the fountain's jets not only makes wavelets in the basin and the canal, it also is dashed upward in countless white drops. Van Eyck even depicts the bubbles created when a stream falls with some force into water, using brown circles similarly juxtaposed to white lines, their volume suggested by two carefully placed specular reflections. A bit farther down the canal, along the outside of the basin, he paints a thin white line with a brown line directly underneath it. Our eyes accordingly fill the canal with water. The precious stones scattered in the small canal surrounding the Fountain of Life are a reference to the gems believed to grow in the fountain in the Garden of Eden, that terrestrial paradise, discussed in chapter 4.[22]

Glass Flasks

The various other transparent materials depicted on the Ghent Altarpiece were painted by the same method used to depict the transparency of water and tears. Van Eyck used it, for example, to paint God the Father's crystal scepter and to paint the large diamond that crowns his gem-studded red cloak. But perhaps the best-studied transparent motif in early Netherlandish painting is the small crystal or glass flask Van Eyck painted behind the Virgin Annunciate (fig. 6.22), which reappears in various of his other works, especially his Madonnas. Such flasks became a popular motif in the works of Van Eyck's contemporaries and followers, both north and south of the Alps. We find it, for instance, on a panel called the *Annunciation* (Musée du Louvre, Paris, fig. 6.23), which is attributed to the workshop of Rogier van der Weyden. And another, rather spectacular example is on the right wing of the Werl Altarpiece (Museo del Prado, Madrid, fig. 6.24), named after its donor, whose name is inscribed on a painted parapet as though it were chiseled in stone: "In the year 1438 Minister Heinrich von Werl, Master of Cologne, had this portrait painted."[23] The two wings of what was once a triptych stand out because, owing to its painted inscription, they are the only works attributed to the Van der Weyden–Flémalle group that can be securely dated. The painters of the Werl and Louvre flasks evoke their transparency by juxtaposing black lines against white; the liquid in the bottle, like that in Van Eyck's flask, casts a colored shadow on the wall behind it.

South of the Alps, Fra Angelico and Fra Filippo Lippi are the first painters to incorporate this northern motif in their work. Fra Angelico (ca. 1395–1455) used the specular reflection to paint a small variety of transparent glassware in his work *The Healing of Justinian by Saint Cosmas and Saint Damian*, which documentary evidence shows he finished between 1438 and 1440. The flasks by Fra Filippo Lippi (ca. 1406–69) have been discussed extensively as an indication of his intimate knowledge of Eyckian or other Flemish examples.[24] By painting transparent flasks, he shows he knew how to use the visual system for depicting transparent materials introduced by the early Netherlandish painters.

To paint the flask on the *Annunciation* kept in the Alte Pinakothek, Munich (fig. 6.25), for example, Filippo Lippi juxtaposed white against black to depict its transparency the same way painters from the North approached the motif.

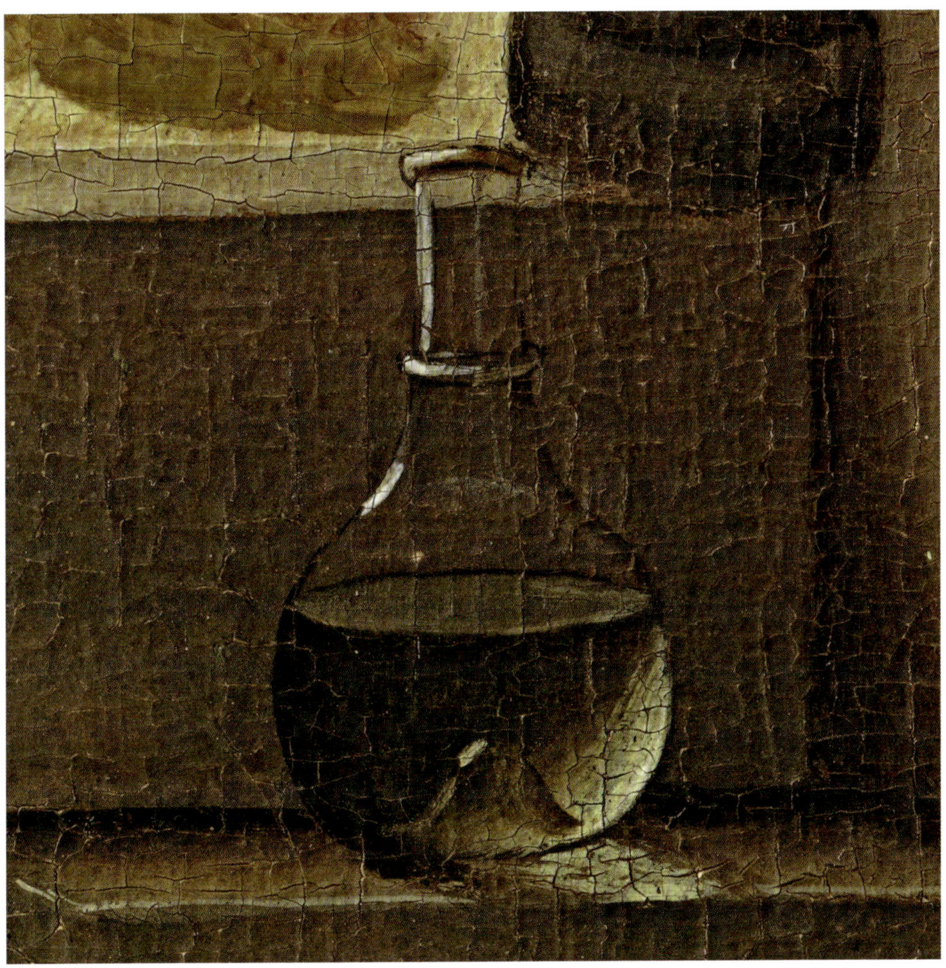

FIGURE 6.22. Jan van Eyck, *Flask* (detail of *The Virgin Annunciate*, the Ghent Altarpiece, after restoration). © Closer to Van Eyck, http://closertovaneyck.kikirpa.be.

The symbolic meaning of these transparent bottles, particularly in paintings depicting the *Annunciation*, has been extensively studied as a reference to the Virgin's immaculate conception.[25] But beyond these religious implications, painters depicted the transparent carafe—by virtue of its reflection only—as a symbol of the new ambition of painting: the mimesis of the visible world. That these early adopters of the specular reflection in Italy still worked in their medium of choice, with pigments bound with egg, shows that at this time the introduction of the Flemish manner of painting in Italy

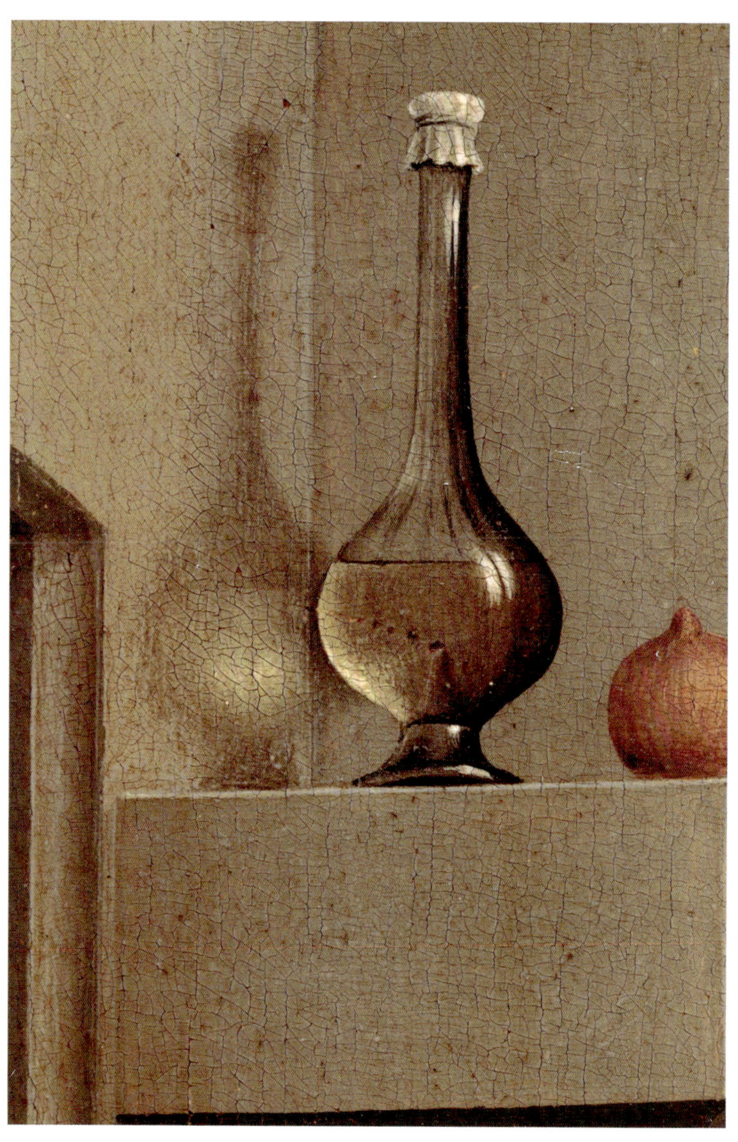

FIGURE 6.23. Rogier van der Weyden (attr.), *Flask* (detail of *The Annunciation*), ca. 1435. Oil on panel. Musée du Louvre, Paris. Photo by Marjolijn Bol.

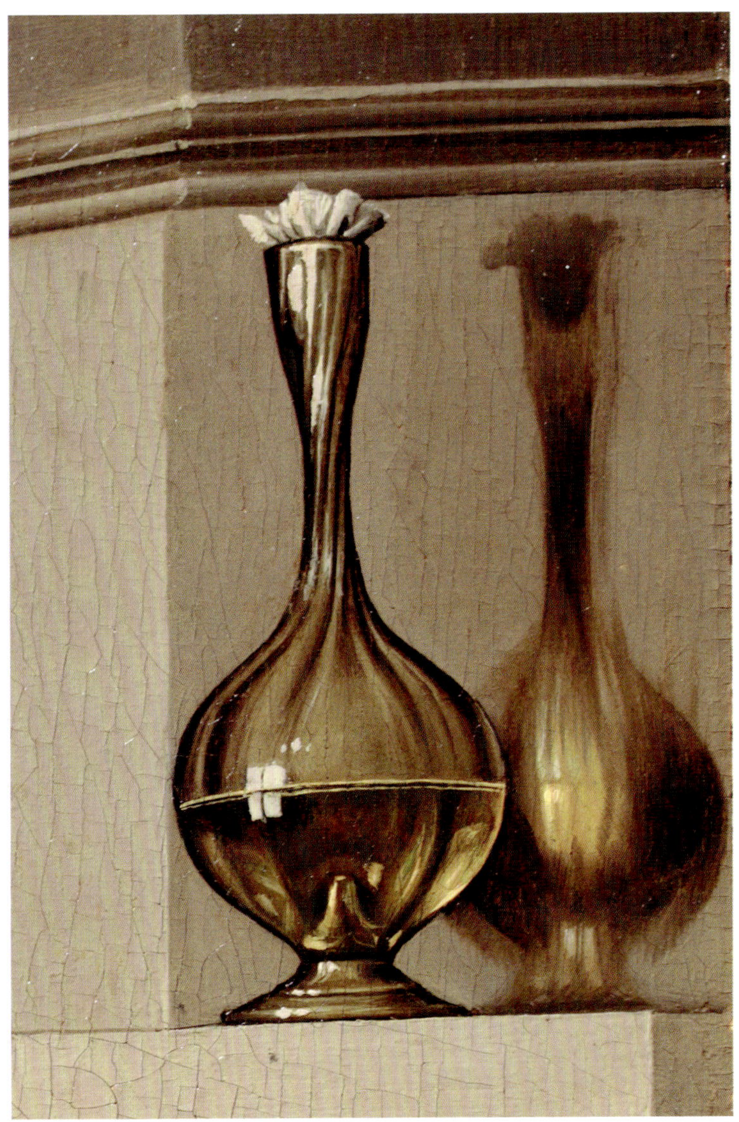

FIGURE 6.24. Master of Flémalle (attr.), *Flask* (detail of *Saint Barbara*). Oil on panel. Museo del Prado, Madrid. Photo by Marjolijn Bol.

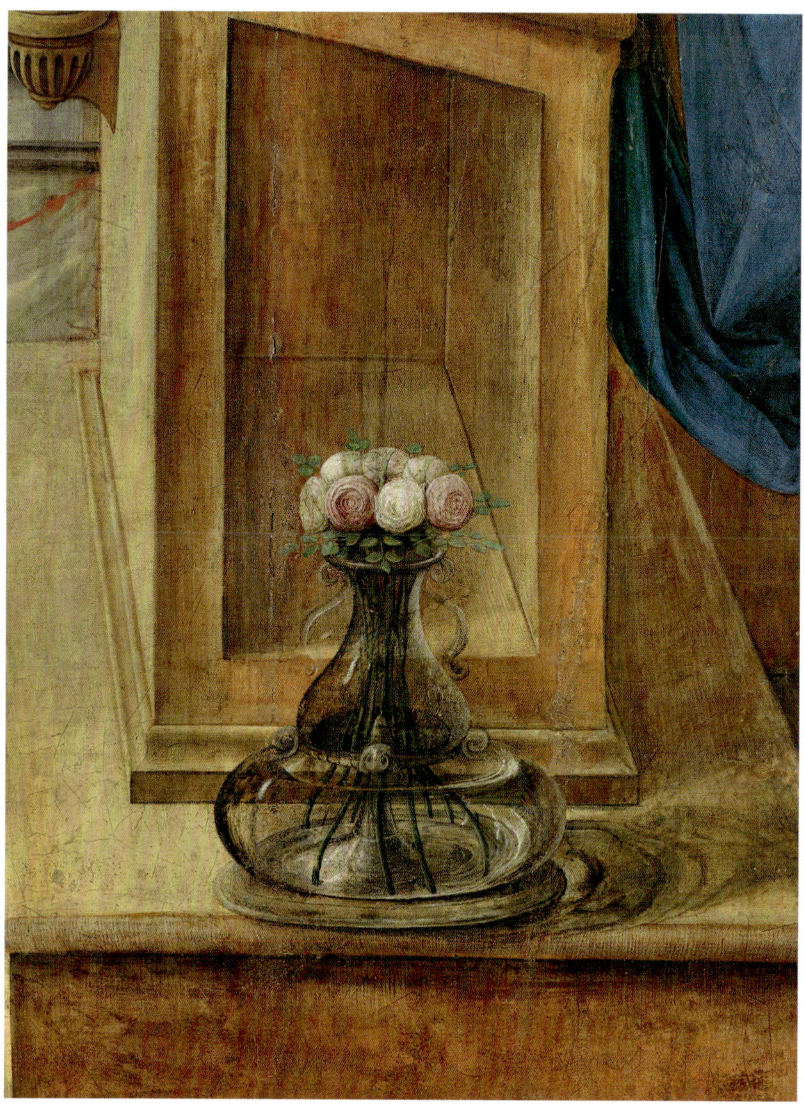

FIGURE 6.25. Fra Filippo Lippi, *Flask* (detail from *The Annunciation*), ca. 1443–45. Tempera. Alte Pinakothek, Munich. © 2023 Photo Scala, Florence / bpk Bildagentur für Kunst, Kultur und Geschichte, Berlin.

had little to do with the "battle" between egg and oil that would ensue decades later. Indeed, the oil medium is not required for painting the actions of light on various surfaces: this can be done perfectly well with an aqueous binding medium.

Alberti's Precious Paint

The impact of the specular reflection on the art of painting was first described by Leon Battista Alberti (1404–72), one of the most influential humanist scholars of the Italian Renaissance. In 1435 Alberti finished *De pictura*, the first treatise to develop a coherent and systematic art theory.[26] Alberti himself was well aware of this and explains in his introduction that he was "not writing a *historiam picturae* like Pliny, but treating of the art in an entirely new way."[27] Alberti acknowledges that whereas he might have had precursors among the ancients such as Apelles (who, according to Pliny, wrote a treatise to Perseus), "none of these exist today." The idea of doing something novel and great without examples was important to Alberti. It was also what he admired in the painters of his own day. He points out that, in contrast to the ancient painters, who did have models they could rely on, the painters of Alberti's time changed their art without precedent. Alberti might in fact be the first to bring the notion of originality to art theory.[28] As we saw in chapter 1 above, this became so important to Renaissance art theory that the wish to surpass the ancients was the reason Vasari introduced the invention of oil paint to the *Vite*; so that the painters of his own day and age could surpass the ancients by virtue of a technique unknown even to Apelles and Zeuxis.[29] And north of the Alps, Dutch art theorists also picked up on this notion of originality to praise Van Eyck for having changed the art of panel painting without examples and without precedent.[30]

Alberti's *De pictura* is of course best known for recording the rules of mathematical perspective. He famously explains the rules of perspective by envisioning the painting as transparent, as a window into the world onto which an image is projected. In this way Alberti came to define our modern notion of the essence of a picture.[31] In *De pictura*, the art of image making is divided into and discussed in three parts learned, according to Alberti, from nature itself: circumscription, composition, and the reception of light. These three things determine how we see and are therefore the province of the painter.[32] Alberti emphasizes the importance of a clear arrangement of figures and the expression of emotions, both of which are part of circumscription and composition.[33] Of special interest to us here, of course, is the last category, in which Alberti discusses how important it is for the painter to understand the nature of the reception of light:

> I would consider of little or no virtue the painter who did not properly understand the effect that every kind of light and shade has on all surfaces [*ego quidem pictorem nullum vel mediocrem putabo eum qui non plane intelligat quam vim umbra omnis lumina in quibusque superficiebus habeant*].[34]

Alberti starts his discussion of the topic by pressing the significance of a careful application of white and black. He explains that only these two "colors" can represent the brightest gleams of the most polished surfaces (*ultimos tersissimarum superficierum fulgores*) and the darkest nights (*ultimas noctis tenebrat*). The passage warrants quoting in full, because Alberti describes, for the first time, the power of painted reflections for rendering polished and see-through materials:

> This composition of white and black has such power that, when skillfully carried out, it can express in painting brilliant surfaces of gold, silver, and glass [*un arte et modo facta aureas argenteasque et vitreas splendissimas superficies demonstret in pictura*]. Consequently those painters who use white immoderately and black carelessly should be strongly condemned. I would like white to be purchased more dearly among painters than precious stones [*pretiosissimas gemmas*]. . . . On this point Zeuxis used to condemn painters because they had no idea what was too much. If some indulgence must be given to error, then those who use black extravagantly are less to be blamed than those who use white imperatively; for by nature, with experience of painting, we learn as time goes by to hate work that is dark and horrid, and the more we learn, the more we attune our hands to grace and beauty. We all by nature love things that are open and bright [*aperta et clara*]. So we must the more firmly block the way in which it is easier to go wrong.[35]

This juxtaposition of white and black, as we have seen, was indeed used by the early Netherlandish painters to paint everything reflective and transparent. And, as Alberti suggests, the power of the specular reflection certainly lies in its careful application. Too much white and the depicted material loses its specularity entirely.

Alberti concludes his treatise on painting by scorning painters who apply real gold to their works "because they think it lends a certain majesty to painting":

> I would try to represent with colours rather than with gold this wealth of rays of gold that almost blinds the eyes of the spectators from all angles.

> Besides the fact that there is greater admiration and praise for the artist in the use of colours, it is also true that, when done in gold on a flat panel, many surfaces that should have been presented as bright and gleaming [*claras et fulgidas*], appear dark to the viewer, while others, that should be darker, probably look brighter.[36]

Together these two passages form an appeal for a new approach to mimesis. In Alberti's theory, painting should not imitate reflection and refraction with materials that reflect or refract but rather, and with paint alone, should imitate the actions of light that can be perceived in the visible world. When this is done with great skill, the result would be the ultimate specular reflection. Alberti asks readers, "What is painting but the art of embracing by means of art the surface of the pool?" (*Quid est enim aliud pingere quam arte superficiem illam fontis amplecti*).[37] It is no coincidence that to reinforce his point Alberti stresses the value of a painting executed by his method by referring to the long medieval tradition of embellishing frames with imitation metalwork. The well-executed painting (*historia*) should not be framed with ornaments imitating metalwork, because the painting that imitates the visible world is worthy of being framed by *real* precious materials:

> Other ornaments done by artificers that are added to painting, such as sculpted columns, bases, and pediments, I would not censure if they were real silver and solid or pure gold [*auro solido vel admodum purissimo*], for a perfect and finished painting is worthy to be ornamented even with precious stones [*gemmarum*].[38]

We know that Alberti finished *De pictura* in 1435 because of a note he added to one of the manuscripts.[39] This is remarkably early for giving such prominence to the power and importance of painted light; the first Italian paintings with specular reflections all postdate Alberti's treatise. This is why it has been suggested that Alberti might have seen works by Van Eyck or other northern painters. This is hard to prove because of Alberti's treatise itself; he is conspicuously silent about naming painters of his own day and explains his theories by referring to ancient painters known from classical texts. When writing about the representation of emotions, only once does Alberti refer by name to a modern painter: Giotto.[40] Paula Nuttall argued that Alberti, as a member of Pope Eugene IV's (1383–1447) entourage, might have seen paintings from the North at the pope's palace, which allegedly was decorated with works by Jan van Eyck.[41] If Alberti had indeed been able to see works by Van Eyck, someone as learned as he was would likely have associated this new way of painting light with what he knew from ancient texts.

Whatever the case, *De pictura* defined a new standard for image making for centuries to come. Next to the mathematical depiction of three-dimensional space on a flat plane, the art of the specular reflection was crucial to capturing the visible world in all its material splendor with paint alone. And as with mathematical perspective, light's refractions and reflections could be depicted convincingly only by adhering to a set of rules. Soon after the early Netherlandish painters depict gems, the human eye, gold, silver, glass, tears, and crystal by imitating their refraction and reflection with paint, artisans from other crafts use the same visual systems to capture the visible world.

To be sure, the oil medium, as we have seen with Filippo Lippi, is not a prerequisite to Van Eyck's method for depicting reflection and refraction. By imitating hue, specular reflection, and internal luster, transparency and translucency can be suggested in any colored medium. This is perhaps demonstrated even more strikingly by some of the miniatures in Jean, duc de Berry's *Très belles heures de Notre-Dame*, better known as the Turin-Milan Hours because of where the various surviving pages of the manuscript are kept today. Several of the illuminations contain specular reflections, with the miniature *The Birth of John the Baptist* (Palazzo Madama, Museo Civico d'Arte Antica, Turin, MS 0467/M, fol. 93v; fig. 6.26) being almost an ode to the specular reflection, with its masterful depiction of a variety of metal objects and transparent glassware.

The left metal jug above the doorway even reflects the red bed hanging to its left. The Turin-Milan Hours has been the subject of much controversy regarding its dating and attribution. Some scholars have argued that several of the miniatures—commonly grouped as "Hand G" since Hulin de Loo's 1911 study—must be of early date, made in the 1420s under the influence of Van Eyck or even by the painter himself (De Loo attributes Hand G to Hubert, Jan's brother).[42] Others attribute Hand G to Van Eyck's followers and argue a date from the 1430s to the 1460s. If the first hypothesis is correct, this would mean that the changes in ambition of painters in the West and, accordingly, the rediscovery of the specular reflection, were born on the painted page rather than on panel. The riddle of the Turin-Milan Hours may never be solved, but for my current argument the main takeaway is that the paint systems as we find them on the Ghent Altarpiece could just as well have been introduced by illuminators. The special translucency of the oil medium when ground with certain pigments is not required to paint everything refractive and reflective. In fact, let us briefly turn to a tapestry

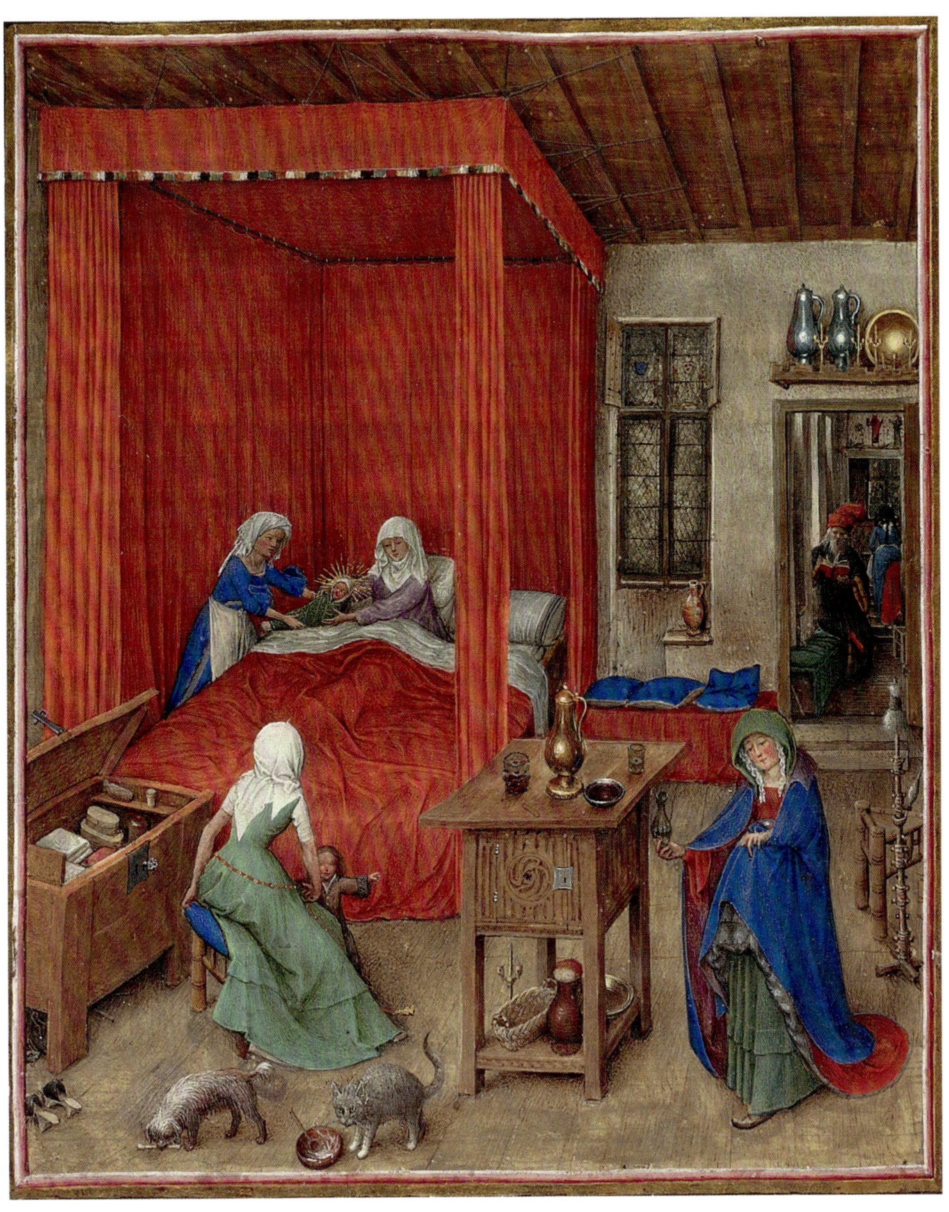

FIGURE 6.26. Jan Van Eyck (attr.), *The Birth of John the Baptist* (detail), *Très belles heures de Notre-Dame*, also known as the Turin-Milan Hours. Tempera and gold leaf on parchment. Palazzo Madama, Museo Civico d'Arte Antica, Turin, inv. MS 0467/M, fol. 93v. Reproduced by permission of the Fondazione Torino Musei, photo by Studio Fotografico Gonella 2012.

series made in the 1450s in a Tournai workshop similar to the two turn-of-the-century tapestry series discussed earlier in this chapter (fig. 6.7). The garments of the figures in this 1450 tapestry series, *Scenes from the Passion of Christ* (Art and History Museum, Brussels), are embellished with an abundance of gems, as are those in *The Nine Heroes* and *The Legend of Saint Piat and Saint Eleutherius* tapestries (fig. 6.27).

But about five decades later, something has changed in how weavers depicted precious stones. Rather than juxtaposing a light and a dark color as the earlier weavers did, they now wove gems by means of the same visual system Van Eyck used; different-colored threads carefully depict each gem by its hue, internal luster, *and* specular reflection.

This is precisely why the recipes in Beurs's *Groote waereld* are representative of Van Eyck's method; Van Eyck's visual schema for rendering refractive and reflective materials by depicting specular and internal reflections was not just efficient but was a true system, a set of rules as it were, quite similar to the rules of perspective first codified by Alberti. And this is why it had not changed when Beurs wrote his manual for oil painters more than two centuries later. Introducing this system, if we may attribute it to him, was Van Eyck's greatest innovation. And like the rules of perspective, part of its beauty was that it also could be adopted by artisans working in materials other than oil paint, from the weavers of tapestries to those painting in enamel or those embellishing the pages of manuscripts.

What had caused this seeming revolution in the art of image making must remain a mystery here as well. The effect of this transformation, however, was to give varnishes and glazes a new purpose in the art of easel painting and to set off new explorations of the oil medium for preparing them. These new explorations were not invented or introduced by Van Eyck; rather, they resulted from new ideas about the nature of painting instigated by his art and that of his contemporaries.

Unlike medieval painters—who relied on glazes to literally evoke refracted light—Van Eyck used fully opaque paint to depict the glow, or internal luster, of light reflected from within translucent materials. Glazes are still important in this system, but they assume a new role. Van Eyck and his contemporaries exploit glazes as part of another imitative faculty of oil paint: its ability to convey the colors of the visible world most convincingly. Glazes are used to imitate the deepest and most saturated blues, greens, and reds, while opaque paints imitate the brightest whites, yellows, and oranges and the darkest blacks. Using the oil medium this way allows painters to summon almost the entire spectrum of colors perceived by the human eye. When Western easel painting changed its ambition from the mimesis of

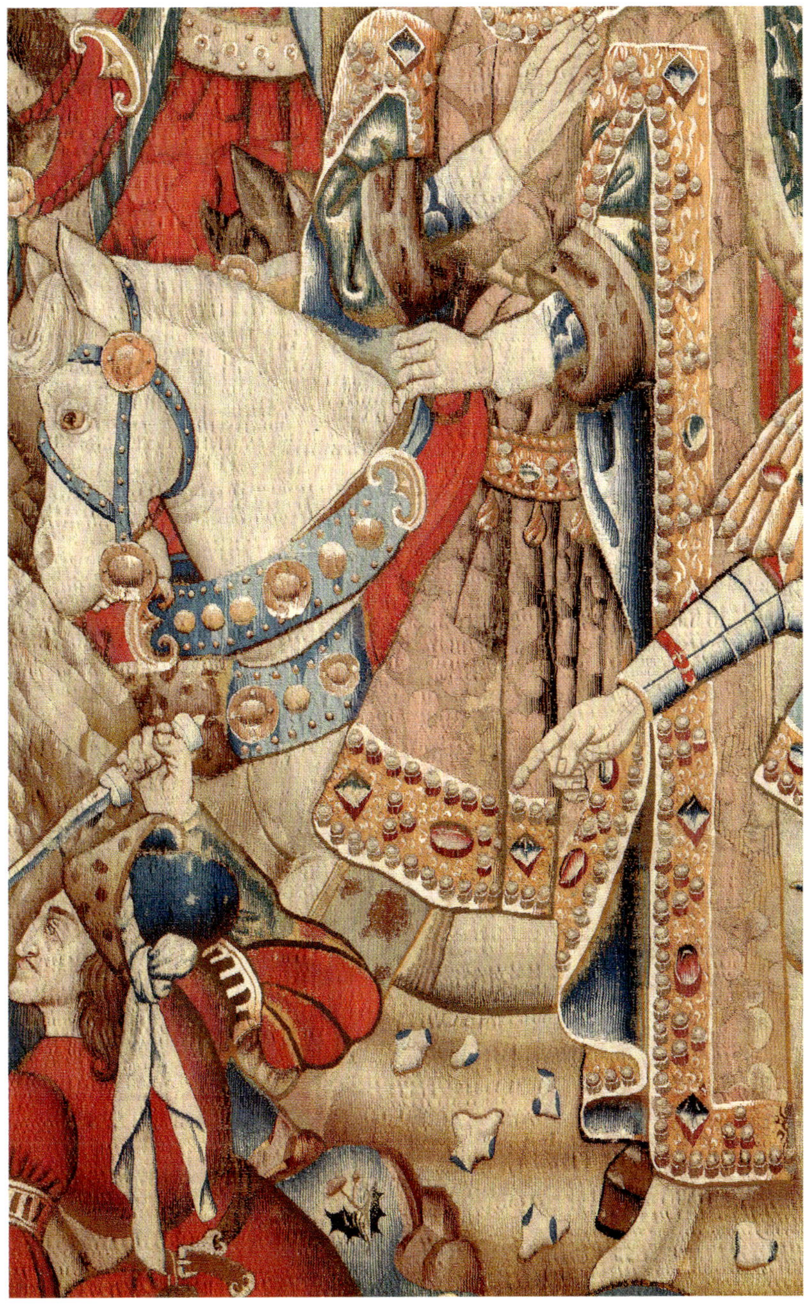

FIGURE 6.27. Tournai workshop, gems on an ornamental border of a robe, detail of *Tapestry with Scenes from the Passion of Christ*, ca. 1445–55. Art and History Museum, Brussels. © KMKG, Brussels.

materials to the mimesis of the visible world, this last faculty of oil paint ensured its permanent place in the toolbox of the easel painter until the discovery of acrylics in the twentieth century.

The varnish—which had functioned together with the glaze as one of the most important material mimetic techniques of the medieval artisan—assumed a new role as well. No longer used to make paintings look like gold and shine like polished enamels, the varnish instead had to become invisible—unifying the complexion of the painting's colors and ensuring its durability by preserving those colors as long as possible. It is safe to say that this new role for varnish, as an invisible agent protecting and bringing out what is underneath, would jeopardize its own survival in later centuries. Few original varnishes have come down to us, because as soon as they stopped fulfilling their previous function they would be removed and replaced.

The new approach to decorating wooden panels, introduced by Van Eyck and his contemporaries, metamorphosed every two-dimensional plane into the artisan's playground for expressing the colors and textures of the materials of the visible world. The depiction of specular and internal reflections helped transform paint into the priceless splendor of precious stones and metals. Or, by such paint systems, as Beurs expressed it, echoing Alberti: "The clever student is made master of great treasures; judging that it is in his power to carefully make carbuncles, turquoises, sapphires, emeralds, jaspers, hyacinths, onyx and sardonyx and even to embellish an entire high priest's breastplate."[43]

Acknowledgments

I would like to acknowledge the financial support of the Dutch Research Council (NWO) for the research project "Deceiving Stuff: Histories, Functions, Techniques, and Effects of Material Mimesis" (with project number 275-54-001) of the NWO Talent program Veni, which was instrumental in realizing this book. The generous financial support of the Young Academy of the Royal Dutch Academy of Sciences (KNAW) helped with the image program. I am grateful also to Utrecht University, the University of Amsterdam, and the Max Planck Institute of the History of Science in Berlin for supporting me and giving me a home for this project in all its stages.

During the years I worked on this book, I was supported by so many kind and generous colleagues: Paul van den Akker, Sven Dupré, Yvonne Elet, Sanli Faez, Peter Hecht, Marieke Hendriksen, Erma Hermens, Paul Hills, Stephan Kemperdick, Jo Kirby, Arjan de Koomen, Friso Lammertse, William R. Newman, Doris Oltrogge, Christian de Pee, Lawrence Principe, Jaya Remond, Erik Rietveld, Marlise Rijks, Dagmar Schäfer, Marc Smith, Pamela Smith, Emma C. Spary, Andreas Speer, Paul Taylor, Jørgen Wadum, Thijs Weststeijn, Tijana Žakula, and the Impact of Oil Team: Jeroen Stumpel, Jan Piet Filedt Kok, Ann-Sophie Lehmann, Arie Wallert, Mark Clarke, Esther van Duijn, and Abbie Vandivere.

I am sincerely grateful to all the team at the University of Chicago Press, especially Karen Merikangas Darling and Tristan Bates for their continued support and expertise. I thank Alice Bennett, whose exceptional copyediting greatly clarified the text, and the anonymous peer reviewers who gave their time and efforts. They helped make the book immeasurably better with their expertise, questions, insights, and suggestions. Any errors remaining are, of course, my own.

Special thanks go to Jeroen Stumpel. Our many conversations were a true inspiration and vital in the coming together of this book.

Finally, I acknowledge my enormous debt to Alexander, who has tirelessly helped and supported me in preparing this book for publication, and to my boys, Hagen and Veit, who joined me as I finished this project.

Notes

Introduction

1. Johann Wolfgang von Goethe, *Über Kunst und Alterthum in den Rhein und Mayn Gegenden*, 3 vols. (Stuttgart: Cottaischen Buchhandlung, 1816), 1:168. Translation from John Gage, *Goethe on Art* (Berkeley: University of California Press, 1980), 141–42. Goethe was a great admirer of Gotthold Ephraim Lessing, who discovered and published parts of the treatise of Theophilus, which cast doubt on Vasari's invention myth because of its discussion of oil paint—also treated later in this introduction.

2. See, for instance, Mojmír Svatopluk Frinta, *Punched Decoration: On Late Medieval Panel and Miniature Painting*, 2 vols. (Prague: Maxdorf, 1998); Jilleen Nadolny, "The Techniques and Use of Gilded Relief Decoration by Northern European Painters, c. 1200–1500" (PhD diss., University of London, Courtauld Institute of Art, 2000).

3. The literature on the history of pictorial representation in art history is too vast to list here, but I should mention Ernst Gombrich's seminal study on the emergence and development of representational skills in the history of painting and sculpture: *Art and Illusion: A Study in the Psychology of Pictorial Representation* (1960; repr., Oxford: Phaidon Press, 1983). For a study of the nature of mimesis in other representational arts, including literature, painting, sculpture, theater, and film, see K. L. Walton, *Mimesis as Make-Believe: On the Foundations of the Representational Arts* (Cambridge, MA: Harvard University Press, 1990). And on the role of materials in mimetic practice see Marjolijn Bol and Emma C. Spary, eds., *The Matter of Mimesis: Studies on Mimesis and Materials in Nature, Art and Science* (Leiden: Brill, 2023).

4. The portraits were likely commissioned during the person's lifetime, and gold leaf—symbolizing immortality and eternity—was occasionally added just before burial, when the portrait would be wrapped with the mummy. According to David L. Thompson, this was the case with the Getty mummy portrait as well: see *Mummy Portraits in the J. Paul Getty Museum* (Malibu, CA: J. Paul Getty Museum, 1982), 8–10.

5. First published as Ernst Gombrich, "Light, Form and Texture in XVth Century Painting," *Royal Society for the Encouragement of Arts* 112, no. 5099 (1964): 826–49, and later in Ernst Gombrich, "Light, Form and Texture in Fifteenth-Century Painting North and South of the Alps," in *The Heritage of Apelles: Studies in the Art of the Renaissance*

(Ithaca, NY: Cornell University Press, 1976), 19–35. Scientists at the National Gallery of London, for instance, concluded that Van Eyck's invention was not one of technique but that "the genius lies in the acute power of observation." In a way, with this conclusion art history made almost a full circle. As early as 1604, Van Eyck's keen observation had already been pointed out by Karel van Mander in his *Schilder-boeck*. See *The Lives of the Illustrious Netherlandish and German Painters . . .* , ed. and trans. Hessel Miedema, 6 vols. (Doornspijk: Davaco, 1994–99), 1:200v: *Desen constighen Schilder heeft gheweest van grooter aendacht.*

6. For of the revival of the interest in Van Eyck during this period see Jenny Graham, *Inventing Van Eyck: The Remaking of an Artist for the Modern Age* (Oxford: Berg, 2007).

7. See Marjolijn Bol, "Technique and the Art of Immortality, 1800–1900," *History of Humanities* 2, no. 1 (2017): 179–99.

8. Horace Walpole, *Anecdotes of Painting in England: With Some Account of the Principal Artists*, 2 vols. (London: J. Dodsley 1762), 1:47. In the same context, Walpole also mentioned an inscription on a fourteenth-century painting known as the Wilton Diptych that appeared to suggest the work was made in oil a few decades before Van Eyck. The inscription reads: "Invention of painting in oil 1410. This was painted before in the beginning of Richard II. 1377." In the twentieth century, researchers at the National Gallery discovered that the Wilton Diptych was painted not in oil, but with an egg-based medium. See Dillian Gordon, Caroline M. Barron, Ashok Roy, and Martin Wyld, *The Wilton Dyptich: Making and Meaning* (London: National Gallery, 1993).

9. Lessing discovered this manuscript in the library in Wolfenbüttel (today Herzog August Bibliothek), where he worked as a librarian. See Gotthold Ephraim Lessing, "Vom Alter der Oelmalerei, aus dem Theophilus Presbyter," in *G. E. Lessing's Gesammelte Werke: In zwei Bänden* (Leipzig: G. J. Göschen, 1855).

10. See Jilleen Nadolny, "The First Century of Published Scientific Analyses of the Materials of Historical Painting and Polychromy, c. 1780–1880," *Reviews in Conservation* 4 (2003): 39–51; Jilleen Nadolny, "A Problem of Methodology: Merrifield, Eastlake and the Use of Oil-Based Media by Medieval English Painters," *14th Triennial Meeting The Hague Reprints* 2 (2005): 1028–33; and Stephen G. Rees-Jones, "Early Experiments in Pigment Analysis," *Studies in Conservation* 35, no. 2 (1990): 93–101.

11. The medieval Old Palace of Westminster was destroyed by a fire in 1834. In its place stands the rebuilt New Palace of Westminster. For an overview of these early attempts at the chemical analysis of binding media, see Nadolny, "First Century of Published Scientific Analyses," 39–51.

12. For details regarding Haslam's chemical experiments to identify the nature of the binding medium, see Rees-Jones, "Early Experiments in Pigment Analyses," 95.

13. John Thomas Smith, *Antiquities of Westminster . . .* (London: T. Bensley, 1807), 223–29.

14. Smith already points to this debate when he tells us that in a 1792 edition of Lessing's *Essay*, the editor—Lessing's brother—remarks that Van Eyck could still be considered the inventor of oil paint because before his time it was not used on figures. See Gotthold Ephraim Lessing, *Sämmtliche Schriften*, ed. Karl Gotthelf Lessing, 30 vols. (Berlin: Nicolaischen Buchhandlung, 1793), 12:360–61.

15. For a thorough overview of the theories that were developed after Lessing's discovery, see Elise Effmann, "Theories about the Eyckian Painting Medium from the Late-Eighteenth to the Mid-Twentieth Centuries," *Reviews in Conservation* 7 (2006): 17–26.

16. An example of the previous is the Maroger medium, developed by and named after the painter and technical director of the laboratory of the Louvre, Jacques Maroger (1884–1962). Maroger believed that the paint medium he developed and made commercially available—it can still be purchased today—held the secret to the technique of the old masters. See Jacques Maroger, "The Secret of Van Eyck Regained: The Long-Lost Emulsions of the First Great Oil Painter," *Magazine of Art* 35 (1943): 219–22.

17. Jacques-Nicolas Paillot de Montabert, *Traité complet de la peinture*, 9 vols. (Paris: Delion, 1829), 9:14–15.

18. Jean François Léonor Mérimée, *De la peinture à l'huile, ou es procédés matériels employés dans ce genre de peinture depuis Hubert et Jan Van Eyck jusqu'à nos jours* (Paris: Huzard, 1830), 6. Mérimée's work was translated into English in 1839 by William Benjamin Sarsfield: see Jean François Léonor Mérimée, *The Art of Painting in Oil, and in Fresco . . .*, ed. and trans. Benjamin Sarsfield (London: Whittaker, 1839), 6.

19. Mérimée, *De la peinture à l'huile*, 7; ed. Sarsfield, 7.

20. Charles Lock Eastlake, *Methods and Materials of Painting of the Great Schools and Masters* (originally *Materials for a History of Oil Painting*), 2 vols. (1947; repr. New York: Dover, 1960), 1:260–67 and 1:412.

21. Since 2012, the Ghent Altarpiece has been undergoing restoration at the Museum of Fine Arts, Ghent. For the detailed documentation of this project, including high-resolution (infrared-) macrophotography and X-radiography, see http://closertovaneyck.kikirpa.be.

22. Paul Coremans, *L'agneau mystique au laboratoire: Examen et traitement*, Les Primitifs flamands, 3, Contributions à l'étude des Primitifs flamands, 2 (Anvers: De Sikkel, 1953); and see also Paul Coremans, R. J. Gettens, and J. Thissen, "La technique des Primitifs flamands: Étude scientifique des matériaux, de la structure et de la technique picturale," *Studies in Conservation* 1, no. 1 (1952): 1–29. In some samples Coremans and his team found another substance *x*, and although its exact nature could not be identified at the time, it is telling that it immediately led to new speculations about Van Eyck's possible use of hard resins. See, for instance, the interpretation of Coremans's research by the American painter Frederic Taubes, *The Mastery of Oil Painting* (London: Thames and Hudson, 1953), 15.

23. Raymond White, "Van Eyck's Technique: The Myth and the Reality, II," in *Investigating Jan Van Eyck*, ed. Susan Foister and Sue Jones (Turnhout: Brepols, 2000), 101–5. This publication also summarizes the state of research regarding Van Eyck's technique; see Ashok Roy, "Van Eyck's Technique: The Myth and the Reality, I," 97–105.

24. Lorne Campbell, Susan Foister, Ashok Roy, and Rachel Billinge, eds., "Methods and Materials of Northern European Painting in the National Gallery, 1400–1550," *National Gallery Technical Bulletin* 18 (1997): 6–55. Because of the National Gallery's representative collection and the extent of the research, this 1997 volume is still a standard reference work for the methods and materials used in early Netherlandish painting.

25. See, for instance, Ann Massing, ed., *The Thornham Parva Retable: Technique,*

Conservation and Context of an English Medieval Painting (Cambridge: Brepols, 2003); Unn Plahter, Erla B. Hohler, Nigel J. Morgan, and Anne Wichstrøm, eds., *Painted Altar Frontals of Norway, 1250–1350*, 3 vols. (London: Archetype, 2004); Paul Binski, Ann Massing, and Marie Louise Sauerberg, eds., *The Westminster Retable: History, Technique, Conservation* (Cambridge: Brepols, 2009); Cyriel Stroo, ed., *Pre-Eyckian Panel Painting in the Low Countries*, 2 vols. (Turnhout: Brepols, 2009).

26. Erwin Panofsky, *Early Netherlandish Painting: Its Origins and Character*, 2 vols. (1953; repr., New York: Harper and Row, 1971), 2:69–70.

27. See Otto Pächt, *Van Eyck and the Founders of Early Netherlandish Painting* (originally *Van Eyck: Die Begründer der Altniederländischen Malerei*) (1989; repr., London: Harvey-Miller, 1994), 15.

28. Even today this old premise resounds in scholarly research and popular scientific publications about Van Eyck and his contemporaries. See, for instance, Fred Kleiner's textbook introducing students to art history, *Gardner's Art through the Ages: Backpack Edition, Book D: Renaissance and Baroque*, 15th ed. (Boston: Wadsworth, 2015), 559. Kleiner explains that although oil paint possibly was used since at least the eighth century, only in the fifteenth century did it become widespread. We learn that in this period "better drying components" were discovered that "enhanced the setting capabilities of oils," and "rather than to apply these oils with the light, flecked brushstrokes that tempera encouraged, artists laid the oils down in transparent glazes over opaque or semi-opaque underlayers."

29. Jill Dunkerton, Susan Foister, Dillian Gordon, and Nicholas Penny, *Giotto to Dürer: Early Renaissance Painting in the National Gallery* (New Haven, CT: Yale University Press, 1991), 194.

30. John Gage, *Colour and Culture: Practice and Meaning from Antiquity to Abstraction* (1993; repr., London: Thames and Hudson, 2009), 131.

31. Jeffrey Chipps Smith, *The Northern Renaissance* (Berlin: Phaidon Press, 2004), 61.

32. Jochen Sander, "*Ars Nova* and European Painting in the Fifteenth Century," in *The Master of Flémalle and Rogier van der Weyden*, ed. Stefan Kemperdick and Jochen Sander (Ostfildern: Hatje Cantz, 2008), 32.

33. See Maximiliaan Martens, Till-Holger Borchert, Jan Dumolyn, Johan De Smet, and Frederica Van Dam, eds., *Van Eyck* (New York: Thames and Hudson, 2020), e.g., 21, 171.

34. On this transformation of the history of science into the history of knowledge, see Peter Burke's historiographical study *What Is the History of Knowledge?* (Cambridge: Polity Press, 2015); Lorraine Daston, "The History of Science and the History of Knowledge," *KNOW: A Journal on the Formation of Knowledge* 1 (2017): 131–54; and the recently established *Journal for the History of Knowledge*, accessed January 11, 2020: www.journalhistoryknowledge.org.

35. See Pamela O. Long, *Artisan/Practitioners and the Rise of the New Sciences, 1400–1600* (Corvallis: Oregon State University Press, 2011); Ursula Klein and Emma C. Spary, eds., *Materials and Expertise in Early Modern Europe: Between Market and Laboratory* (Chicago: University of Chicago Press, 2010).

36. See Pamela Smith, *The Body of the Artisan: Art and Experience in the Scien-*

tific Revolution (Chicago: University of Chicago Press, 2004); William R. Newman, *Promethean Ambitions: Alchemy and the Quest to Perfect Nature* (Chicago: University of Chicago Press, 2004).

37. Thompson R. Campbell, "A Survey of the Chemistry of Assyria in the Seventh Century BC," *Ambix* 2, no. 1 (1938): 5–6.

38. See A. L. Oppenheim, R. H. Brill, D. Barag, and A. von Saldern, eds., *Glass and Glassmaking in Ancient Mesopotamia: An Edition of the Cuneiform Texts Which Contain Instructions for Glassmakers with a Catalogue of Surviving Objects*, Corning Museum of Glass Monographs, vol. 3 (Corning, NY: Corning Museum of Glass, 1970); Mark Clarke, "The Earliest Technical Recipes: Assyrian Recipes, Greek Chemical Treatises and the Mappae Clavicula Text Family," in *Craft Treatises and Handbooks: The Dissemination of Technical Knowledge in the Middle Ages*, ed. Ricardo Córdoba (Turnhout: Brepols, 2013), 9–31.

39. For a good introduction to this subject, I refer to chapter 1, "Art Technological Source Research," in *The Conservation of Easel Paintings*, ed. Joyce Hill Stoner and Rebecca Anne Rushfield, Routledge Series in Conservation and Museology (London: Routledge, 2012). The proceedings of the symposia of the Art Technological Source Research Working Group (ICOM–Committee for Conservation) also address the issues brought up in this paragraph; see especially Mark Clarke, Joyce H. Townsend, and Ad Stijnman, eds., *Art of the Past-Sources and Reconstructions* (London: Archetype, 2005).

40. Stefanos Kroustallis, "Theophilus Matters: The Thorny Question of the 'Schedula diversarium artium' Authorship," in *Zwischen Kunsthandwerk und Kunst: Die "Schedula diversarium artium,"* ed. Andreas Speer, Maxime Mauriège, and Hiltrud Westermann-Angerhausen (Berlin: De Gruyter, 2013), 52–71. Kroustallis proposes we date the *Schedula* to the eleventh century.

41. William Eamon, *Science and the Secrets of Nature: Books of Secrets in Medieval and Early Modern Culture* (Princeton, NJ: Princeton University Press, 1996), 94.

42. See Pamela O. Long, *Openness, Secrecy, Authorship: Technical Arts and the Culture of Knowledge from Antiquity to the Renaissance* (Baltimore: Johns Hopkins University Press, 2003), esp. chap. 4, "Authorship on the Mechanical Arts in the Last Scribal Age," 102–42.

43. For an in-depth discussion of some of the topics below, I refer to Thomas B. Brill, *Light: Its Interaction with Art and Antiquities* (New York: Plenum Press, 1980); R. W. G. Hunt and M. R. Pointer, *Measuring Colour*, 4th ed. (Chichester, UK: John Wiley, 2011); M. H. Freeman, C. C. Hull, and W. N. Charman, *Optics*, 11th ed. (New York: Butterworth-Heinemann, 2003).

44. Hunt and Richter, *Measuring Colour*, 11–13 (on saturation); see also 44–46 (on color purity).

45. Hunt and Richter, *Measuring Colour*, 157.

46. See Ronald W. Fleming, Henrik Wann Jensen, and Heinrich H. Bülthoff, "Perceiving Translucent Materials," in *APGV '04 Proceedings of the 1st Symposium on Applied Perception in Graphics and Visualization* (2004), 127–34.

47. Freeman, Hull, and Charman, *Optics*, 15–22.

48. For a discussion of these terms in the work of Theophilus, see Marjolijn Bol, "Seeing through the Paint: The Dissemination of Technical Terminology between Three

Métiers: *Pictura Translucida*, Enameling and Glass Painting," in *Zwischen Kunsthandwerk und Kunst: Die "Schedula Diversarum Artium,"* ed. Andreas Speer (Berlin: De Gruyter, 2013), 145–62.

49. In the premodern period varnishes were sometimes also colored; see chapter 2 below.

50. In Paul Hills, *Venetian Colour: Marble, Mosaic, Painting and Glass, 1250–1550* (New Haven, CT: Yale University Press: 1999), Hills investigates a similar problem. He considers the special color palette of the arts of the Venetian lagoon in relation to social, cultural, and environmental factors. In his search for the colors of Venice, material culture in its broadest sense plays an important role, since its manufacture, marketing, and consumption affect cognitive skill.

Chapter One

1. For the 1550 text I refer to Giorgio Vasari, *Le vite de' più eccellenti architetti, pittori, et scultori italiani, da Cimabue, insino a' tempi nostri: Nell'edizione per i tipi di Lorenzo Torrentino, Firenze 1550*, ed. Luciano Bellosi and Aldo Rossi, 2 vols. (Turin: Einaudi, 1991). For the 1568 text I refer to *Le vite de' più eccellenti pittori, scultori, e architettori scritte da Giorgio Vasari*, ed. Gaetano Milanesi, *Le opere di Giorgio Vasari*, 9 vols. (Florence: Sansoni, 1878–85). For a critical edition that is particularly useful for side-by-side comparisons of the 1550 and 1568 editions of the *Vite*, see *Le vite de' più eccellenti pittori, scultori, e architettori nelle redazioni del 1550 e 1568*, ed. Paola Barocchi and Rosanna Bettarini, 6 vols. (Florence: Sansoni, 1966–87). For the English translation (based on the 1568 edition), I refer to *Lives of the Painters, Sculptors, and Architects: Giorgio Vasari*, ed. and trans. Gaston du C. De Vere, intro. David Ekserdjian, Everyman's Library, 2 vols. (1912–15; repr., New York: Alfred A. Knopf, 1996).

2. See the introduction to this book for an overview of the long historiography of this debate.

3. Vasari, *Vite*, ed. Bellosi and Rossi, 1:94; ed. Vere, 1:32–33.

4. Vasari, *Vite*, ed. Bellosi and Rossi, 1:7; ed. Vere, 1:13.

5. Vasari, *Vite*, ed. Bellosi and Rossi, 1:102; ed. Vere, 1:46–47.

6. The *Vite* is thus organized as follows: 1. Dedicatory letter to Cosimo I de' Medici; 2. *Proemio* to the whole work; 2. Technical *introduzione* to architecture, sculpture, and painting; 3. *Proemio* to the first age (followed by the biographies of this period); 2. *Proemio* to the second age plus biographies; 3. *Proemio* to the third age plus biographies.

7. Vasari, *Vite*, ed. Bellosi and Rossi, 1:207; ed. Vere, 1:245;.

8. Vasari, *Vite*, ed. Bellosi and Rossi, 1:207–8; ed. Vere, 1:246.

9. This is pointed out by David Frangenberg, who uses it to question Vasari's authorship of parts of the *Vite*. See Frangenberg, "Bartoli, Giambullari and the Prefaces to Vasari's Lives (1550)," *Journal of the Warburg and Courtauld Institutes* 65, no. 15 (2002): 255–56. Because of their erudition, both in language and in concept, the four *proemi* to the *Vite* are sometimes thought to have been written by, or influenced by, Vasari's learned collaborators—an issue that Charles Hope has also discussed in detail; see, e.g., "Vasari's *Vite* as a Collaborative Project," in *The Ashgate Research Companion to Giorgio Vasari*, ed. David. J. Cast (London: Routledge, 2016), 11–23. For a different

interpretation, see Eliana Carrara, "Reconsidering the Authorship of the 'Lives': Some Observations and Methodological Questions on Vasari as a Writer," *Studi di Memofonte* 15 (2015): 53–90.

10. Vasari, *Vite*, ed. Bellosi and Rossi, 1:359; ed. Vere, 1:424.

11. Vasari, *Vite*, ed. Bellosi and Rossi, 1:359; ed. Vere, 1:424.

12. Vasari, *Vite*, ed. Bellosi and Rossi, 1:360; ed. Vere, 1:427.

13. Vasari, *Vite*, ed. Bellosi and Rossi, 1:67–68 (chaps. 20–21). And for the English translation of Vasari's technical treatise, see G. Baldwin Brown and Louisa S. Maclehose, eds., *Vasari on Technique: Being the Introduction to the Three Arts of Design, Architecture, Sculpture and Painting, Prefixed to the Lives of the Most Excellent Painters, Sculptors and Architects* (London: Dent, 1907), 225.

14. See Paula Nuttall, *From Flanders to Florence: The Impact of Netherlandish Painting, 1400–1500* (New Haven, CT: Yale University Press, 2004).

15. An interesting discussion of several early authors who criticize Vasari's idea that Van Eyck was the inventor of oil paint can be found in Jan Emmens, "De Uitvinding van de Olieverf: Een kunsthistorisch probleem in de zestiende eeuw," in *Kunsthistorische Opstellen* (Amsterdam: Van Oorschot, 1981), 1:191–203. See also Pieter Brinkman, *Het Geheim van Van Eyck: Aantekeningen bij de uitvinding van het olieverven* (Zwolle: Waanders, 1993), 62–77.

16. For Borghini's letter, see Karl Frey and Herman Walther Frey, *Il Carteggio di Giorgio Vasari*, 2 vols. (Munich: Georg Müller, 1923), 2:26–28.

17. Vasari, *Vite*, ed. Milanesi, 1:644–45.

18. Vasari, *Vite*, ed. Bellosi and Rossi, 1:644–45; ed. Vere, 1:200.

19. Vasari, *Vite*, ed. Milanesi, 1:644–45; ed. Vere, 1:200.

20. Chapter 5 of this book shows that most of Cennini's remarks about using oil paint on panel concern the practice of glazing.

21. Vasari, *Vite*, ed. Milanesi, 2:593; ed. Vere, 1:430.

22. For an excellent discussion of the influence of Flemish paintings on Florentine art, see Nuttall, *From Flanders to Florence*. And for the impact of Flemish paintings on Venetian art, see Bernard Aikema, Beverly Louise Brown, and Giovanna Nepi Sciré, eds., *Renaissance Venice and the North: Crosscurrents in the Time of Bellini, Dürer, and Titian* (New York: Rizzoli, 2000). See also the various essays on this topic in Ingrid Alexander-Skipnes, ed., *Cultural Exchange between the Low Countries and Italy (1400–1600)* (Turnhout: Brepols, 2007).

23. Brown and Maclehose, *Vasari on Technique*, 227–28. This information is repeated in the life of Antonello da Messina.

24. Vasari, *Vite*, ed. Bellosi and Rossi, 1:361.

25. Vasari, *Vite*, ed. Bellosi and Rossi, 1:362.

26. Vasari, *Vite*, ed. Bellosi and Rossi, 1:362.

27. See Bernard Aikema and Beverly Louise Brown, "Painting in Fifteenth-Century Venice and the *Ars Nova* of the Netherlands," in *Renaissance Venice and the North*, ed. Aikema, Brown, and Nepi Sciré, 176–85. See also chapter 6 of this book.

28. Cited from Panofsky, *Early Netherlandish Painting*, 1:2. See also Nuttall, *From Flanders to Florence*, 32; and Brinkman, *Het Geheim van Van Eyck*, 53–52.

29. Baxandall translates *colorum* as "colors," but taking into account ideas about Van

Eyck's presumed special technique in other sources south of the Alps, it is also possible that Facius tried to convey that Van Eyck discovered properties of "paints" recorded by the ancients. In Italian, *colore* was (and is) often used to mean paint more generally. I am grateful to Jeroen Stumpel for bringing this to my attention.

30. Michael Baxandall, "Bartholomaeus Facius on Painting: A Fifteenth-Century Manuscript of the *De Viris Illustribus*," *Journal of the Warburg and Courtauld Institutes* 27 (1964): 102–3.

31. Baxandall, "Bartholomaeus Facius on Painting," 96. Maximiliaan Martens takes the passage from Facius on Van Eyck as the starting point for his argument that Van Eyck might have been familiar with Pliny's encyclopedia, and book 35 in particular; see "Jan van Eyck's Optical Revolution," in *Van Eyck*, ed. Martens et al., 141–79.

32. For the English translation, see Nuttall, *From Flanders to Florence*, 32; for the original Latin text, see Massimo Miglio, *Storiografia pontificia del quattrocento* (Bologna: Patron, 1975), 141n31.

33. Antonio di Pietro Averlino (Filarete), *Antonio Averlino detto il Filarete: Trattato di Architettura*, ed. Anna Maria Finoli and Liliana Grassi, 2 vols., Classici Italiani di scienze tecniche e arti (Milan: Polifilo, 1972), 2:668.

34. Filarete, *Trattato di Architettura*, ed. Finoli and Grassi, 2:668.

35. In passing, Vasari often refers to Pliny as his source.

36. For the influence of Pliny's *Natural History* on Vasari's art history, see Sarah Blake McHam, *Pliny and the Artistic Culture of the Italian Renaissance: The Legacy of the "Natural History"* (New Haven, CT: Yale University Press, 2013), 273–85.

37. Vasari and his advisers also turned to other ancient sources for the *Vite*'s model of art history and for the division of the *Lives* into three ages. Gombrich, for instance, argued that Vasari's evolutionary hypothesis of art history's development may have come from Cicero's *Brutus*; see Ernst Gombrich, "Vasari's *Lives* and Cicero's Brutus," *Journal of the Warburg and Courtauld Institutes* 23, no. 3–4 (1960): 309–11. For the differences in Vasari's concept of history between the 1550 and the 1568 editions of the *Vite*, see Zygmunt Waźbiński, "L'idée de l'histoire dans la première et la seconde édition des 'Vies' de Vasari," in *Il Vasari storiografo e artista: Atti del congresso internazionale nel IV centenario della morte . . .* (Florence: Istituto nazionale di studi sul Rinascimento, 1976), 1–25.

38. For an overview of the extensive literature concerning Vasari's idea of progress, see Matteo Burioni, "Vasari's *Rinascita*: History, Anthropology or Art Criticism," in *Renaissance?: Perceptions of Continuity and Discontinuity in Europe, c. 1300–c. 1550*, ed. Alexander Lee, Pit Péporté, and Harry Schnitker (Leiden: Brill, 2010), note 43.

39. Vasari, *Vite*, ed. Bellosi and Rossi, 1:207; ed. Vere, 1:245.

40. Pliny, *Natural History, Volume 9: Books 33–35*, ed. and trans. H. Rackham, Loeb Classical Library 394 (Cambridge, MA: Harvard University Press, 1961), 332–33 [35.36.97].

41. It is perhaps significant that in the second edition of the *Vite* the remark that Jan van Eyck invented a varnish that "that all the painters in the world desired" was changed into the varnish "that he [Jan van Eyck] desired." Was this intended to slightly downplay the importance of an invention by a Flemish painter?

42. Vasari, *Vite*, ed. Bellosi and Rossi, 1:360–61.

43. Pliny, *Natural History* [35.36.79], ed. Rackham, 9:319.

44. Vasari, *Vite*, ed. Bellosi and Rossi, 2:543; ed. Vere, 1:622.

45. Vasari, *Vite*, ed. Bellosi and Rossi, 1:364; ed. Vere, 1:429.

46. Francisco de Hollanda, *On Antique Painting*, ed. Helmut Wohl, trans. Alice Sedgwick Wohl (University Park: Pennsylvania State University Press, 2015), xi. Hollanda's treatise was not published until the nineteenth century, and the earliest surviving manuscript is a 1563 Castilian translation kept in the Real Academia de Bellas Artes de San Fernando, Madrid (MS 361/13). The Portuguese text (possibly the original or a copy made for the purpose of sending the manuscript to Spain), from which the 1563 Castilian translation was made, was transcribed in the eighteenth century. This transcription is also kept in the Academia de Bellas Artes de San Fernando (MS 651-azul). The fifteenth-century Portuguese text unfortunately disappeared soon after and has not been rediscovered. Sedgwick's English translation is based on the eighteenth-century transcription.

47. De Hollanda, *On Antique Painting*, ed. Wohl, trans. Sedgwick, 128.

48. De Hollanda, *On Antique Painting*, ed. Wohl, trans. Sedgwick, 147–48.

49. Vasari, *Vite*, ed. Bellosi and Rossi, 2:539; ed. Vere, 1:621–22.

50. Vasari, *Vite*, ed. Bellosi and Rossi, 2:542; ed. Vere, 1:620.

51. Vasari, *Vite*, ed. Milanesi, 4:493.

52. Vasari uses the same rhetoric to prove that Michelangelo surpassed the ancients in all three arts; see Vasari, *Vite*, ed. Bellosi and Rossi, 2:543: "Ma quello che fra i morti e' vivi porta la palma, e trascende e ricuopre tutti, è il divino Michel Agnolo Buonarroti, il qual non solo tien il principato di una di queste arti, ma di tutte tre insieme. Costui supera e vince non solamente tutti costoro che hanno quasi che vinto già la natura, ma quelli stessi famosissimi antichi che sì lodatamente fuor d'ogni dubbio la superano, et unico giustamente si trionfa di quegli, di questi e di lei, [. . .]. Il che medesimamente per consequenzia si può credere le sue pitture; le quali, se per adventura ci fussero di quelle famosissime greche o romane da poterle a fronte a fronte paragonare, tanto resterebbono in maggior pregio e più onorate, quanto più appariscono le sue sculture superiori a tutte le antiche."

53. Giovanni Battista Armenini, *De' veri precetti della pittura . . .* (Ravenna: Francesco Tebaldini, 1586), 123; and for the English translation, see *On the True Precepts of the Art of Painting by Giovanni Battista Armenini*, ed. and trans. Edward J. Olszewski (New York: Burt Franklin, 1977), 190–91.

54. Giovan Paolo Lomazzo, *Idea del tempio della pittura* (Milan: Paolo Gottardo Pontio, 1590), 49; and for the English translation, see *Idea of the Temple of Painting*, ed. and trans. Jean Julia Chai (University Park: Pennsylvania State University Press, 2013), 84.

55. Lucas D'Heere, *Den hof en boomgaerd der poësien door Lucas d'Heere*, ed. W. Waterschoot (Ghent, 1565; repr., Zwolle: Tjeenk Willink, 1969), 29–32.

56. Jochen Becker, "Zur niederländischen Kunstliteratur des 16. Jahrhunderts: Lucas De Heere," *Simiolus-Netherlands Quarterly for the History of Art* 6 (1972–73): 118–20.

57. Karel van Mander makes a similar comparison when he writes about Van Eyck; see *Het Schilder-Boeck* (Haarlem: printed by Jacob de Meester for Passchier van Westbusch, 1604), 200v–201r; and for a side-by-side parallel translation, see *The Lives of*

the Illustrious Netherlandish and German Painters, from the First Edition of the Schilder-boeck (1603–1604): Preceded by the Lineage, Circumstances and Place of Birth, Life and Works of Karel van Mander, Painter and Poet and Likewise His Death and Burial, from the Second Edition of the Schilder-boeck (1616–1618), ed. and trans. Hessel Miedema, 6 vols. (Doornspijk: Davaco, 1994–99), 1:200v–201r.

58. De Heere, *Den hof en boomgaerd der poësien*, ed. Waterschoot, 31–32. Becker points out that in the last verse, De Heere echoes Vasari's biography of Giotto, where a similar remark can be found. See "Zur Niederländischen Kunstliteratur," 120n45. Such comments, pressing the importance of having surpassed the ancients without precedent, can also be found in the writings of Marcus van Vaernewyck and Karel van Mander discussed later in this chapter. Chapter 6 of this book discusses the previous matter in relation to Leon Battista Alberti's *De pictura*.

59. Translation by the author. See Marcus van Vaernewyck, *Den spieghel der Nederlandscher audtheyt* (Ghent: Gheeraert van Salenson, 1568), fol. 118.

60. Van Vaernewyck, *Den spieghel*. fol. 118.

61. Van Vaernewyck, *Den spieghel*, fol. 118.

62. Van Vaernewyck, *Den spieghel*, fol. 119.

63. Van Vaernewyck, *Den spieghel*, fol. 119. The same point is repeated by Karel van Mander in his general introduction to the *Schilder-boeck*; see ed. Miedema, 1:198v: *Twee Maeseycksche doorluchtige ghebroeders, Hubert en Ian, die alree in onse Const wonder gheschickt, en op een redelijcke schoon wijse de verwen gehandelt hebben, en met geenen onheblijcken aerdt van teyckeninghe, dat het te verwonderen is, datse in soo vroeghen tijdt soo volcomen en blinckende zijn voort-gecomen: Want ick vinde niet, dat in hoogh oft neder Duytsch-landt eenige vroeger Schilders bekent zijn, oft ghenoemt worden.*

64. Dominicus Lampsonius, *Les effigies des peintres célèbres des Pays-Bas*, ed. and trans. Jean Puraye and Marie Delcourt (Paris: Éditions Desclée de Brouwer, 1956), 23–26. And for the English translation of this verse, see "Picturing the Netherlandish Canon," accessed August 10, 2017: https://sites.courtauld.ac.uk/netherlandish-canon/artist/jan-van-eyck/.

65. Van Mander, *Schilder-boeck*, ed. Miedema, 1, fol. 203r.

66. Van Mander, *Schilder-boeck*, ed. Miedema, 1, fol. 71v. This anecdote, as Brinkman points out, seems to run parallel to the story of the cleaning of the Predella of the Ghent Altarpiece that was first recorded by Van Vaernewyck and is repeated by Van Mander; Brinkman, *Het Geheim van Van Eyck*, 93.

67. Van Mander, *Schilder-boeck*, ed. Miedema, 1, fol. 82r.

68. For a detailed comparison between Vasari's and Van Mander's accounts of Van Eyck's invention of oil paint, see Brinkman, *Het Geheim van Van Eyck*, 90–98.

69. Van Mander, *Schilder-boeck*, ed. Miedema, 1, fol. 199r.

70. Van Mander, *Schilder-boeck*, ed. Miedema, 1, fols. 199v–200r.

71. Judith Bryce, *Cosimo Bartoli (1503–1572): The Career of a Florentine Polymath* (Geneva: Librairie Droz, 1983), 55–71.

72. Since the publication of Polydore Vergil's history of ancient and modern inventions and discoveries in his 1499 treatise *De inventoribus rerum*, Italians had been increasingly fascinated with the phenomenon of invention; see Uta Bernsmeier, "Die *Nova Reperta* des Jan Van Der Straet-Ein Beitrag zur Problemgeschichte der Entdeckun-

gen und Erfindungen im 16. Jh." (Diss., University of Hamburg, 1984), and Catherine Atkinson, *Inventing Inventors in Renaissance Europe: Polydore Vergil's "De inventoribus rerum"* (Tübingen: Mohr Siebeck, 2007).

73. See, e.g., Ernst van de Wetering, *Rembrandt: The Painter at Work* (Amsterdam: Amsterdam University Press 1997), 136–39, who argued that the print shows Van Eyck in a workshop representative of sixteenth-century artistic practice. Compare also Emmens, who places Vasari's invention myth in the history of *nova reperta*, but only because later authors speak about the oil painting myth in this context. He concludes that the invention myth represents the discovery of a secondary technique (as opposed to fresco) introduced by Vasari in the life of Antonello da Messina as an expression of his anti-Venetian disposition. See "De uitvinding van de olieverf," 191–203. More recently, Lucy Davis argued that Stradanus's print should be studied from the perspective of Stradanus working in Italy as a Flemish artist. Through the print Stradanus imagined the historical legacy of Van Eyck in Florence. See "Renaissance Inventions: Van Eyck's Workshop as a Site of Discovery and Transformation in Jan van der Straet's *Nova Reperta*," *Nederlands Kunsthistorisch Jaarboek* 59 (2010): 223–48.

74. Translation by the author. Samuel van Hoogstraten, *Inleyding tot de hooge schoole der schilderkonst: Anders de zichtbaere werelt* (Rotterdam: Francois van Hoogstraten, 1678), 338.

Chapter Two

1. Pliny, *Natural History* [35.36.97], ed. Rackham, 9:332–33.

2. For an analysis and overview of the context and nature of these cleaning controversies as they emerged since the nineteenth century, see David Bomford, "Picture Cleaning: Positivism and Metaphysics," in *The Conservation of Easel Paintings*, ed. Joyce Hill Stoner and Rebecca Anne Rushfield (London: Routledge, 2012), 481–91.

3. In his account of the various black pigments, Pliny explains that one type of ink was made from burned pitch resin (resins were also a key ingredient in varnish) mixed with soot (a black pigment); see Pliny, *Natural History* [35.25.41–43], ed. Rackham, 9:290–93. The Greek medical writer Pedanius Dioscorides (ca. AD 40–90), author of the *Materia Medica* (AD 50–70), also mentions of ink prepared in a similar manner. Dioscorides, *De Materia Medica: Being an Herbal with Many Other Medicinal Materials Written in Greek in the First Century of the Common Era*, ed. and trans. Tess Anne Osbaldeston and Robert P. A. Wood (Johannesburg: Ibidis Press, 2000), 88–89 [1–86], 94–95 [1–93], 830 [5–183].

4. For the 1950s controversy, which largely took place in the *Burlington Magazine*, see (a selection) Cesare Brandi, "The Cleaning of Pictures in relation to Patina, Varnish, and Glazes," *Burlington Magazine* 91, no. 556 (1949): 183–88; Neil Maclaren and Anthony Werner, "Some Factual Observations about Varnishes and Glazes," *Burlington Magazine* 92, no. 568 (1950): 189–92; Helmut Ruhemann, "Leonardo's Use of Sfumato," *British Journal of Aesthetics* 1 (1961): 231–37; Ernst Gombrich, "Dark Varnishes: Variations on a Theme by Pliny," *Burlington Magazine* 104, no. 707 (1962): 51–55; and Joyce Plesters, "Dark Varnishes-Some Further Comments," *Burlington Magazine* 104, no. 716 (1962): 452–60.

5. See Bomford, "Picture Cleaning," 481–91.

6. In addition to varnish layers, *craquelure* also frequently occurs in the ground layer of a painting and in the paint film. See, e.g., David Bucklow, "The Description of Craquelure Patterns," *Studies in Conservation* 43, no. 3 (1997): 129–40; Ludovic Pauchard and Frédérique Giorgiutti-Dauphiné, "Craquelures and Pictorial Matter," *Journal of Cultural Heritage* 46 (2020): 361–73.

7. On the aging of varnishes see Robert L. Feller, Nathan Stolow, and Elizabeth H. Jones, *On Picture Varnishes and Their Solvents*, rev. and enl. ed. (Cleveland: Press of Case Western Reserve University, 1971), 154–67. And for an introduction to the subject of aging in the art of painting, see Paul Taylor, *Condition: The Ageing of Art* (London: Paul Holberton, 2015).

8. For a general introduction to the history of art restoration and conservation (mostly of paintings), see Alessandro Conti, *History of the Restoration and Conservation of Works of Art* (1973; repr., Oxford: Butterworth-Heinemann, 2007), and David Bomford and Mark Leonard, eds., *Issues in the Conservation of Paintings* (Los Angeles: Getty Conservation Institute, 2004).

9. A notable exception is the early (possibly even original) sandarac and walnut varnish found on a painting (ca. 1470–75) by Carlo Crivelli, now kept in the National Gallery of London; see Jill Dunkerton and Raymond White, "The Discovery and Identification of an Original Varnish on a Panel by Carlo Crivelli," *National Gallery Technical Bulletin* 21 (2000): 70–76.

10. Stoner and Rushfield, *Conservation of Easel Paintings*, 635–57.

11. For the ecological roles of resins and the most comprehensive work on plant resins to date, see Jean H. Langenheim, *Plant Resins: Chemistry, Evolution, Ecology, and Ethnobotany* (Portland, OR: Timber Press, 2003), 197–253.

12. In addition, resins were used for many more purposes than making glue and varnish. Where relevant, some of these will be discussed later on in this book.

13. Langenheim, *Plant Resins*, 275; and see F. N. Howes, "Age-Old Resins of the Mediterranean Region and Their Uses," *Economic Botany* 4, no. 4 (1950): 307–16.

14. Frank N. Jones, Mark E. Nichols, and Socrates Peter Pappas, eds., *Organic Coatings: Science and Technology* (Hoboken, NJ: John Wiley, 2017), 6–8; and see also Feller, Stolow, and Jones, *On Picture Varnishes and Their Solvents*, 125–27.

15. Gerald W. R. Ward, ed., *The Grove Encyclopedia of Materials and Techniques in Art* (New York: Oxford University Press, 2008), 729–32.

16. The optical characteristics of varnishes have been extensively studied by E. René de la Rie; see, for instance, "The Influence of Varnishes on the Appearance of Paintings," *Studies in Conservation* 32 (1987): 1–13. See also Thomas B. Brill, *Light: Its Interaction with Art and Antiquities* (New York: Plenum Press, 1980), 99–105.

17. Langenheim, *Plant Resins*, 45–47.

18. Howes, "Age-Old Resins," 307.

19. Almost all of the medieval Norwegian frontals investigated by Plahter et al. were found to have original glair varnishes. The authors point out that this might have been more a local exception than the rule; see Plahter et al., *Painted Altar Frontals of Norway*, 2:61. Andrea Kirsh and Rustin S. Levenson list a few more examples where glair varnishes were discovered on Italian and Spanish paintings; see *Seeing through Paintings:*

Physical Examination in Art Historical Studies (New Haven, CT: Yale University Press, 2000), 230–31.

20. For a discussion of glair varnishes and their recipes, see Renate Woudhuysen-Keller and Paul Woudhuysen, "A Short History of Eggwhite Varnishes," in *Firnis: Material-Ästhetik-Geschichte; Internationales Kolloquium, Braunschweig, 15.–17. Juni 1998*, ed. A. Harmssen (Stegen: AdR-Arbeitsgemeinschaft der Restauratoren, 1999), 80–86.

21. Pliny the Elder, *Natural History, Volume 7: Books 24–27*, ed. and trans. W. H. S. Jones and A. C. Andrews, Loeb Classical Library 393 (Cambridge, MA: Harvard University Press, 1956), 28–31 [24.22.34–35]; *Natural History, Volume 4: Books 12–16*, ed. and trans. H. Rackham, Loeb Classical Library 370 (Cambridge, MA: Harvard University Press, 1968), 266–67 [14.25.122–23].

22. For the English translation cited here, see Eastlake, *Materials for a History of Oil Painting*, 21–22. Aëtius's complete work has never been fully published in Greek, but it was published in a sixteenth-century Latin translation by Janus Cornarius; see Aetius Amidenus, *Aetii medici Graeci contractae ex veteribus medicinae sermones XVI*, ed. and trans. Janus Cornarius (Basel: Johannes Froben, 1542), 21.

23. For a recent discussion on the *Mappae clavicula* and *Compositiones lucencis* being two separate traditions (with an overview of relevant literature), see Guido Frison and Giulia Brun, "*Compositiones lucenses* and *Mappae clavicula*: Two Traditions or One? New Evidence from Empirical Analysis and Assessment of the Literature," *Heritage Science* 6, no. 1 (2018): 1–24, https://doi.org/10.1186/s40494-018-0189-y.

24. For a critical edition with German translation of codex Lucensis 490, see Hjalmar Hedfors, *Compositiones ad tingenda musiva [Codex lucensis 490]* (Uppsala: Almqvist och Wiksells, 1932).

25. Cyril Stanley Smith and John G. Hawthorne, "*Mappae clavicula*: A Little Key to the World of Medieval Techniques," *Transactions of the American Philosophical Society* 64, no. 4 (1974): 1–128. For the Latin transcription I refer to Thomas Phillipps, "Letter from Sir Thomas Phillipps . . . Communicating a Transcript of a MS Treatise . . . Entitled *Mappae clavicula*," in *Archaeologia, or Miscellaneous Tracts relating to Antiquity* (London: Society of Antiquaries of London, 1770). Smith and Hawthorne use the same recipe numbers as Phillipps does.

26. The catalog entry was titled *Mappae clavicula de efficiendo auro volumen 1*; see Rozelle P. Johnson, "Notes on Some Manuscripts of the *Mappae clavicula*," *Speculum* 10 (1935): 72.

27. "*Mappae clavicula*," ed. Smith and Hawthorne, 4–14.

28. For an introduction to craft writing in the Latin West, see Bernard Bischoff, "Die Überlieferung der technischen Literatur," in *Mittelalterliche Studien: Ausgewählte Aufsätze zur Schriftkunde und Literaturgeschichte* (Stuttgart: A. Hiersemann, 1981), 3:277–97; and for their seminal discussion of the origin of the *Mappae clavicula* and its relation to alchemical traditions, see Robert Halleux and Paul Meyvaert, "Les origines de la *Mappae clavicula*," *Archives d'histoire doctrinale et littéraire du Moyen-Âge* 54 (1987): 7–58.

29. "*Mappae clavicula*," ed. Smith and Hawthorne, 15.

30. Halleux and Meyvaert, "Origines de la *Mappae clavicula*," 25.

31. *Compositiones ad tingenda musiva*, ed. Hedfors [no. M 10–20]; compare with "*Mappae clavicula*," ed. Smith and Hawthorne, 66–67 [no. 247]. Because multiple recipes appear on any given page in the editions cited, I will refer to the numbers assigned to specific recipes in these editions by putting them in square brackets.

32. *Compositiones ad tingenda musiva*, ed. Hedfors, 29 [no. M 10–20]. The equivalent recipe in the *Mappae clavicula* compilation does not mention that the varnished object has to be dried in the sun; compare "*Mappae clavicula*," ed. Phillipps, 234 [no. 247]: *illucidare poteris, lucidata autem siccare.*

33. Smith and Hawthorne translated it this way: "How transparent varnishes ought to be put over pigments"; see "*Mappae clavicula*," [no. 247], ed. Phillipps, 23; ed. Smith and Hawthorne, 66–67.

34. In Sélestat MS 17 this recipe is titled *Ad pingendam picturam hudis* (To paint a picture against dampness); see "*Mappae clavicula*," ed. Smith and Hawthorne, 43 [no. 109] and n. 64.

35. Dioscorides mentions that the oil called *cicinum* is called *ricinus* by the Romans. *De Materia Medica*, ed. and trans. Osbaldeston and Wood, 719–20 [4–164]; on oil of *cicinum* see also pages 37–38 [1–38].

36. "*Mappae clavicula*," [no. 109], ed. Phillipps, 211; ed. Smith and Hawthorne, 43.

37. Andrew Dalby, *Food in the Ancient World from A to Z* (London: Routledge, 2003), 76.

38. Dehydrated castor oil, usually heated in a vacuum with certain catalysts added, is in fact an important drying oil in today's surface coatings industry; it is resistant to water, gives a flexible film, and does not easily yellow compared with other drying oils such as linseed oil.

39. *Compositiones ad tingenda musiva*, ed. Hedfors, 45 [U11–16].

40. *Compositiones ad tingenda musiva*, ed. Hedfors, 45 [U8–11].

41. Bernardo Oderzo Gabrieli, "L'inventario della spezieria di Pietro Fasolis e il commercio dei materiali per la pittura nei documenti piemontesi (1332–1453): Parte Seconda," *Bollettino della Società Storica Pinerolese* 30 (2013): 19.

42. For more information about dyestuffs and how they were made into pigments suitable for painting with, see chapter 5 of this book.

43. *Compositiones ad tingenda musiva*, ed. Hedfors, 25 [K32–L11], and the notes on pages 137–38; and see "*Mappae clavicula*," ed. Smith and Hawthorne, 64–65 [246].

44. The ability of mastic to make an oil-resin mixture more siccative is also mentioned in a recipe for making Greek glue; see "*Mappae clavicula*," ed. Smith and Hawthorne, 41 [98]. See also chapter 3 below for the apparent increase in the use of mastic in varnishes in the fifteenth and sixteenth centuries.

45. *Compositiones ad tingenda musiva*, ed. Hedfors, 25 [K-32–L 11].

46. This recipe is included twice in MS5 (Corning Museum of Glass); see "*Mappae clavicula*," ed. Smith and Hawthorne, [no 116, 208]. The *Mappae clavicula* also contains various recipes for coating tin foil without an oil-resin varnish (compare 60, 61, 207, 115). These recipes likely reflect the making of "golden tin" for manuscript illumination, which was not compatible with oily binders. See also the instructions for coating tin foil from the treatise of Theophilus discussed later in this chapter.

47. Rainwater was most likely added while cooking the varnish so as to purify it.

Purifying oil with water is first mentioned in more detail in the sixteenth century and will be discussed in chapter 3.

48. See "*Mappae clavicula*" [nos. 116, 208 refer to the same recipe included twice in the manuscript], ed. Phillipps, 212, 227; ed. Smith and Hawthorne, 44, 59.

49. Josephine Darrah, "White and Golden Tin Foil in Applied Relief Decoration: 1240–1530," in *Looking through Paintings: The Study of Painting Techniques and Materials in Support of Art Historical Research*, ed. Erma Hermens, Annemiek Ouwekerk, and Nicola Costaras (Baarn: Archetype, 1998), 49–79, and especially table 2. Unfortunately it is difficult to identify yellow dyestuffs such as saffron and aloe by instrumental means. For recent advances in the identification of aloe in oil-resin varnishes, see Adriana Rizzo, Nobuko Shibayama, and Daniel P. Kirby, "A Multi-analytical Approach for the Identification of Aloe as a Colorant in Oil-Resin Varnishes," *Analytical and Bioanalytical Chemistry* 399, no. 9 (2011): 3093–94, doi: 10.1007/s00216-010-4402-4.

50. Lucretia Kargère and Adriana Rizzo, "Twelfth-Century French Polychrome Sculpture in the Metropolitan Museum of Art: Materials and Techniques," *Metropolitan Museum Studies in Art, Science, and Technology* 1 (2010): 39–72. Analysis of the varnish on the tin foil decorations on a twelfth-century sculpture of the *Torso of Christ*, also in the collection of the Metropolitan Museum of Art, yielded similar results; see p. 53.

51. See, e.g., the analysis published in Plahter et al., *Painted Altar Frontals of Norway*, 2:168–74, and fig. III.2.23 and table 48; and Massing, ed., *Thornham Parva Retable*, 54. For results from yellow varnishes on sculptures and other polychromy, see also various contributions in Katja Kollandsrund and Jilleen Nadolny, eds., *Medieval Painting in Northern Europe: Techniques, Analysis, Art History, Studies in Commemoration of the 70th Birthday of Unn Plahter* (London: Archetype, 2006).

52. For an edited volume devoted to Theophilus's *Schedula*, see Andreas Speer, Maxime Mauriège, and Hiltrud Westermann-Angerhausen, eds., *Zwischen Kunsthandwerk und Kunst: Die "Schedula Diversarium Artium"* (Berlin: De Gruyter, 2013).

53. To compare a representative selection of the manuscripts containing the treatise of Theophilus and the various critical editions of the *Schedula*, see *The "Schedula diversarum artium" . . . a Digital Critical Edition*, ed. Andreas Speer and Hiltrud Westermann-Angerhausen (Cologne: Thomas-Institute, University of Cologne, 2010), accessed June 11, 2021: http://schedula.uni-koeln.de.

54. Theophilus, *"On Divers Arts": The Foremost Medieval Treatise on Painting, Glassmaking and Metalwork*, ed. and trans. John G. Hawthorne and Cyril Stanley Smith (1963; repr., New York: Dover, 1979), and *Theophilus Presbyter und das mittelalterliche Kunsthandwerk: Gesamtausgabe der Schrift "De diversis artibus" in zwei Bänden*, ed. and trans. Erhard Brepohl, 2 vols. (Cologne: Böhlau, 1999). Unless stated otherwise, citations are from the Dodwell edition, which contains a critical edition of the Latin text side-by-side with an English translation: Theophilus, *The Various Arts: "De diversis artibus,"* ed. and trans. Charles Reginald Dodwell (Oxford: Clarendon Press, 1986).

55. Theophilus, *Schedula*, ed. Smith and Hawthorne, xxx.

56. Theophilus, *Schedula*, ed. Dodwell, 19 [no. XXI].

57. The handle, of course, would have prevented burned hands when the metal becomes hot.

58. On the vocabulary Theophilus uses to describe transparent and translucent effects, see Bol, "Seeing through the Paint," 145–62.

59. Theophilus, *Schedula*, ed. Dodwell, 19 [XXI].

60. *Compositiones ad tingenda musiva*, ed. Hedfors, 28 [M 10–20].

61. Theophilus, *Schedula*, ed. Dodwell, 25–26 [XXVIII].

62. *Original Treatises Dating from the XIIth to XVIIIth Centuries on the Arts of Painting*, ed. and trans. Mary Philadelphia Merrifield, 2 vols. (London: John Murray, 1849), 1:2.

63. See Anne D. Hedemann, "Making the Past Present: Visual Translation in Jean Lebègue's 'Twin' Manuscripts of Sallust," in *Patrons, Authors and Workshops: Books and Book Production in Paris around 1400*, ed. Godfried Croenen and Peter F. Ainsworth (Louvain: Peeters, 2006), 173, and note 3 for an overview of the literature on LeBègue's relation to the arts. See also Inès Villela-Petit, "La peinture médiévale vers 1400 autour d'un manuscrit de Jean Lebègue: Édition des *Libri colorum*" (Diss., École nationale des Chartes, 1995).

64. *Aultres receptes en Latin et en François per Magistrum Johannem dit Le Begue . . .*, ed. Merrifield, 1:313–15 [no. 341].

65. Cennino Cennini, *Cennino Cennini's "Il libro dell'arte": A New English Translation and Commentary with Italian Transcription*, ed. and trans. Lara Broecke (London: Archetype, 2015), 1–5.

66. This manuscript has been digitized; see MS Ott. Lat. 2974, *Digital Vatican Library*, accessed June 11, 2021: https://digi.vatlib.it/view/MSS_Ott.lat.2974.

67. See Bernardo Oderzo Gabrieli, "L'inventario della spezieria di Pietro Fasolis e il commercio dei materiali per la pittura nei documenti piemontesi (1332–1453): Parte prima," *Bollettino della Società Storica Pinerolese* 29 (2012): 7–43. See especially the entry "vernicis liquida" in the apothecary's inventory on page 26; Gabrieli, "L'inventario della spezieria di Pietro Fasolis: Parte seconda," 7–53.

68. Cennini, *Il libro dell'arte*, ed. Broecke, 201–2 [no. 169].

69. "*Mappae clavicula*," [no. 247], ed. Phillipps, 234; ed. Smith and Hawthorne, 65–66.

70. Theophilus, *Schedula*, ed. Dodwell, 22–24 [no. XXIV].

71. Compare this recipe with a similar recipe titled *De deauratura* from the treatise of Eraclius, *De coloribus*, ed. Merrifield, book 3, 220–21 [no. 276], in which tin foil is similarly varnished two or three times and afterward immersed in a mixture of a yellow dye and beer. This recipe does not mention the final layer of varnish that Theophilus advises.

72. Compare with a similar recipe for coating tin foil with a "golden" color that can be found in Eraclius's *De coloribus*, ed. Merrifield, 1:220–21 [no. 276]. The recipe differs from Theophilus's in three respects: a different yellow dyestuff is used; Eraclius does not instruct one to cover the golden tin with varnish one last time; and he does not explain if or how one should paint on the golden tin.

73. A colophon in the text states that it was finished in 1262, but it has been proposed that, based on the script used, the manuscript is most likely fifteenth century. The recipes nevertheless seem to indicate earlier traditions; see Devon L. Strolovitch, "Old Portuguese in Hebrew Script" (PhD diss., Cornell University, 2005), 117; and see also

António João Cruz and Luís Urbano Afonso, "On the Date and Contents of a Portuguese Medieval Technical Book on Illumination: *O livro de como se fazem as cores*," *Medieval History Journal* 11, no. 1 (2008): 1–28.

74. This is difficult to translate, but it is likely related to one of the other terms used to describe the resin *vernix*: *gressa*. See also the section "Sweat and Tears: *Vernix* in Medico-botanical Treatises" below in this chapter.

75. *O livro de como se fazem as cores*, ed. Strolovitch, 141 [no. 42].

76. The recipes on art technology contained in Amplonius fol. 49 were edited by Ernst Ploss in "Studien zu den deutschen Maler und Färberbüchern des Mittelalters: Ein Beitrag zur deutschen Altertumskunst und Wortforschung" (Diss., Ludwig-Maximilian Universität, 1952), 177.

77. Unn Plahter, "*Líkneskjusmíð*: 14th-Century Instructions for Painting from Iceland, Norwegian Medieval Altar Frontals and Related Material," in *Norwegian Medieval Altar Frontals and Related Material: Papers from the Conference in Oslo 16th to 19th December 1989*, ed. Magne Malmanger, Laszlo Berczelly, and Signe Horn Fuglesang (Rome: Giorgio Bretschneider, 1995), 163.

78. "*Líkneskjusmíð*," ed. Plahter, 160. This might be a reference to how the varnish is bleached by the sun. See also chapter 5 of this book for recipes describing how painter's oil becomes paler in the sun. Plahter also points out that silver, a white metal, cannot be heated in the sun because it reflects the sun's rays.

79. *Mediaeval Painter's Materials and Techniques: The Montpellier "Liber diversarum arcium,"* ed. and trans. Mark Clarke (London: Archetype, 2011), 1–2.

80. *Montpellier "Liber diversarum arcium,"* ed. Clarke, 139–40 and 280 [nos. 2.8.1 and 2.8.2].

81. *Rasa* may be sandarac, from its Sanskrit name *chanda-rasa*; see *Montpellier "Liber diversarum arcium,"* ed. Clarke, 139–40 and 280 [no. 2.8.2].

82. Jonathan Thornton, "Use of Dyes and Colored Varnishes in Wood Polychromy," in *Painted Wood: History and Conservation . . .* , ed. Valerie Dorge and F. Carey Howlett (Los Angeles: Getty Conservation Institute, 1998), 238.

83. *Montpellier "Liber diversarum arcium,"* ed. Clarke, 139–40 and 280 [no. 2.8.1].

84. Eraclius, *De coloribus*, ed. Merrifield, 1:182 and 183 [no. I].

85. See Eraclius, *De coloribus*, ed. Merrifield, 1:166–67. The British Library manuscript of Eraclius has also been published by Rudolph Erich Raspe, *A Critical Essay on Oil-Painting: Proving That the Art of Painting in Oil Was Known before the Pretended Discovery of John and Hubert van Eyck; to Which Are Added Theophilus, "De arte pigendi," Eraclius, "De artibus Romanorum," and a Review of Farinator's "Lumen animae"* (London: Printed for the author by H. Goldney, 1781).

86. This advice is also congruent with Cennini's warning not to varnish gold leaf. Eraclius, *De coloribus*, ed. Merrifield, 1:224–25 [no. 267].

87. Eraclius, *De coloribus*, ed. Merrifield, 1:240–41 [no. 274].

88. Note that the two main resins used to make this yellow varnish are the same as those mentioned in the varnish recipe of Theophilus.

89. Eraclius, *De coloribus*, ed. Merrifield, 1:224–25 [no. 267].

90. Binski, Massing, and Sauerberg, *Westminster Retable*; see especially the contributions to chapter 6, "Materials and Techniques."

91. Raymond White and Jo Kirby, "Medium Analysis," in *Westminster Retable*, ed. Binski, Massing, and Sauerberg, 255–59; and from the same publication, see also Marie Louise Sauerberg, Ashok Roy, Marika Spring, Spike Bucklow, and Mary Kempski, "Materials and Techniques," 244–45.

92. Peter of Saint-Omer, *De coloribus*, ed. Merrifield, 1:158–65 [nos. 202–9]. See also recipe 165, in which Peter of Saint-Omer explains how to stain wood with saffron and wine. We learn that if you want to "make the wood shine," you should first let the saffron-wine mixture dry and then coat it with one more layer of saffron with oil.

93. Peter of Saint-Omer, *De coloribus*, ed. Merrifield, 1:158–59 [no. 202].

94. Because dyes dissolve in their medium (so refractive index is irrelevant), animal glue can also be used to create a translucent coating. Such a translucent yellow layer of animal glue would nevertheless be less lustrous than a yellow oil-resin varnish.

95. Peter of Saint-Omer, *De coloribus*, ed. Merrifield, 1:160–63 [no. 205].

96. Theophilus, *Schedula*, ed. Dodwell, 29 [no. XXX].

97. Peter of Saint-Omer, *De coloribus*, ed. Merrifield, 1:160–63 [no. 205].

98. Peter of Saint-Omer, *De coloribus*, ed. Merrifield, 162–65 [no. 208].

99. See also Thornton, "Use of Dyes and Colored Varnishes in Wood Polychromy," 238. As we will see in the next chapter, adding alum might have helped clear the oil-resin varnish of impurities.

100. Theophilus similarly refers to this coating in other places in the *Schedula* as *glutine vernition* or simply as *gluten*.

101. Theophilus, *Schedula*, ed. Dodwell, 16 [no. XVII] for the recipe for casein glue, and 145 [no. LXXXII] for the recipe for hide glue.

102. For various recipes that describe how to make a mixture of linseed oil and various gums and resins for applying gold leaf, see "*Mappae clavicula*," ed. Smith and Hawthorne, 44 [nos. 113, 113-A, and 114] and the various recipes in Cennini, *Il libro dell'arte*, ed. Broecke, 127–36 [nos. 91–102].

103. Theophilus, *Schedula*, ed. Dodwell, 19 [no. XXI].

104. See also Theophilus, *Schedula*, ed. Smith and Hawthorne, 28n1.

105. That these terms are often interpreted this way becomes especially clear from the many editions that translate it as such. See, for instance, Theophilus, *Schedula*, ed. Dodwell, and "*Mappae clavicula*," ed. Smith and Hawthorne, who consistently translate this ingredient as "varnish." This is probably because *vernix* and *vernice* became general terms to denote varnishes from the fifteenth century onward.

106. As *veronice* and *vernicis*.

107. "*Mappae clavicula*" [no. 98], ed. Phillipps, 208; ed. Smith and Hawthorne, 41.

108. It is noteworthy that by the fourteenth century the roles seem to be reversed. In Alcherius's recipe collection, for instance, one of the main ingredients for a glue that mends crystal, glass, and gems is "painter's liquid varnish" (*vernice liquida pictorem*). Significantly, this glue also has to be dried in the sun; see ed. Merrifield, 1:74–77 [no. 60].

109. It has, for instance, been discussed in numerous artist's handbooks in the context of the search for a durable paint medium (briefly discussed in the introduction to this book). See Mérimée's identification of the resin in Theophilus's varnish with copal in *De la peinture à l'huile* (1830), 3–6. (Copal came from the New World or the African

continent. There is no evidence that copal from Africa was known or used during the time of Theophilus.)

110. *Theophili qui est Rugerus, presbyteri et monachi, libri III. de diversis artibus: Seu Diversarium artium schedula . . .* , ed. and trans. Robert Hendrie (London: Johannes Murray, 1847), 63–74.

111. Theophilus, *Schedula*, ed. Hendrie, 65.

112. Theophilus, *Schedula*, ed. Hendrie, 68.

113. Eastlake, *Materials for a History of Oil Painting*, 230.

114. Compare also with Merrifield's identification of *vernix* with sandarac in her English translation of Cennino Cennini, *A Treatise on Painting by Cennino Cennini Written in the Year 1437 . . .* , ed. and trans. Mary Philadelphia Merrifield (London: Edward Lumley, 1844), 161–62.

115. Ps. Bartholomaeus Mini de Senis, *Tractatus de herbis* (London, British Library, MS Egerton 747), ed. Iolanda Ventura (Florence: Sismel Edizioni del Galluzzo, 2009), 288–89 [no. 66].

116. London, British Library, MS Harley 270, fol. 128v, accessed June 11, 2021: http://www.bl.uk/manuscripts/Viewer.aspx?ref=harley_ms_270_f123r.

117. Elizabeth Keen, *Journey of a Book: Bartholomew the Englishman and the Properties of Things* (Canberra: ANU E Press, 2007), 5.

118. Bartholomaeus Anglicus, *De proprietatibus rerum* (Lyon: Petrus Ungarus, 1482), [book 17, *De Resina*]. An oleo-gum resin consists of water-soluble gum, resin, and essential oil.

119. Bartholomaeus Anglicus, *De proprietatibus rerum*, [book 17, *Bdellium*].

120. In an entry in the glossary of the treatise that Jehan LeBègue finished compiling in 1431, *vernicem liquidem* and *glutine* are considered synonyms. See Merrifield, 1849, 1:358: *vernicem liquidam id est glutinam pro pictoribus*.

121. Andrea Ubrizsy Savoia and Luís Ramón-Laca, "The *Libri Picturati* Watercolours and Early Botanical Illustrations," in *Drawn after Nature: The Complete Botanical Watercolours of the 16th-Century "Libri Picturati,"* ed. J. de Koning, Gerda van Uffelen, Alicja Zemanek, and Bogdan Zemanek (Zeist: KNV, 2008), 60–67.

122. *Ortus sanitatis* (Mainz: Jakob Meydenbach, 1491), Cambridge University Library (Inc.3.A.1.8[37]), fol. 232v, accessed June 11, 2021: https://cudl.lib.cam.ac.uk/view/PR-INC-00003-A-00001-00008-00037/1.

123. Langenheim, *Plant Resins*, 382–84.

124. Significantly, the chemical composition of the two resins is so similar that even with instruments it is difficult to distinguish them. See Clara Azémard, Matthieu Ménager, and Cathy Vieillescazes, "On the Tracks of Sandarac: Review and Chemical Analysis," *Environmental Science and Pollution Research* 24, no. 36 (2017): 27746–54.

125. Compare the entries on these materials in the *Breviarium Bartholomei* (1387) composed by John Mirfeld and a similar untitled glossary of the mid-fifteenth century, included in *Sinonoma Bartholomei: A Glossary from a Fourteenth-Century Manuscript in the Library of Pembroke College, Oxford*, ed. and trans. J. L. G. Mowat (Oxford: Clarendon Press, 1882), 12, 14, 16, 26; and *Alphita, a medico-botanical glossary from the Bodleian manuscript, Selden B.35*, ed. and trans. J. L. G. Mowat (Oxford: Clarendon Press, 1887), 22, 33, 71, 85, 88, 173.

126. *Sinonoma Bartholomei*, ed. Mogwat, 12 and *Alphita, a medico-botanical glossary*, ed. Mowat, 22. The two entries are similar; I cite from *Aphita*.

127. Compare the entry on *Carabre* in which Mirfeld explains that this resin is also used in pulverized form to prepare parchment: *Sinonoma Bartholomei*, ed. Mogwat, 14: *Carabre vel cacabre, i. verm[n]icium id s. pulverizatum cum quo paratur percamenum, classis idem*. This indeed is one of the other known uses of sandarac resin: to remove the grease from vellum so it is easier to write on. For an explanation of this practice, see Michelle P. Brown, *The British Library Guide to Writing and Scripts* (Toronto: University of Toronto Press, 1998), 67.

128. Today amber is classified not as an organic mineral but as an organic-derived gemstone (like pearl, coral, and ivory); see Gilberto Artioli, *Scientific Methods and Cultural Heritage: An Introduction to the Application of Materials Science to Archaeometry and Conservation Science* (Oxford: Oxford University Press, 2010), 372–73.

129. See Faya Causey, *Amber and the Ancient World* (Los Angeles: Getty Publications, 2011), 52–58; Gage, *Colour and Culture*, 26.

130. Eraclius, *De coloribus*, ed. Merrifield, 1:225 [no. 267].

131. Merrifield suggests that it was transcribed from an older Eraclius manuscript that might have been previously owned, and perhaps worked on, by Jehan Alcherius; see *Original Treatises*, 1:168–69.

132. Its earliest occurrence may be in the 1269 Paris Painters Statutes, which uses variations of the appellation *vernicie* to describe the act of varnishing; François Bonnardot and René de Lespinasse, eds., *Les métiers et corporations de la ville de Paris, XIIIe siècle: Le livre des métiers* (Paris: Imprimerie national, 1879), 168–73 [e.g., vi, xxxv, vi].

133. *Mededeelingen van de vereeniging ter beoefening der geschiedenis van 's-Gravenhage* (The Hague, 1863), 1:298 [R. 143 7/8, fol. 83].

134. Wilhelmus P. van Stockum, ed., *'S-Gravenhage in den loop der tijden*, 2 vols. (The Hague: W. P. van Stockum, 1889), 1:70.

135. See the 1485 Middle Dutch translation of Bartholomew of England's encyclopedia, Bartholomeus Engelsman, *Van den proprieteyten der dinghen* (Haarlem: Jacob Bellaert, 1485), fol. 312v: "Ende hier bij ouerdraecht een gomme die men vernijs heet ende is een gomme van enen boem ende byndt oeck zeer op ghescoerde ende het maect een dinck claer ende behoudet lange ende daer om oerbarent die malers."

136. Translation by author. This manuscript has been digitized and transcribed and can be found on the website of the Bodleian Library; see Jan Van Boendale, MS Marshall 29, fols. 45v–46r, accessed June 11, 2021: http://dutch.clp.ox.ac.uk/?q=chw0096: "Maer si (de vrouwen) smeren ende salven / Haere aensichten, omdet si / Te scoenre scinen, ende bedi / Dat si te meer selen sijn besien / Beide van desen ende van dien. / Maer als een meester heeft vernist / Een beelde met al siere list / Dat scone blict als gout / Nochtan soo eest binnen hout: / alsoo ghelikerwijs es .i. wijf: / Alse si heeft vernist haer lijf / Dat scone blicket ende scijnt / Dat es al om niet ghepijnt / Dattere was, dat moet daer bliven / Dat en canse niet verdriven."

137. Ludo Jongen, Cees Schotel, and Josephine Franken, eds., *Het leven van Liedewij, de maagd van Schiedam: De Middelnederlandse tekst naar de bewaarde bronnen*

uitgegeven, vertaald en van commentaar voorzien (Schiedam: Fonds Historische Publikaties Schiedam, 1989).

138. *Het leven van Liedewij*, ed. Jongen, Schotel, and Franken, 100 [XLII].

139. There is an early reference to this practice in the *Etymologies* of the archbishop of Seville, *Isidorus hispalensis* (ca. 560–636); see *The Etymologies of Isidore of Seville*, ed. and trans. Stephen A. Barney, W. J. Lewis, J. A. Beach, and O. Berghof (Cambridge: Cambridge University Press, 2006), 349 [viii. Aromatic Trees, no. 7]: "*mastix* is the sap of the mastic tree [*Ientiscus*]. It is also called *granomastix* because it looks like grains [*granum*]. The better kind grows on the island of Chios and is good smelling, with the whiteness of Punic wax-hence it beautifies the glow of one's skin."

140. Later, *vernis* would become known as *blanketsel* in Dutch. See, e.g., *Het christelick huys-wijf* (1625) by Jacob Cats, who also mentions that gum/resin was an ingredient of face paint. Like Van Boendale en Cennini, Cats had nothing positive to say about the practice, which he considered deceptive; see Jacob Cats, *Alle de wercken, so ouden als nieuwen* (Amsterdam: Jan Jacobsz. Schipper, 1658), 144.

141. Cennini, *Il libro dell'arte*, ed. Broecke, 248–49 [no. 233].

142. Cennini, *Il libro dell'arte*, ed. Broecke, 248–49 [no. 234].

Chapter Three

1. Medieval recipes suggest that artisans did sometimes want their varnish ingredients to be of a certain clarity. This likely was about the quality of materials they wanted to use, because the method of preparation described in these recipes would have still produced a relatively dark substance.

2. Cennini, *Il libro dell'arte*, ed. Broecke, 200–201 [no. 168]. I have slightly adapted the translation from Broecke, who translates *lucida* as "transparent."

3. See J. D. J. van den Berg, "Analytical Chemical Studies on Traditional Oil Paints" (Diss., University of Amsterdam, 2002), 14–16; see also Renate Woudhuysen-Keller, "Leinöl als Malmittel. Rekonstruktionsversuche nach Rezepten aus dem 13. bis 19. Jahrhundert," *Maltechnik Restauro* 2 (1973): 75–76.

4. For Eastlake's notes on the manuscript, see *Materials for a History of Oil Painting*, 1:126–40. Eastlake's transcription formed the basis for the 1966 English translation by Rosamund Borradaile and Viola Borradaile, *The Strasburg Manuscript: A Medieval Painter's Handbook/Das Strassburger Manuskript: Handbuch für Maler des Mittelalters* (1966; repr., Munich: Callwey 1976). I cite the recent critical edition by Sylvie Neven, *The Strasbourg Manuscript: A Medieval Tradition of Artists' Recipe Collections* (London: Archetype, 2016).

5. For two comprehensive studies into these treatises, see William Eamon, "Arcana Disclosed: The Advent of Printing, the Books of Secrets Tradition and the Development of Experimental Science in the Sixteenth Century," *History of Science* 22, no. 2 (1984): 111–50, and more recently Ernst Striebel, "Das Augsburger Kunstbuechlin von 1535: Eine kunsttechnologische Quellenschrift der deutschen Renaissance" (Diss., Technische Universität München, 2007).

6. Valentin Boltz, *Illuminier Buoch . . .* (Basel: Jacob Kündig, 1549), xvii–xviii, accessed June 11, 2021: http://doi.org/10.3931/e-rara-5578. See also the edition by C. J.

Benziger, which also contains a biography of Boltz: Valentin Boltz, *Illuminierbuch: Wie man allerlei Farben bereiten, mischen und auftragen soll*, ed. C. J. Benziger (Munich: Georg D. W. Callwey, 1913). I cite the original 1549 treatise.

7. Boltz, *Illuminier Buoch*, iii–v.

8. A hemp oil-mastic varnish from the *Liber illuministarum*, for instance, recurs literally in Boltz's treatise; compare *Der "Liber illuministarum" aus Kloster Tegernsee: Edition, Übersetzung und Kommentar der kunsttechnologischen Rezepte*, ed. and trans. Anna Bartl, Christoph Krekel, Manfred Lautenschlager, and Doris Oltrogge (Stuttgart: Franz Steiner Verlag, 2005), 264–65 [no. 438]; Boltz, *Illuminier Buoch*, xx–xxi.

9. On the German tradition of *Kunstbüchlein* and the history of books of secrets in the medieval and early modern period, see Eamon, *Science and the Secrets of Nature*, 112–33, and Doris Oltrogge, "Writing on Pigments in Natural History and Art Technology," in *Early Modern Color Worlds*, ed. Tawrin Baker, Sven Dupré, Sachiko Kusukawa, and Karin Leonhard (Leiden: Brill, 2015), 47–69.

10. *Strasbourg Manuscript*, ed. Neven, 132–34 [no. 92].

11. This is, in fact, a rather surprising result. It is well known that paint films darken, particularly when they are not exposed to light. Further research is required to find out why these oil-resin mixtures lightened while in the dark. One theory is that it was perhaps not so much the relative darkness as their age that made them lighter. This is also observed by Armenini, cited later in this chapter.

12. *Strasbourg Manuscript*, ed. Neven, 132–34 [no. 92].

13. Compare the varnish recipe in the treatise of Theophilus discussed in chapter 2.

14. *Strasbourg Manuscript*, ed. Neven, 132–34 [no. 92]. Be aware that hot varnish is very dangerous, and that the experiment I refer to here was executed with a small amount of varnish that was allowed to cool (as the recipe also prescribes) before using a glove to attempt to pull a thread.

15. *Strasbourg Manuscript*, ed. Neven, 134–35 [no. 93].

16. Boltz, *Illuminier Buoch*, xi.

17. These are the most relevant genera of Pinaceae in the current context. See John S. Mills and Raymond White, *The Organic Chemistry of Museum Objects* (Oxford: Butterworth-Heinemann, 1994), 87–89, for a more complete overview. Mills and White also suggest that the etymology of turpentine comes from yet another oleoresin exuded by *Pistacia terebinthus*, known as the terebinth tree; see page 95.

18. It may be significant that Leonhart Fuchs, in his 1543 *New Kreüterbuch*, explains that the *Lörchenbaum* is said to be called *Larix* in Latin and Greek, and that its resin is sold everywhere in the pharmacies and other places as *terbentin*, "a substance that they call *Gloret*." Fuchs continues that "without a doubt" the name *Gloret* is derived from the tree name *Lörch*. See Leonhart Fuchs, *New Kreüterbuch* (Basel: Michael Isingrin, 1543), n.p. [Cap. CLXXXIX]: *Dises verkaufft man allenthalben in den Apotheken / und sonst auch / für Terbentin / das sie Gloret nennen.*

19. Langenheim, *Plant Resins*, 306. In his *New Kreüterbuch* Fuchs also points out that larch resin never becomes thick (*wuürt nimmer dick*) after being exuded by the tree. Fuchs, *New Kreüterbuch*, n.p. [Cap. CLXXXIX].

20. Boltz, *Illuminier Buoch*, xi.

21. *Strasbourg Manuscript*, ed. Neven, 134–35 [no. 94].

22. In Eastlake's copy of the manuscript there was a lacuna in the first line of the recipe. Since this recipe recurs literally in Boltz von Ruffach's 1549 *Illuminierbuch*, it is likely that the missing ingredient is indeed the *bimstein* Eastlake proposed; compare Boltz, *Illuminier Buoch*, xx–xxi.

23. See Woudhuysen-Keller, "Leinöl als Malmittel," 76–80. Compare also the recipes for clarifying oil in eighteenth- and nineteenth-century painter's handbooks discussed by Leslie Carlyle in *The Artist's Assistant: Oil Painting Instruction Manuals and Handbooks in Britain, 1800–1900, with Reference to Selected Eighteenth-Century Sources* (London: Archetype, 2001), 31–34.

24. Boltz, *Illuminier Buoch*, xvii–xix.

25. See Charles S. Tumosa and Marion F. Mecklenburg, "The Influence of Lead Ions on the Drying of Oils," *Reviews in Conservation* 6 (2005): 39–47. This article uses modern theories of dryers to understand and classify the effects of lead compounds on the drying behavior and physical properties of oil paints.

26. See Woudhuysen-Keller, "Leinöl als Malmittel," 82–85. She suggests that the main drying effect of calcium in recipes that do not mention other dryers may have been due to the relatively high water content of untreated cold-pressed drying oils. She established that when oils are mixed with burned chalk (dissolved in water to turn it into calcium hydroxide), a clear drying effect could be established after heating the oil to 100°C.

27. Manfred Lautenschlager, "Kodikologische Angaben zur Handschrift," in *"Liber illuministarum,"* ed. Bartl et al., 49–54.

28. *"Liber illuministarum,"* ed. Bartl et al., 328–29 [no. 1081B].

29. *"Liber illuministarum,"* ed. Bartl et al., 284–85 [no. 683].

30. Linus Pauling, *General Chemistry* (1947; repr. from 1970 ed., Newburyport, MA: Dover, 2014), 634.

31. For the ability of alum to coagulate mucilaginous material in painter's oil, see Raymond White and Jo Kirby, "Rembrandt and His Circle: Seventeenth-Century Dutch Paint Media Re-examined," *National Gallery Technical Bulletin* 15 (1994): 68–69.

32. The *Liber illuministarum* contains one recipe that explains how to glue parchment ornaments to a varnished panel painting, but it does not provide any details about what kind of varnish it was or how it was applied; see *"Liber illuministarum,"* ed. Bartl et al., 180–81 [no. 247].

33. See, for instance, a varnish made with egg white, gum, and honey in the *Strasbourg Manuscript*, ed. Neven, 135–37 [97], and in *"Liber illuministarum,"* ed. Bartl et al., 230–31 [no. 345]. According to Boltz this varnish (he calls it *Hußfürnüß*) gives shine (*glantz*) to everything painted on parchment and paper; see Boltz, *Illuminier Buoch*, xiv–xvi.

34. *Strasbourg Manuscript*, ed. Neven, 104–5 [33] and 134–35 [no. 96].

35. *Strasbourg Manuscript*, ed. Neven, 104–5 [no. 33].

36. The mastic varnish described in the manuscript is made with linseed oil, hempseed oil, or nut oil. It is praised as the best varnish possible and very good and clear; see *Strasbourg Manuscript*, ed. Neven, 104–5 [no. 33].

37. *Strasbourg Manuscript*, ed. Neven, 104–5 [no. 33].

38. *Strasbourg Manuscript*, ed. Neven, 104–5 [no. 33].

39. In Boltz's treatise, one recipe for making an oil-mastic varnish explains that it is suitable for application to *perment oder läder*. Here *perment*, of course, refers to the parchment windows Boltz explains how to varnish at the end of his treatise. Note also that all Boltz's varnishes are mastic varnishes.

40. Hemp oil has not been demonstrated on easel paintings through scientific analysis and is rarely mentioned in recipe collections outside the Strasbourg tradition. Hempseed oil is mentioned in LeBègue's manuscript in a recipe for painter's varnish (ed. Merrifield, 1:313–15 [no. 341]), and the manuscript by the physician Theodore Turquet de Mayerne (1573–1655) includes several recipes for varnish made with hemp oil that likely derive from the Strasbourg tradition. De Mayerne mentions that hempseed oil is rather green, but that it is drying and therefore can be used for making varnish. More than two hundred years after the first Strasbourg recipes were recorded, De Mayerne might not have realized they did not deal with picture varnishes per se; see Theodore Turquet de Mayerne, *Pictoria, sculptoria, tinctoria et quae subalternarum atrium spectantia* (1620–46), London, British Library, Sloane MS 2052, fols. 44v–45r.

41. For a discussion of how these recipes fit within a larger cultural context concerned with relieving eyestrain by using emeralds and the color green, see Marjolijn Bol, "The Emerald and the Eye: On Sight and Light in the Artisan's Workshop and the Scholar's Study," in *Perspective as Practice: Renaissance Cultures of Optics*, ed. Sven Dupré (Turnhout: Brepols, 2019), 71–97; and see Marjolijn Bol, Henk de Groot, and Arie Wallert, "Glass and Parchment with a View: Oil Paint and the Imitation of (Stained) Glass Windows, 1400–1600," in *Making and Transforming Art: Changes in Artists' Materials and Practice*, ed. Hélène Dubois, Sigrid Eyb-Green, and Joyce H. Townsend (London: Archetype, 2014), 129–30.

42. Amberg, Staatlichen Provinzial Bibliothek, Codex 77 (Amberger Malerbüchlein), fol. 224r. See Emil Ernst Ploss, "Das Amberger Malerbüchlein: Zur Verwandtschaft der spätmhd. Farbrezepte," in *Festschrift für Hermann Hempel zum 70. Geburtstag* . . . , 3 vols. (Göttingen: Vandenhoeck und Ruprecht, 1972), 3:693–703. Compare with similar recipes in Leiden, Bibliotheek der Rijksuniversiteit, Voss. Chymicus octavo 6, fol. 23v; and Prague, Národní knihovna České republiky, Codex XI D 10 (Prager Malerbuch, ca. 1452), fol. 84v.

43. For a discussion of several of these recipes, see Bol, "Emerald and the Eye," 86–93.

44. Hugh Platt, *The Jewell House of Art and Nature Containing Divers Rare and Profitable Inventions, Together with Sundry New Experimentes in the Art of Husbandry, Distillation, and Moulding / Faithfully and Familiarly Set Downe, according to the Authors Owne Experience, by Hugh Platte* (London: Peter Short, 1594), 76–77.

45. *Artliche kunste—mancherley weyse Dinten und allerhand farben zubereyten* (Mainz: Peter Jordan, 1531), xxii.

46. The recipe that describes how to make colored, see-through parchment is identical to that in the *Artliche Künste*. Compare *Schreyberey: Mancherley weyse Dinten und allerhand farben zubereyten* (Mainz: Peter Jordan, 1532), xxii, and *Allerhand Farben/und mancherlay weyse/Dünten zůbereyten* (Augsburg: Heynrich Steyner, 1533), xix.

47. On the interpretation of variations of the term *glas* in German varnish recipes, see Johannes Alexander van de Graaf, "The Interpretation of Old Painting Recipes,"

Burlington Magazine 104, no. 716 (1962): 473–75. Van de Graaf considers all later recipes that mention adding Venetian glass to paints as a dryer or optical filler a misinterpretation of earlier varnish recipes that use it to refer to resin. Marika Spring, however, has demonstrated through the scientific analysis of paint samples that crushed glass was added to paints as early as the fifteenth century. She argues that it is unlikely that using glass in paints should be entirely attributed to a misunderstanding of earlier recipes; see Marika Spring, "Colourless Powdered Glass as an Additive in Fifteenth- and Sixteenth-Century European Paintings," *National Gallery Technical Bulletin* 33 (2012): 4–26.

48. The earliest reference to this ingredient is in a varnish recipe in the *Liber illuministarum*. Here the ratio of oil to resin is three to two (common for varnishes); see *"Liber illuministarum,"* ed. Bartl et al., 212–13 [no. 316]: *Item wiltu gutten fürnes machen. so nÿm venedigisch glas stos das klain ze buluer*. Bartl et al. translate it to mean "Venetian glass"; see 593.

49. Much has been published about the use of lead compounds as dryers in painter's oil. A good overview of this research can be found in Charles S. Tumosa and Marion F. Mecklenburg, "The Influence of Lead Ions on the Drying of Oils," *Reviews in Conservation* 6 (2005): 39–47.

50. *Segreti d'arti diverse nel regno di Napoli: Il manoscritto It. III.10 della Biblioteca Marciana di Venezia*, ed. Fabio Frezzato and Claudio Seccaroni (Saonara: Prato, 2010), 20–22; Merrifield, *Original Treatises*, 2:606. The numbering of the recipes is the same in both editions, but the 2010 edition by Frezzato and Seccaroni also includes recipes (many of them medical and cosmetic) that Merrifield chose not to transcribe.

51. Paris, Bibliothèque nationale de France, An., MS Fr. 640. The manuscript has recently been published in an online digital critical edition; see *Secrets of Craft and Nature in Renaissance France: A Digital Critical Edition and English Translation of BnF MS Fr. 640*, ed. and trans. by the Making and Knowing Project, Pamela H. Smith, Naomi Rosenkranz, Tianna Helena Uchacz, Tillmann Taape, Clément Godbarge, Sophie Pitman, Jenny Boulboullé, Joel Klein, Donna Bilak, Marc Smith, and Terry Catapano (New York: Making and Knowing Project, 2020), accessed February 15, 2022: https://doi.org/10.7916/78yt-2v41.

52. Armenini, *De' veri precetti della pittura*, ed. Olszewski, 70–78.

53. MS It. III.10, ed. Frezzato and Seccaroni, 173 [no. 402]; ed. Merrifield, 2:632–35.

54. MS It. III.10, ed. Frezzato and Seccaroni, 172–73 [no. 399]; ed. Merrifield, 2:630–33.

55. BnF. MS Fr. 640, ed. Making and Knowing Project, fol. 3v.

56. MS It. III.10, ed. Frezzato and Seccaroni, 173 [no. 402]; ed. Merrifield, 2:632–35.

57. Armenini, *De' veri precetti della pittura*, 129; ed. Olszewski, 195–96.

58. MS. It. III.10, ed. Frezzato and Seccaroni, 172–73 [no. 399]; ed. Merrifield, 2:630–33.

59. MS It. III.10, ed. Frezzato and Seccaroni, 172 [no. 398]; ed. Merrifield, 2:630–31.

60. This is also where the manuscript containing Sansovino's varnish recipe is kept today (coincidentally, because the codex has been part of the Marciana Library's collections only since the eighteenth century). Jacopo Sansovino is also mentioned in a stucco recipe in the same manuscript [no. 393].

61. MS It. III.10, ed. Frezzato and Seccaroni, 156 [no. 339]; ed. Merrifield, 2:620–21.

The mordant is made by grinding roasted mastic, *coperosa* (sulfate of zinc, a dryer in oil), varnish in grains (sandarac), and burned roche alum with purified linseed oil to the consistency of ink.

62. MS It. III.10, ed. Frezzato and Seccaroni, 172 [no. 398]; ed. Merrifield, 2:630–31.

63. Merrifield translates this as "dew," which is indeed an alternative translation of *al sereno*. In this meaning, the recipe would explain a type of water washing of the oil, similar to cleaning it by boiling it with water.

64. MS It. III.10, ed. Frezzato and Seccaroni, 172 [no. 398]; ed. Merrifield, 2:630–31.

65. "*Mappae clavicula*" [nos. 116, 208 refer to the same recipe included twice in the manuscript], ed. Phillipps, 212, 227; ed. Smith and Hawthorne, 44, 59.

66. William Eamon, "Arcana Disclosed: The Advent of Printing, the Books of Secrets Tradition and the Development of Experimental Science in the Sixteenth Century," *History of Science* 22, no. 2 (1984): 128.

67. Eamon, "Arcana Disclosed," 129.

68. For the English translation of the introduction to Ruscelli's *Secreti nuovi*, see William Eamon and Françoise Paheau, "The Accademia Segreta of Girolamo Ruscelli: A Sixteenth-Century Italian Scientific Society," *Isis* 75, no. 2 (1984): 339–40. Girolamo Ruscelli, *Secreti nuovi di maravigliosa virtù* (Venice: Gli Heredi Marchiò Sessa, 1567), fols. 3v–4r.

69. According to Ernesto Terzi, the author of an 1871 work on *prodotti delle conifere*, this material refers to the turpentine oleoresin dripping from pines that has hardened during the transition from summer to autumn. We learn that *ragia di pino*, when freed of impurities, turns into a bright, hard whitish transparent mass that easily softens when exposed to heat. Its only difference from turpentine is "the small quantity of essential oil that it contains" (*si distingue dalla trementina solo per la poca quantita d'olio etereo che contiene*). See Ernesto Terzi, *I prodotti delle conifere memoria di Ernesto Terzi*, ed. C. Molinari (Milan: Galleria Vittorio Emanuele, 1871), 28.

70. *De' secreti del reverendo donno Alessio Piemontese, prima parte, divisa in sei libri* (Milan: Valerio, e Hieronymo fratelli da Meda, 1557), fol. 61v. The English translations are from a 1595 English edition that contains all four books for the first time (the first book was published in English in 1558, translated from the 1557 French edition by Plantin Moretus): *The Secrets of the Reverend Maister Alexis of Piemont . . . : Translated out of French into English by William Ward* (London: Peter Short for Thomas Wight, 1595), 172v.

71. *De' secreti del reverendo donno Alessio Piemontese*, 62v; *Secrets of Alexis of Piemont*, ed. William Ward, 171v.

72. Woudhuysen-Keller, "Leinöl als Malmittel," 75–77. Woudhuysen-Keller observes that cooking oil with water may work better to separate the mucilaginous matter from the oil.

73. BnF. MS Fr. 640, ed. Making and Knowing Project, fol. 97v. Compare also with a recipe on folio 60v for making mastic varnish that "dries in an hour."

74. In another recipe, the author of MS Fr. 640 explains something similar when he writes that some make varnish by mixing mastic with oil of turpentine, which after heating has to be "put in another vessel to purge it of its dregs," see BnF. MS Fr. 640, ed. Making and Knowing Project, fol. 99v.

75. BnF. MS Fr. 640, ed. Making and Knowing Project, fol. 97v.

76. A note on folio 3r suggests that *la colophoine* is *rousine* "cooked again" (*recuite*) to make it liquid and to purify it (sometimes also called *resine* in the same context). The author of MS Fr. 640 does not specify what kind of resin is purified in this process, but that the resin is said to be "reboiled" could mean that it refers to the hard residue left after distilling one of the turpentine oleoresins. MS Fr. 640 explains on folio 3v that this resin is known as *la colophoine*, or *pix græca*.

77. BnF. MS Fr. 640, ed. Making and Knowing Project, fol. 3r. Compare also fol. 31r and fol. 97v.

78. Armenini, *De' veri precetti della pittura*, 128–29; ed. Olszewiski, 195–96.

79. Robert James Forbes, *Studies in Early Petroleum History*, 2 vols. (Leiden: Brill, 1958), 1:85–86.

80. Mills and White, *Organic Chemistry*, 7.

81. Armenini, *De' veri precetti della pittura*, 128–29; ed. Olszewiski, 195–96.

82. Armenini, *De' veri precetti della pittura*, 129; ed. Olszewiski, 195.

83. Cennini, *Il libro dell'arte*, ed. Broecke, 200–201 [no. 169].

84. BnF. MS Fr. 640, ed. Making and Knowing Project, fol. 3v.

85. Armenini, *De' veri precetti della pittura*, 128–29; ed. Olszewiski, 195–96.

86. Armenini, *De' veri precetti della pittura*, 128–29; ed. Olszewiski, 195–96.

87. BnF. MS Fr. 640, ed. Making and Knowing Project, f. 97v.

88. Cennini, *Il libro dell'arte*, ed. Broecke, 199–200 [no. 168].

89. Albrecht Dürer, *Dürer: Schriftlicher Nachlaß*, ed. Hans Rupprich, 3 vols. (Berlin: Deutscher Verlag für Kunstwissenschaft, 1956–69), 1:72–73 [no. 19].

90. BnF. MS Fr. 640, ed. Making and Knowing Project, fol. 3r.

91. BnF. MS Fr. 640, ed. Making and Knowing Project, fol. 66r. Compare with Mayerne, *Pictoria, sculptoria, tinctoria*, fol. 9v: *Les mort des colours est quand l'huyle nageante au dessus, se seiche & faict une peau, qui noircit a l[']air. Il y a quelques couleurs, & les esmaulx entre aultres qui ne se meslent pas aisement avec l'huyle, ains[i] vont tousjours à fonds se lier, et ainsi meurent facilement, i.e. noircissent.* For the transcription, see *Beiträge zur Entwicklungs-Geschichte der Maltechnik. Folge 4. Quellen für Maltechnik während der Renaissance und deren Folgezeit (16.–18. Jahrhundert) in Italien, Spanien, den Niederlanden, Deutschland, Frankreich und England nebst dem de Mayerne Manuskript*, ed. and trans. Ernst Berger (Munich: G. D. W. Callwey, 1901), 112–15 [no. 11].

92. BnF. MS Fr. 640, ed. Making and Knowing Project, fol. 3r.

93. BnF. MS Fr. 640, ed. Making and Knowing Project, fol. 3r.

94. BnF. MS Fr. 640, ed. Making and Knowing Project, fol. 4r.

95. BnF. MS Fr. 640, ed. Making and Knowing Project, fol. 4r.

96. BnF. MS Fr. 640, ed. Making and Knowing Project, fols. 3r–3v.

97. Because many of the earlier recipe sources are German and very little evidence survives from Southern Europe, it is of course entirely possible that walnut oil, which was more commonly available in the South, played a prominent role in varnishing and glazing at an earlier date there as well. This is hard to prove, however, because of the lack of early recipes from this part of Europe. But even if this was the case, the sources that do survive suggest that specifically avoiding linseed oil because it tended

to produce a yellower coating (and thus replacing it with walnut oil), began to concern painters only in the fifteenth and sixteenth centuries.

Chapter Four

1. The English translation cited here is from Erwin Panofsky, *Abbot Suger on the Abbey Church of St.-Denis and Its Art Treasures* (Princeton, NJ: Princeton University Press, 1979), 23–24. For the entire poem in a recent critical edition, see Abbot Suger von Saint-Denis, *Ausgewählte Schriften: Ordinatio, De consecratione, De administratione*, ed. and trans. Andreas Speer and Günther Binding (2000; repr., Darmstad: Wissenschaftliche Buchgesellschaft, 2008), 325 [*De administratione*, 174], and compare also 345 [*De administratione*, 224].

2. Suger, *Ausgewählte Schriften*, ed. Speer and Binding, 345 [*De administratione*, 224]. The beginning of this passage is based on Ezekiel 28:13; ed. Panofsky, 62–65.

3. Precious stones are also mentioned in Exodus 28:15–31, Exodus 39:10–21, Ezekiel 28:13, and elsewhere.

4. See Gage, *Colour and Culture*, esp. chap. 4, "A Dionysian Esthetic."

5. Marjolijn Bol, "Coloring Topaz, Crystal and Moonstone: Gems and the Imitation of Art and Nature, 300–1500," in *Fakes!?: Hoaxes, Counterfeits and Deception in Early Modern Science*, ed. Marco Beretta and Maria Conforti (Sagamore Beach, MA: Science History Publications, 2014), 108–12.

6. Marjolijn Bol, "*Polito et Claro*: The Art and Knowledge of Polishing, 1200–1500," in *Gems in the Early Modern World: Materials, Knowledge and Global Trade, 1450–1800*, ed. Sven Dupré and Michael Bycroft (Cham: Palgrave Macmillan, 2018), 223–57.

7. See also the introduction to this book, where I explain the specular reflection as well.

8. For the influence of Pliny during the medieval period, see Arno Borst, *Das Buch der Naturgeschichte: Plinius und seine Leser im Zeitalter des Pergaments* (1994; repr., Heidelberg: Universitätsverlag C. Winter, 1995), and Marjorie Chibnall, "Pliny's Natural History and the Middle Ages," in *Empire and Aftermath: Silver Latin II*, ed. Thomas A. Dorey (London: Routledge and Kegan Paul, 1975), 57–78.

9. John F. Healy, *Pliny the Elder on Science and Technology* (Oxford: Oxford University Press, 1999), 83–86.

10. It is not known whether Isidore cites Pliny directly or through intermediate sources; see Isidore of Seville, *Etymologies*, ed. Barney et al., 14.

11. For the Latin text, see *Isidori Hispalensis episcopi etymologiarum sive originum libri XX*, ed. Wallace Martin Lindsay, Scriptorum classicorum bibliotheca Oxoniensis, 2 vols., vol. 1 (Oxford: Oxford University Press, 1911), [I.xxix], and for the English translation see Isidore of Seville, *Etymologies*, ed. Barney et al., 55 [1.29.1].

12. This number is mentioned in Isidore of Seville, *Etymologies*, ed. Barney et al., 24.

13. Isidore of Seville, *Etymologiarum* [16.15.28], ed. Lindsay, 2:[n.p.]; ed. Barney et al., 328.

14. Isidore of Seville, *Etymologiarum* [16.6.1], ed. Lindsay, 2:[n.p.]; ed. Barney et al., 322.

15. For a modern interpretation of the etymology of *gemma*, see Michiel de Vaan,

Etymological Dictionary of Latin and the Other Italic Languages (Leiden: Brill, 2008), 257. De Vaan argues that *gemma* originally meant bud or eye (in trees).

16. Isidore of Seville, *Etymologiarum* [16.6.1], ed. Lindsay, 2:[n.p.]; ed. Barney et al., 322.

17. Translation slightly adapted from Barney et al.; see Isidore of Seville, *Etymologiarum* [3.31.1], ed. Lindsay, 2:[n.p.]; ed. Barney et al., 100.

18. Translation slightly adapted from Barney et al.; see Isidore of Seville, *Etymologiarum* [16.25.1], ed. Lindsay, 2.[n.p.]; ed. Barney et al., 328.

19. Marjolijn Bol, "Gems in the Water of Paradise: The Iconography and Reception of Heavenly Stones in the Ghent Altarpiece," in *Van Eyck Studies: Papers Presented at the Eighteenth Symposium for the Study of Underdrawing and Technology in Painting, Brussels, 19–21 September 2012*, ed. Christina Currie and Bart Fransen (Leuven: Peeters, 2017), 34–48.

20. For the Latin text, see *Augustinus: De Genesi contra Manichaeos*, ed. Dorothea Weber, vol. 91 (Vienna: Austrian Academy of Sciences, 1998), 116 [2.1.1]; and for the English translation see Augustine, *On Genesis: A Refutation of the Manichees; Unfinished Literal Commentary on Genesis; The Literal Meaning of Genesis*, ed. and trans. Edmund Hill, vol. 13 (New York: New City Press, 2002), part 1, 69 [2.1.1].

21. For an extensive work on the sources for the allegorical meanings of gems, see Christel Meier, *Gemma spiritalis: Methode und Gebrauch der Edelsteinallegorese vom frühen Christentum bis ins 18.* (Munich: Wilhelm Fink Verlag, 1977), 1:62.

22. *Augustinus: De Genesi contra Manichaeos* [2.10.13–14], ed. Weber, 135; Augustine, *On Genesis*, ed. Hill, 80–81.

23. See Jean Delumeau, *History of Paradise: The Garden of Eden in Myth and Tradition*, trans. Matthew O'Connell (1992; repr., Urbana: University of Illinois Press, 2000), 18–19; Alessandro Scafi, *Mapping Paradise: A History of Heaven on Earth* (London: British Library, 2006), 61.

24. An ancient precedent for the idea that gems were carried by paradisal rivers can be found in Pliny the Elder: "The rivers that produce gems are the Chenab and the Ganges, and of all the lands that produce them India is most prolific." The Ganges and the Pishon were considered to be the same river. *Natural History, Volume 10: Books 36–37*, ed. and trans. D. E. Eichholz, Loeb Classical Library 419 (Cambridge, MA: Harvard University Press, 1962), 326–27 [37.76.200–201].

25. See Delumeau, *History of Paradise*, 18–19; Scafi, *Mapping Paradise*, 62.

26. Isidore of Seville, *Etymologiarum* [13.21.9–10], ed. Lindsay, 2:[n.p.]; ed. Barney et al., 280–81.

27. Bol, "Gems in the Water of Paradise," 34–48.

28. Richard Jones, *The Medieval Natural World* (London: Routledge, 2013), 12–18.

29. All citations in this section are from the preface to Hildegard's book on stones. For the Latin text, see Hildegard von Bingen, *Physica: Liber subtilitatum diversarum naturarum creaturarum*, ed. Reiner Hildebrandt and Thomas Gloning (Berlin: De Gruyter, 2010), 2:228–29 [IV *Prefatio*]; for the English translation, see *Hildegard von Bingen's "Physica": The Complete English Translation of Her Classic Work on Health and Healing*, ed. and trans. Priscilla Throop (Rochester, VT: Healing Arts Press, 1998), 137–38.

30. Cantimpré lists Saint Augustine, Pliny, Solinus (third-century Latin author of

De mirabilis mundi), Isidore, and Jacques de Vitry (French theologian ca. 1160/70–1240) as his authorities, see Thomas of Cantimpré, *Thomas Cantimpratentsis "Liber de natura rerum": Editio princeps secundum codices manuscriptos*, ed. Helmut Boese (Berlin: Walter de Gruyter, 1973), 351.

31. Thomas of Cantimpré, *Liber de natura rerum*, ed. Boese, 351.

32. Thomas of Cantimpré, *Liber de natura rerum*, ed. Boese, 355.

33. Thomas of Cantimpré, *Liber de natura rerum*, ed. Boese, 355.

34. For the ways Aristotelian thought shaped Albertus Magnus's mineralogical theories, see Albertus Magnus, *Book of Minerals*, ed. and trans. Dorothy Wyckoff (Oxford: Oxford University Press, 1967), especially xxx–xxxv; for the Latin text I refer to Albertus Magnus, *De mineralibus* [*Mineralium libri quinque*], ed. Auguste Borgnet, *Opera omnia Alberti Magni*, vol. 5 (Paris: Apud Ludovicum Vives, 1890).

35. Albertus Magnus, *De mineralibus* [1.1.1], ed. Borgnet 5:4; ed. Wyckoff, 10.

36. Albertus Magnus, *Book of Minerals*, ed. Wyckoff, xxx–xxxi.

37. Albertus Magnus, *De mineralibus* [1.1.1], ed. Borgnet 5:4; ed. Wyckoff, 10.

38. Albertus Magnus, *De mineralibus* [1.1.2], ed. Borgnet 5:4; ed. Wyckoff, 14.

39. Albertus Magnus, *De mineralibus* [1.2.3], ed. Borgnet 5:4; ed. Wyckoff, 44–45.

40. Albertus Magnus uses *translucidus* three times to describe green stones and once when he writes about a very clear type of amber; see *De mineralibus*, ed. Borgnet, 5:39, 41, 43, and 46.

41. Sometime between 1240 and 1260, and before he wrote his book on minerals, Albert had written a commentary on *De sensu*; see Cemil Akdogan, "Optics in Albert the Great's *De sensu et sensato*: An Edition, English Translation, and Analysis" (PhD diss., University of Wisconsin–Madison, 1978). See also Wyckoff, who dates *De mineralibus* after Albert's *De sensu*; Albertus Magnus, *Book of Minerals*, ed. Wyckoff, xxxv–xli.

42. David C. Lindberg, *Theories of Vision from Al-Kindi to Kepler* (Chicago: University of Chicago Press, 1976), 6–9, and see pages 104–7 for a discussion of Albertus Magnus and the Aristotelian tradition in the thirteenth century.

43. Albertus Magnus, *De mineralibus* [1.1.2], ed. Borgnet, 5:3; ed. Wyckoff, 12.

44. Albert's theory of *congelatio* and *conglutinatio* as formative processes for precious stones was based on Avicenna's (*Abu Ali ibn Sina*, 980–1037) *De congelatione et conglutinatione lapidum*; see Albertus Magnus, *Book of Minerals*, ed. Wyckoff, 14, 283–85.

45. Compare also Aristotle, *Physics* [2.8.199a15–17].

46. Albertus Magnus, *De mineralibus* [1.1.3], ed. Borgnet 5:4; ed. Wyckoff, 14–17.

47. Albertus Magnus, *De mineralibus* [1.1.3], ed. Borgnet 5:4; ed. Wyckoff, 14–17.

48. Albertus Magnus, *De mineralibus* [1.2.2], ed. Borgnet, 5:15; ed. Wyckoff, 39.

49. Albertus Magnus, *De mineralibus* [1.2.2], ed. Borgnet 5:16; ed. Wyckoff, 40.

50. Albertus Magnus, *De meteoris, lib. IV* [2.1.7], ed. Auguste Borgnet, *Opera omnia Alberti Magni*, vol. 4 (1890), 534.

51. Albertus Magnus, *De mineralibus* [1.2.2], ed. Borgnet, 5:15–17; ed. Wyckoff, 38–43. The idea that transparent minerals such as rock crystal were petrified ice can be found with many ancient writers; see, e.g., Pliny, *Natural History* [36.45.161], ed. Eichholz, 10:128–29.

52. Albertus Magnus, *De mineralibus* [1.2.2], ed. Borgnet, 5:17; ed. Wyckoff, 38–43.

53. R. J. Charleston, "Lead in Glass," *Archeometry* 3, no. 1 (1960): 1–4.

54. Albertus Magnus, *De mineralibus* [1.3.2], ed. Borgnet, 5:50; ed. Wyckoff, 132–33.

55. Dodwell does not think these rings were worn on the finger, but Smith and Hawthorne argue that the diameter of the ring and the fact that a simulated gem was added suggest otherwise. *"On Divers Arts,"* ed. Smith-Hawthorne, 73–74 [no. 31].

56. Eraclius, *De coloribus*, ed. Merrifield, 1:216 [no. 271].

57. Eraclius, *De coloribus*, ed. Merrifield, 1:208–12 [no. 255]. This passage is also interesting because Eraclius refers to Isidore of Seville when he writes *"vitrum dictum, ut ait Ysidorus, quod visui perspecuitate transluceat"* (Glass is so called, as Isidorus says, because with its translucency it transmits light to one's sight.) He subsequently brings up the Plinian myths of the invention of glass and unbreakable glass, both of which can also be found in Isidore's *Etymologies* [16.16.1–6].

58. See Bol, "Coloring Topaz, Crystal and Moonstone," 108–29; Bol, "Emerald and the Eye," 71–97.

59. Seneca the Younger, *Seneca: Epistles 1–65*, ed. and trans. Richard M. Gummere, Loeb Classical Library 75, 3 vols. (Cambridge, MA: Harvard University Press, 1917), 1:421 [90.33].

60. Pliny, *Natural History* [37.20.79], ed. Eichholz, 10:227.

61. Pliny, *Natural History* [37.75.196–98], ed. Eichholz, 10:325.

62. Isidore of Seville, *Etymologiarum* [16.15.27], ed. Lindsay, 2:[n.p.]; ed. Barney et al., 328.

63. Pliny, *Natural History* [37.20.79 and 37.75.196–98], ed. Eichholz, 10:226–27 and 10:324–27.

64. The Stockholm Papyrus was discovered with another papyrus—Papyrus Leiden X—written in the same hand, which today is kept at Leiden University. For the transcription and translation of these papyri, see *Les alchimistes grecs: Papyrus de Leyde–Papyrus de Stockholm–Recettes*, ed. and trans. Robert Halleux, Collection des Universités de France, vol. 1 (Paris: Belles Lettres, 1981). For an English translation of the two papyri, see Earle Radcliffe Caley, "The Leyden Papyrus X: An English Translation with Brief Notes," *Journal of Chemical Education* 3 (1926): 1149–66, and Earle Radcliffe Caley, "The Stockholm Papyrus: An English Translation with Brief Notes," *Journal of Chemical Education* 4 (1927): 979–1002. Caley based his translation on Otto Lagercrantz's German translation, *Papyrus graecus holmiensis (P. holm.): Recepte für Silber, Steine und Purpur*, ed. and trans. Otto Lagercrantz, Arbeten utgifna med understöd af Vilhelm Ekmans Universitetsfond, Uppsala, vol. 13 (Uppsala, 1913).

65. Berthelot first pointed out that the papyri were preserved in the mummy case of an Egyptian chemist, and Lagercrantz argues that the papyri may have been a luxury copy made for the entombment (internal evidence shows the manuscripts were copied from another source). This would explain why the papyri were preserved after Diocletan's AD 296 decree banning all treatises dealing with alchemy. See Marcellin Berthelot, *Introduction à l'étude de la chimie des anciens et du Moyen Âge* (Paris: G. Steinheil, 1889), 5; and *Papyrus graecus holmiensis*, ed. Lagercrantz, 13:55. For the transmission of these Greek alchemical texts into various languages and cultures, see also Matteo Martelli, "Translating Ancient Alchemy: Fragments of Graeco-Egyptian Alchemy in Arabic Compendia," *Ambix* 64, no. 4 (2017): 326–42.

66. Compare with a Greek collection of recipes for the imitation of precious stones, which contains recipes similar to those found in the Stockholm Papyrus. This Byzantine recipe treatise, titled *Deep Tincture of Stones, Emeralds, Rubies and Jacinths from the Book Taken Out from the Sancta Sanctorum of Temples*, likely echoes the work of the Greek atomist Democritus. For the Greek text with a French translation, see *Collection des anciens alchimistes Grecs, texte et traduction*, ed. and trans. Marcellin Berthelot and Charles Émile-Ruelle, 3 vols. (Paris: Georges Steinheil, 1888), 3:333–49; and for the reconstruction of the original four books of Democritus, see Matteo Martelli, *The Four Books of Pseudo-Democritus* (Leeds: Maney, 2013).

67. See also Bol, "Coloring Topaz, Crystal and Moonstone," 108–29; Bol, "Emerald and the Eye," 71–97.

68. Caley, "Stockholm Papyrus," XX [no. 74]; Halleux, *Alchimistes grecs* [no. 74]. The production of verdigris was also ancient, and similar methods were described by Theophrastus and Pliny.

69. In another recipe in the same papyrus that explains how to dye textiles we learn that "by celandine one means a plant root. It dyes [a] gold color by cold dyeing. Celandine is costly, however. You should accordingly use the root of the pomegranate tree, and it will act the same. And if wolf's milk [possibly *Euphorbia virgata* Waldst.& Kit., also known as "leafy spurge"] is boiled and dried it produces yellow. If, however, a little verdigris is mixed with it, it produces green; and safflower blossom likewise." "Stockholm Papyrus," ed. Caley, 997 [no. 139].

70. This may be mastic, which is known to have been used since antiquity as a type of chewing gum for its beneficial effect on the teeth; see, e.g., Pliny, *Natural History* [24.74.121–22], ed. and trans. Jones and Andrews, 7:88–89. Halleux, however, points out that the passage can also be understood as resin flowing directly from a cut in the tree; see Halleux, *Alchimistes grecs* [no. 76], and p. 196 [no. 5].

71. Caley, "Stockholm Papyrus," 989–90 [no. 76]; Halleux, *Alchimistes grecs*, 130 [no. 76].

72. The alum is likely dissolved in vinegar, as other recipes in the same papyrus suggest.

73. Robert C. Kammerling et al., "Fracture Filling of Emeralds: Opticon and Traditional Oils," *Gems and Gemology* 27, no. 2 (1991): 70–85. The authors experimented with dyeing quench-cracked quartz with the synthetic resin Opticon to demonstrate its efficiency in the practice of emerald oiling. They observe that if a suspension rather than a solution was used to color the crystal, the color was filtered out of the Opticon at the surface of the fractures. I observed something similar. Experiments using oil ground with copper green were much less effective at producing color in quench-cracked crystals than immersing them in copper green dissolved in resin.

74. See also chapter 5 below.

75. Note that indigoids (indigo, Tyrian purple) are "vat dyes." They are the only natural dye not directly soluble in water. Before they can be used in dyeing, they have to go through fermentation so that the insoluble dye becomes soluble. See, e.g., Jo Kirby, Maarten van Bommel, and André Verhecken, *Natural Colorants for Dyeing and Lake Pigments: Practical Recipes and Their Historical Sources* (London: Archetype, 2014), 26–28.

76. See Kirby, Bommel, and Verhecken, *Natural Colorants*, 25–26.

77. Albertus Magnus, *De mineralibus* [2.2.7], ed. Borgnet, 5:45–46; ed. Wyckoff, 119.

78. Al-Beruni, *Al-Beruni's Book on Mineralogy: The Book Most Comprehensive in Knowledge on Precious Stones*, ed. and trans. Hakim Mohammad Said, One Hundred Great Books of Islamic Civilization (Islamabad: Pakistan Hijra Council, 1989), 66, 141.

79. Arab writers on stones appear to have developed slightly different ideas about the origin of emeralds and malachite. Compare the entries on "emerald" and "malachite" in an Arab lapidary attributed to Aristotle. Here emeralds are said to be found in gold mines, whereas malachite is found in copper mines. Like translucent emeralds of the beryl variety, malachite is said to grow from the green tarnish on copper, i.e., verdigris; see Pseudo-Aristotle, *Das Steinbuch des Pseudo-Aristotle nach der arabische Handschrift*, ed. and trans. Julius Ruska (Heidelberg: Carl Winter's Universitätsbuchhandlung, 1912), 134 and 146.

80. Vannoccio Biringuccio, *De la "Pirotechnia" Libri X . . . Composti Per il Vannoccio Biringuccio Sennese* (Venice: Venturino Roffinello, 1540), 41; for an English translation, see Vannoccio Biringuccio, *The "Pirotechnia" of Vannoccio Biringuccio: The Classic Sixteenth-Century Treatise on Metals and Metallurgy*, ed. and trans. Cyril Stanley Smith and Martha Teach Gnudi (1942; repr., Cambridge, MA: MIT Press, 1959), 12.

81. See Birgit Arrhenius, "Garnet Jewelry of the Fifth and Sixth Centuries," in *From Attila to Charlemagne: Arts of the Early Medieval Period in the Metropolitan Museum of Art*, ed. Katharine Reynolds Brown, Dafydd Kidd, and Charles T. Little (New York: Metropolitan Museum of Art, 2000), 214–15.

82. Noël Adams, "The Garnet Millennium: The Role of Seal Stones in Garnet Studies," in *"Gems of Heaven": Recent Research on Engraved Gemstones in Late Antiquity, AD 200–600*, ed. Christopher Entwistle and Noël Adams (London: British Museum, 2011), 10–24.

83. Noël Adams, "Rethinking the Sutton Hoo Shoulder Clasps and Armour," in *Intelligible Beauty: Recent Research on Byzantine Jewellery* (London: British Museum Press, 2010), 83–112.

84. Herbert Maryon, *Metalwork and Enamelling: A Practical Treatise on Gold and Silversmiths*, 5th rev. ed. (1912; repr., London: Dover, 1971), 166–99.

85. David Buckton, "Enamelling on Gold: A Historical Perspective," *Gold Bulletin* 15, no. 3 (1982): 102.

86. David Buckton, "The Oppenheim or Fieschi-Morgan Reliquary in New York, and the Antecedents of Middle Byzantine Enamel," *Byzantine Studies Conference Abstracts* 8 (1982): 35–36.

87. Theophilus, ed. Dodwell, 98–109 [nos. L-LVII], and for the description of how to make the enamel, see 105–7 [no. LIIII, *De electro*].

88. According to Buckton it was more common to bend the edges of the base plate, since this would hold up better in the furnace, "Enamelling on Gold," 103.

89. Theophilus, ed. Dodwell, 104–5 [no. LIII, *De imponendis gemmis et margaritis*].

90. Theophilus, ed. Dodwell, 44–45 [no. XII, *De diversis vitri coloribus non translucidus*].

91. Buckton, "Enamelling on Gold," 105.

92. Theophilus, ed. Dodwell, 105–7 [no. LIIII, *De electro*].

93. Bol, "*Polito et Claro*," 232–37.

94. Theophilus, ed. Dodwell, 105–7 [no. LV, *De poliendo electro*]; compare [no. XCV, *De poliendis gemmis*].

95. Theophilus, ed. Dodwell, 25 [no. XXVII].

96. Rudolf Distelberger and Manfred Leithe-Jasper, *Kunsthistorisches Museum, Vienna: The Imperial und Ecclesiastical Treasury* (Munich: C. H. Beck / Scala, 2009), 49.

97. Buckton, "Enamelling on Gold," 102.

98. Neil Stratford, *Catalogue of Medieval Enamels in the British Museum*, vol. 2, *Northern Romanesque Enamel* (London: British Museum Press, 1993), [no. 3]. These plaques share some similarities with Mosan illuminated manuscripts.

99. Stratford, *Catalogue of Medieval Enamels*, vol. 2, [no. 3]. A similar technique is also found on the famous twelfth-century Stavelot Triptych (Pierpoint Morgan Library, New York); see Wilhelm Voelkle and Charles Ryskamp, *The Stavelot Triptych: Mosan Art and the Legend of the True Cross* (London: Pierpont Morgan Library, 1980).

Chapter Five

1. See also the introduction for a detailed explanation of this definition.

2. For an overview of natural film-forming materials used in the past, see Liliane Masschelein-Kleiner, *Ancient Binding Media, Varnishes and Adhesives* (Rome: ICCROM, 1995), especially chap. 2.

3. Masschelein-Kleiner, *Ancient Binding Media*, 15–17.

4. Theophilus, *Schedula*, ed. Dodwell, 18–19 [no. 20].

5. Water needs to be reintroduced because it is difficult to press oil from seeds that are too dry. Compare also the various recipes for pressing linseed and walnut oil from the late sixteenth-century Marciana manuscript, which use a method similar to that of Theophilus. As with Aëtius, these instructions can be found grouped with medicinal and skin-care recipes; see MS It. III.10, ed. Frezzato and Seccaroni, 87–93 (especially nos. 119, 120).

6. Another option is that the oil is darker because the linseed was heated over fire; high temperatures turn the oil browner; see also figure 3.3.

7. Theophilus, *Schedula*, ed. Dodwell, 24 [XXV]. The recipe on preparing varnish discussed in chapter 4 shows that Theophilus must at least have been aware of some of the methods for creating a heat-bodied oil.

8. On clearing oils used to make varnishes, see chapters 2 and 3 above.

9. The process of polymerizing in a drying oil is extremely complex and may be further influenced by the presence of light, certain metals (i.e., from the pigments or added dryers), film thickness (relative oxygen intake), and the extent to which an oil has been "polymerized" or "bodied" before use. See Mills and White, *Organic Chemistry of Museum Objects*, 30.

10. When these oils are removed from the light, they eventually regain their yellow or brown hue. Science still does not perfectly understand how this yellowing and bleaching oil works. See Jacky Mallégol, Jacques Lemaire, and Jean-Luc Gardette, "Yellowing of Oil-Based Paints," *Studies in Conservation* 46, no. 2 (2001): 121–31; Jacky Mallégol, Jean-Luc Gardette, and Jacques Lemaire, "Long-Term Behavior of Oil-Based

Varnishes and Paints: Photo- and Thermooxidation of Cured Linseed Oil," *Journal of the American Oil Chemists' Society* 77, no. 3 (2000): 257–63; and Jones et al., *Organic Coatings*, 208.

11. On the increase of the refractive index of oils during autoxidative drying, see, e.g., De la Rie, "Varnishes," 8, and White, "Van Eyck's Technique," 103 (see also p. 105 for the literature listed in notes 5–7).

12. Since book 3 of Eraclius survives only in a fifteenth-century manuscript, we cannot be entirely sure of the age of this recipe.

13. See Woudhuysen-Keller, "Leinöl als Malmittel," 82–85, and my discussion of her findings in the context of preparing varnish with lime in note 26 of chapter 3 above.

14. Eraclius, *De coloribus*, ed. Merrifield, 1:232–33 [no. 260]. Compare also the recipe on page 225 [no. 267], where Eraclius advises using a thick oil (*crasso oleo*) for varnish. This thick oil likely was similarly prepared by exposing it to sun and air.

15. Book 3 of *De coloribus* also mentions *cerosium* in a recipe for preparing orpiment [no. 261] (when you want to use orpiment on paper, you "distemper it like *cerosium*"). The treatise of Alcherius, also included in the manuscript of LeBègue, suggests a white lead-based pigment for *cerosium*; see *Experimenta de coloribus*, ed. Merrifield, 1:270–71 [no. 293, "to make a good rose color"]: *cum bracha pulverizata quae aliter dicitur album plumbum aliter cerusa, atque aliter Hispaniae, et dimittatur incorporari cum ipsa creta vel cerosio* [compare also no. 294 *blacha seu cerosio*]. Matters are confused, however, by a recipe in the treatise of Peter of Saint-Omer [no. 159], which speaks about adding *succum cerosium* to temper verdigris for use on paper. Here it is unlikely that lead white is implied. Merrifield suggests cherry juice or *cervisia* (beer).

16. Alcherius, *Experimenta de coloribus*, ed. Merrifield, 1:302–3 [no. 319]. As early as Dioscorides (ca. AD 40–90) it was recommended that oil be placed in the sun in open earthenware vessels at least for several days to "whiten it." Dioscorides, *De Materia Medica*, ed. and trans. Osbaldeston and Wood, 35 [1–32].

17. Yet to keep a skin from forming, stirring would have been important.

18. Cennini, *Il libro dell'arte*, ed. Broecke, 126–27 [no. 91].

19. Cennini, *Il libro dell'arte*, ed. Broecke, 128 [no. 92].

20. *Strasbourg Manuscript*, ed. Neven, 120–21 [no. 66].

21. *Strasbourg Manuscript*, ed. Neven, 120–21 [no. 66].

22. *"Liber illuministarum,"* ed. Bartl et al., 184–85 [no. 260].

23. Plahter, "*Líkneskjusmíð*," 160.

24. For an overview of the types of copper acetate compounds (and the various other copper compounds such as copper carbonate and copper chloride), see Nicholas Eastaugh, Valentine Walsh, Tracey Chaplin, and Ruth Siddall, *Pigment Compendium: A Dictionary and Optical Microscopy of Historical Pigments* (2008; repr., Abingdon, UK: Routledge, 2013), 430–34.

25. Hermann Kühn, "Verdigris and Copper Resinate," in *Artists' Pigments: A Handbook of Their History and Characteristics*, ed. Ashok Roy (Washington, DC: National Gallery of Art, 1997), 2:131–47; for reconstructions with historical recipes see also David A. Scott, *Copper and Bronze in Art: Corrosion, Colorants, Conservation* (Los Angeles: Getty Publications, 2002), esp. 279–93.

26. Alcherius, *Experimenta de coloribus*, ed. Merrifield 1849, 1:284–85 [300].

27. For an excellent and close to encyclopedic work on the historical sources of dyes, see Dominique Cardon, *Natural Dyes: Sources, Tradition, Technology and Science* (London: Archetype, 2007).

28. For an extensive discussion of these colorants, see Kirby, Van Bommel, and Verhecken, *Natural Colorants for Dyeing and Lake Pigments*, esp. 9–14; and see also Jo Kirby and Raymond White, "The Identification of Red Lake Pigment Dyestuffs and a Discussion of Their Use," *National Gallery Technical Bulletin* 17 (1996): 56–80.

29. Kirby, Bommel, and Verhecken, *Natural Colorants for Dyeing and Lake Pigments*, 71–74, and Kirby and White, "Identification of Red Lake Pigment Dyestuffs," 66–69.

30. For a few examples, see Jo Kirby, Marika Spring, and Catherine Higgit, "The Technology of Red Lake Pigment Manufacture: Study of the Dyestuff Substrate," *National Gallery Technical Bulletin* 26 (2005): 83n1. Yellow lake pigments were sometimes mixed with copper green to "improve" its color.

31. Kirby, Spring, and Higgit, "Technology of Red Lake Pigment Manufacture," 71–87.

32. Kirby, Spring, and Higgit mention that lime and calcium carbonates, such as chalk, marble dust, eggshells, or cuttlefish bone, could have been used as well, but that these more often appear in recipes for making yellow lake pigments and red or rose pink lake pigments from brazilwood; see "The Technology of Red Lake Pigment Manufacture," 71 and 83n1.

33. LeBègue, *Autre recepte*, ed. Merrifield, 1:292–93 [304].

34. LeBègue, *Autre recepte*, ed. Merrifield, 1:292–93 [304]. According to Kirby et al., dyes prepared with a large amount of calcium carbonate have some degree of translucency but less than dyes prepared with hydrated alumina substrates. Using white lead (RI 1.94–2.09) as a substrate would indeed have produced an opaque pigment; see Kirby, Bommel, and Verhecken, *Natural Colorants*, 77–78.

35. *Segreti d'arti diverse*, ed. Frezzato and Seccaroni, 144–45 [no. 301]; ed. Merrifield, 2:608–10 [no. 301]. Another recipe from the same manuscript [no. 309] provides another list of colors without "body." It mentions verdigris (*verderame*), red lake (*la laccha*), ochre (*l'ocria*), and green earth (*verde terra*), which "are very proper for mixing with those colors which have body." Since ochre and green earth are also considered part of this list of colors without body, the Marciana manuscript appears to have here compiled a more general list of pigments with relatively little covering power, not necessarily tailored to creating glazes of deep and saturated color.

36. Ferrante Imperato, *Dell'historia naturale di Ferrante Imperato napolitano. Libri XXVIII. Nella quale ordinatamente si tratta della diversa condition di miniere, e pietre. Con alcune historie di piante et animali; sin' hora non date in luce* (Naples: Costantino Vitale, 1599), 95. See also p. 96, where he describes the technique of oil painting and mentions that the ancients did not know about this technique: *è cosi di pochi anni introdotta, e del tutto a gli antichi incognita* (see chap. 1 above).

37. BnF. MS Fr. 640, ed. Making and Knowing Project, fol. 65v.

38. BnF. MS Fr. 640, ed. Making and Knowing Project, fol. 57r.

39. BnF. MS Fr. 640, ed. Making and Knowing Project, fol. 58r (compare fol. 56v and fol. 65r).

40. See also the discussion of "refractive index" in my introduction.

41. See Joyce Plesters, "Ultramarine Blue, Natural and Artificial," *Studies in Conservation* 11, no. 2 (1966): 62–91, and Eastaugh et al., *Pigment Compendium*, 381–82 and 582–83. Significantly, ultramarine was used in the Norwegian Altar Frontals discussed later in this chapter; see Unn Plahter, "The Trade in Painters' Materials in Norway in the Middle Ages: Part 2, Materials, Techniques and Trade from the Twelfth Century to the Mid-Fourteenth Century," in *Trade in Artists' Materials: Markets and Commerce in Europe to 1700*, ed. Jo Kirby, Susie Nash, and Joanna Cannon (London: Archetype, 2010), 64–73.

42. Eastaugh et al., *Pigment Compendium*, 39–40 and 590–91; see also Rutherford J. Gettens and Elisabeth West Fitzhugh, "Azurite and Blue Verditer," *Studies in Conservation* 11, no. 2 (1966): 54–61. Another color that can be produced from a copper carbonate mineral is the green pigment malachite; see Rutherford J. Gettens and Elisabeth West Fitzhugh, "Malachite and Green Verditer," *Studies in Conservation* 19, no. 1 (1974): 2–23. Malachite is often found with azurite, shares much of its chemical composition, and can be extracted in a similar manner. Both Harley and Thompson point out that while malachite has greater natural abundance than azurite, it is not often found on medieval panel paintings, and its preparation is rarely discussed in historical recipes. One reason for this, they suggest, is that verdigris may have been preferred over malachite because it has better working properties in oil: it could be ground to a much finer particle size. For azurite there was no such alternative except the much more costly ultramarine. See Thompson, *Materials and Techniques of Medieval Painting*, 161, and Rosamond D. Harley, *Artists' Pigments c. 1600–1835* (London: Butterworths, 1970), 72.

43. Plesters, "Ultramarine Blue," 67.

44. *Montpellier "Liber diversarum arcium,"* ed. Clarke, 140 and 281 [no. 2.9.5].

45. Daniel V. Thompson, *The Materials and Techniques of Medieval Painting* (1936; repr., New York: Dover, 1956), 130–35.

46. *Montpellier "Liber diversarum arcium,"* ed. Clarke, 141 and 281 [no. 2.9.7.A].

47. On the thirteenth-century Thornham Parva Retable, for instance, most colors are applied in their pure state, but to create a purple glaze azurite is mixed with a red lake; see Massing, *Thornham Parva Retable*, 56.

48. *Montpellier "Liber diversarum arcium,"* ed. Clarke, 140–41 and 280–81 [no. 2.9.7A].

49. Theophilus, *Schedula*, ed. Dodwell, 25 [no. XXVII].

50. See chapter 2 above; Theophilus, *Schedula*, ed. Dodwell, 25 [no. XXVII]. For varnishing tin foil yellow, see 22–23 [no. XXIV].

51. See chapters 2 and 4 above.

52. Marie Louise Sauerberg, Ashok Roy, Marika Spring, Spike Bucklow, and Mary Kempski, "Materials and Techniques," in *The Westminster Retable*, ed. Binski, Massing, and Sauerberg, 241.

53. Sauerberg et al., "Materials and Techniques," 241 and 265.

54. For the technical analysis of the enamels on the Westminster Retable recounted here, see Binski, Massing, and Sauerberg, *Westminster Retable*, 236–37.

55. As with an oil-resin varnish, close contact between the glass and the painted surface is crucial for the convincing material imitation of enamel. Any air between the glass

and the painted surface would have caused desaturation of the colors and decreased the visibility of the enamel. When attempting to reconstruct the retable's imitation enamel, Clare Heard and Spike Bucklow discovered that it took tremendous skill to establish the necessary close contact between glass and paint; see "Reconstruction of the Westminster Retable," in *Westminster Retable*, ed. Binski, Massing, and Sauerberg, 396–98; and from the same publication, see White and Kirby, "Medium Analysis," 237.

56. *The Guelph Treasure Shown at the Art Institute of Chicago: Catalogue of the Exhibition* (Chicago, 1931), 33.

57. See also chapter 6 below for a discussion of porphyry and other stone imitations on the frames and backs of some of Jan van Eyck's paintings.

58. *Montpellier "Liber diversarum arcium,"* ed. Clarke, 142 and 281 [no. 2.12.1A].

59. The recipe in the treatise of Theophilus reads: *Et inde cooperies locum, quem ita pingere uolueris.* The recipe in the Montpellier manuscript reads: *Et inde cohoperies crocum quam ita pingere volueris.*

60. Unn Plahter et al., *Painted Altar Frontals of Norway*, 1:102–6; for the technical information, see 2:221–31.

61. For the technical information on the Kaupanger frontal, see Plahter et al., *Painted Altar Frontals of Norway*, 2:229–31 [cat. Kaupanger]. In the Hauge frontal (see 2:221–24 [cat. Hauge]), the figural scenes are similarly outlined in black and executed with yellow varnish over silver to suggest gold. A few details such as the linings and some of the foliate ornaments have been decorated with green and red glazes, here applied directly over silver leaf.

62. Plahter et al., *Painted Altar Frontals of Norway*, 2:224–28 [cat. Heddal].

63. Plahter et al., *Painted Altar Frontals of Norway*, 2:106–7, 2:163.

64. Plahter et al., *Painted Altar Frontals of Norway*, 2:98.

65. See, for instance, Dillian Gordon, "A Sienese Verre Eglomisé and Its Setting," *Burlington Magazine* 123 (1981): 148–53; and Georg Swarzenski, "The Localisation of Medieval Verre Eglomisé in the Walters Collection," *Journal of the Walters Art Gallery* 3 (1940): 54–68.

66. Cennini, *Il libro dell'arte*, ed. Broecke, 226–30 [no. 198–203].

67. *Segreti d'arti diverse*, ed. Merrifield, 2:614–17 [no. 325].

68. *Montpellier "Liber diversarum arcium,"* ed. Clarke, 142 and 281 [no. 2.13.1A–B].

69. See chapter 2 above.

70. Krishan K. Chawla, *Fibrous Materials* (Cambridge: Cambridge University Press, 2016), 59.

71. For a detailed overview of the materials and methods used to make these "applied brocades," see Ingrid Geelen and Delphine Steyaert, *Imitation and Illusion: Applied Brocade in the Art of the Low Countries in the Fifteenth and Sixteenth Centuries* (Brussels: Royal Institute for Cultural Heritage, 2011), and Esther van Duijn, "All That Glitters Is Not Gold: The Depiction of Gold-Brocaded Velvets in Fifteenth- and Early Sixteenth-Century Netherlandish Paintings" (Diss., University of Amsterdam, 2013).

72. Geelen and Steyaert, *Imitation and Illusion*, 65–73; see also Pamela Betts and Glenn Gates, "Dressed in Tin: Analysis of the Textiles in the *Abduction of Helen* Series," *Journal of the Walters Art Museum* 74 (2019), accessed February 14, 2022: https://journal.thewalters.org/volume/74/essay/pressbrokat/.

73. *"Liber illuministarum,"* ed. Bartl et al., 176–77 [no. 239].

74. *"Liber illuministarum,"* ed. Bartl et al., 176–77 [no. 240].

75. *"Liber illuministarum,"* ed. Bartl et al., 188–89 [no. 268]. Another recipe in the same manuscript includes a variation of the previous recipe. It details how to paint on a gilded *"pilt,"* a German term used to refer to "image." In this case "all pigments should be ground with oil" (*ryb sÿ al mit öl an*). The manuscript mentions *paris rott* ("lac lake"), *zinober* ("cinnabar"), *ringen lasur* ("simple azure") and *spangrün* (verdigris); see *"Liber illuministarum,"* ed. Bartl et al., 184–85 [no. 257].

76. Cennini, *Il libro dell'arte*, ed. Broecke, 180 [no. 147].

77. Cennini, *Il libro dell'arte*, ed. Broecke, 181 [no. 148].

78. Cennini, *Il libro dell'arte*, ed. Broecke, 182 [no. 150].

79. Next to the more technical studies mentioned above, see also, e.g., Anne E. Wardwell, "The Stylistic Development of 14th- and 15th-Century Italian Silk Design," *Kunstblätter* 47 (1976): 177–226; Lisa Monnas, *Merchants, Princes and Painters: Silk Fabrics in Italian and Northern Paintings, 1300–1550* (New Haven, CT: Yale University Press, 2008); Rembrandt Duits, *Gold Brocade and Renaissance Painting: A Study in Material Culture* (London: Pindar Press, 2008); Anke Koch, *Darstellung von Seidenstoffen in der Kölner Malerei der ersten Hälfte des 15. Jahrhunderts* (Weimar: Verlag und Datenbank für Geisteswissenschaften, 2010).

80. The origin and meaning of these terms as descriptors for color modeling have been the subject of some discussion. See Eleanor Webster Bulatkin, "The Spanish Word 'Matiz': Its Origin and Semantic Evolution in the Technical Vocabulary of Medieval Painters," *Traditio* 10 (1954): 459–527; Hans Roosen-Runge, "Einführung: Entwicklung der Forschung über die Technik der Buchmalerei," in *Farbmittel, Buchmalerei, Tafel- und Leinwandmalerei des Mittelalters*, ed. Hermann Kühn, Heinz Roosen-Runge, Rolf E. Straub, and Manfred Koller (1984; repr., Stuttgart: Reclam, 2002), 1:66–68; for practical reconstructions of the modeling prescriptions, see Annette Scholtka, "Theophilus Presbyter-Die maltechnischen Anweisungen und ihre Gegenüberstellung mit naturwissenschaftlichen Untersuchungsbefunde," *Zeitschrift für Kunsttechnologie und Konservierung* 6, no. 1 (1992): 1–54; for a summary of the discussion in the context of the Montpellier manuscript, see *Montpellier "Liber diversarum arcium,"* ed. Clarke, 190–98.

81. Theophilus, *Schedula*, ed. Dodwell, 50 [no. XX].

82. The appellation *lumina/illuminare* and the accompanying bright paints are not to be confused with the application of specular reflections such as the Fayum painter used to depict the gold, gems, and eyes portrayed (fig. I.2). As we will see in chapter 6, this technique for rendering the optics of reflective and refractive materials was rediscovered only in the fifteenth century.

83. M. A. van Lokeren, B. de Saint-Genois, P. C. van der Meersch, and K. de Volkaersbeke, eds., *Messager des sciences historiques, ou Archives des arts et de bibliographie de Belgique* (Ghent: L. Hebbelynck, 1859), 128–30.

84. For the identification of *sinopere* with red lake see Susie Nash, "'Pour couleurs et autres choses prise de lui . . .': The Supply, Acquisition, Cost and Employment of Painters' Materials at the Burgundian Court, c. 1375–1419," in *Trade in Artists' Materials: Markets and Commerce in Europe to 1700*, ed. Joanna Cannon, Jo Kirby, and Susie Nash (London: Archetype, 2010), 141–48, 178n75.

85. Jan Baptist van der Straelen, *Jaerboek der vermaerde en kunstryke gilde van Sint Lucas binnen de stad Antwerpen* (Antwerp: Pieter-Theodor Moons-Van der Straelen, 1855), 15 [no. XI].

86. The 1391 ordinances of the *Tailleurs d'images, sculpteurs, peintres et enlumineurs de la ville de Paris*, for instance, do not yet use the term. Compare a similar rule ordering that fine gold or polished silver covered with a "golden tint" ought to be used for what should be golden (*et ce qui devra estre doré, soit de fin or ou d'argent bruny, et doré teinte*). See "Statuts des tailleurs d'images, sculpteurs, peintres et elumineurs de la ville de Paris," in *Collection des meilleurs dissertations, notices et traités relatifs a l'histoire de France*, ed. Jean Michel Constant Leber, vol. 19 (Paris: G. A. Dentu, 1838), 454 [no. 6].

87. Alphonse Goovaerts, "Les ordonnances données en 1480, à Tournai, aux métiers des peintres et des verriers," *Compte-rendu des séances de la Commission royale d'histoire* 65, no. 6 (1896): 163–64 [nos. 24, 25].

88. Hans Rott, *Quellen und Forschungen zur südwestdeutschen und schweizerischen Kunstgeschichte im XV. und XVI. Jahrhundert (Band 3, 2): Quellen II (Schweiz)*, 3 vols. (Stuttgart: Strecker und Schröder Verlag, 1936), 2:50.

89. A similar color scheme can be found on the silvered floor tiles of the Ghent Altarpiece; see chapter 6 below.

90. Hans Rott, *Quellen und Forschungen zur südwestdeutschen und schweizerischen Kunstgeschichte im XV. und XVI. Jahrhundert (Band 3, 1): Quellen I (Baden, Pfalz, Elsass)*, 3 vols. (Stuttgart: Strecker und Schröder Verlag, 1936), 1:221.

91. Ernst Ploss, "Studien zu den deutschen Maler und Färberbüchern des Mittelalters: Ein Beitrag zur deutschen Altertumskunst und Wortforschung" (Diss., Ludwig-Maximilian Universität, 1952), 102. *Glasieren* eventually became the German term for making vitreous glazes, and *lasieren* was used for making glazes with oil paint. In Dutch, on the other hand, the appellation *glacis* is now used to describe glazes made with oil paint, and *glazuur* describes the technique of applying vitreous coatings to metal or ceramics.

92. Van Mander, *Schilder-boeck*, ed. Miedema, 1, fol. 44r.

93. Van Mander, *Schilder-boeck*, ed. Miedema, 1, fol. 286r.

94. Translation by author. Giovan Paolo Lomazzo, *Trattato dell'arte de la pittura, scoltura, et architettura* (Milan: Paolo Gottardo Pontio, 1584), 197–98 [book 3, chap. 9].

95. Translation by author. Lomazzo, *Trattato dell'arte de la pittura*, 191 [book 3, chap. 4]; see also Barbara Tramelli, *Giovanni Paolo Lomazzo's "Trattato dell'arte della pittura"* (Leiden: Koninklijke Brill NV, 2017), 112.

96. Eastaugh et al., *Pigment Compendium*, 174 [*giallo santo*]. Note that *giallo santo* was sometimes also used to refer to yellow lakes extracted from other plants such as dyer's broom (*Genista tinctoria* L.) and weld (*Reseda luteola* L.).

97. Tramelli, *Lomazzo's Trattato*, 112.

98. Lomazzo, *Trattato*, 197–98 [bk. 3, chap. 9].

99. Armenini, *De' veri precetti*, 126 [bk. 2], ed. Olszewski, 193–94.

100. See also note 97 above.

101. In the Low Countries a separate term has described scumble since the seventeenth century: *schommelen*; see, e.g., Gerard de Lairesse, *Groot Schilderboek*, 2 vols. (Amsterdam: Erfgenamen van Willem de Coup, 1707), 1:12: *Want wy konnen het*

eeven sterk hoogen en schaduwen, waar door dezelve gloed en verheevenheid zal bewerkt worden, is het niet aanstonds door de bloote kracht der Verw, men schommel en lakseer het zo lang, tot dat het die eigenschap bezit.

102. My translation. See Filippo Baldinucci, *Vocabolario toscano dell'arte del disegno* (Florence: Per Santi Franchi al segno della Passione, 1681), 174.

103. Note that paintings show that scumbles were also made with little medium, using a dry brush technique.

104. Baldinucci also uses the term *velare* to describe glaze paint. In his entry on gold leaf (*oro in foglia*), for example, he explains that silver is used as a ground for colors that have no body and are see-through (*non aver corpo trampariscono*), and that this method of coloring on silver is called *velare*; see *Vocabolario toscano*, 115; also compare Baldinucci's entry on *verde eterno*, where he explains that "eternal green" is "nothing more" than a *velatura* made of silver leaf covered with verdigris; *Vocabolario toscano*, 178.

Chapter Six

1. Giovanni Santi, *La vita e le gesta di Federico di Montefeltro, duca d'Urbino: Poema in terza rima (codice Vat. Ottob. lat. 1305)*, ed. Luigi Michelini Tocci (Vatican City: Biblioteca Apostolica Vaticana, 1985).

2. For a recent discussion and technical examination, see Cristina Currie, "Genesis of a Pre-Eyckian Masterpiece: Melchior Broederlam's Painted Wings for the Crucifixion Altarpiece," in Stroo, *Pre-Eyckian Panel Painting*, 1:2–61.

3. See Currie, "Genesis of a Pre-Eyckian Masterpiece," 1:63, and fig. 32.

4. Timothy Husband, *The Art of Illumination: The Limbourg Brothers and the "Belles Heures" of Jean de France, Duc de Berry* (New York: Metropolitan Museum of Art, 2008), 205–6 and 259–60.

5. Save for the rare exception such as on the diadem of one of the dancing girls tempting Saint Jerome (fol. 186r), even in the sumptuous *Très riches heures* (Chantilly, Musée Condé, MS 65), commissioned by John of Berry (1340–1416), Duke of Burgundy, this approach toward imitating light has not changed. The manuscript was left uncompleted when all three Limbourg brothers died unexpectedly in 1416, and several other painters worked on it in the course of the fifteenth century—as a whole therefore, it is less representative for the Eyckian turning point.

6. Laura Weigert, *Weaving Sacred Stories: French Choir Tapestries and the Performance of Clerical Identity* (Ithaca, NY: Cornell University Press, 2004), esp. chap. 2, "The Lives of Piat and Eleutherius."

7. The authenticity of the quatrain has long been debated, but the recent restoration campaign of the Ghent Altarpiece found that it was likely contemporary with the altarpiece; see Susan Frances Jones, Anne-Sophie Augustyniak, and Hélène Dubois, "The Authenticity of the Frames of the Quatrain and the other Frame Inscriptions," in *The Ghent Altarpiece: Research and Conservation of the Exterior*, ed. Bart Fransen and Cyriel Stroo (Turnhout: Brepols, 2020), 173–308; and see also Joris C. Heyder, "Further to the Discussion of the Highlighted Chronogram on 'The Ghent Altarpiece,'" *Simiolus: Netherlands Quarterly for the History of Art* 38, no. 1/2 (2015): 5–16.

8. Willem Beurs, *De groote waereld in't kleen geschildert, of Schilderagtig tafereel*

van 's weerelds schilderyen . . . (Amsterdam: Johannes en Gillis Jansonius van Waesberge, 1692), 123. A critical edition of this treatise is forthcoming; see *Willem Beurs' "The Big World Painted Small,"* ed. Jeroen Stumpel and Ann-Sophie Lehmann, trans. Myra Scholz (Los Angeles: Getty Publications, forthcoming).

9. Beurs, *Groote waereld,* 120.

10. See, for instance, Lorne Campbell, "Robert Campin, the Master of Flémalle and the Master of Mérode," *Burlington Magazine* 116, no. 860 (1974): 634–46; and Stefan Kemperdick, "Robert Campin, Jacques Daret, and Rogier van der Weyden: The Written Sources," in *The Master of Flémalle and Rogier van der Weyden,* ed. Kemperdick and Jochen Sander (Ostfildern: Hatje Cantz Verlag, 2008), 53–61.

11. Compare also with the halos on a painting attributed to the Master of Flémalle depicting the *Madonna lactans* in the Städel Museum, Frankfurt, which has similar glazed gems on gold with specular reflections.

12. Kemperdick and Sander, *Master of Flémalle and Rogier van der Weyden,* 246–57 [no. 13].

13. See Cathy Metzger and Griet Steyaert, "Painting a Distinct Profession," in *Rogier van der Weyden, 1400–1464: Master of Passions,* ed. Lorne Campbell and Jan Van der Stock (Leuven: Davidsfonds, 2009), 174. Figure 86 shows a paint sample of the background.

14. See chapter 5 above.

15. See J. R. J. van Asperen de Boer, "A Scientific Re-examination of the Ghent Altarpiece," *Oud Holland* 93, no. 3 (1979): 167–75 and 21, and see also the report on the conservation treatment of the closed Ghent Altarpiece (in Dutch), Bart Devolder, Hélène Dubois, Jochen Ketels, and Simon Laevers, eds., *Behandeling van het Lam Godsveelluik. Verslag van fase 1: Het gesloten altaarstuk (2012–2016)* (Brussels: Koninklijk Instituut voor het Kunstpatrimonium, 2017), 37–49.

16. See also Millard Meiss, "Light as Form and Symbol in Some Fifteenth-Century Paintings," *Art Bulletin* 27, no. 3 (1945): 175–81; and E. Melanie Gifford, "Van Eyck's Washington Annunciation: Technical Evidence for Iconographic Development," *Art Bulletin* 81, no. 1 (1999): 109–16.

17. Anne-Sophie Augustyniak and Laure Mortiaux, "The Frames: In Search of Lost Unity," in *Ghent Altarpiece,* ed. Fransen and Stroo, 169–87.

18. Philippi Villani, *De origine civitatis Florentie et de eiusdem famosis civibus,* ed. Giuliano Tanturli (Padua: Antenore, 1997), 154. Translation from Laurie Schneider, *Giotto in Perspective* (Englewood Cliffs, NJ: Prentice-Hall, 1974), 38.

19. Baxandall, "Facius on Painting," 104.

20. This painting was in the private apartments of the prince of Ferrara; see Baxandall, "Facius on Painting," 104.

21. Beurs, *Groote waereld,* 87.

22. See also Bol, "Gems in the Water of Paradise," 34–48.

23. Kemperdick and Sander, *Master of Flémalle and Rogier van der Weyden,* 285–90 [no. 22].

24. Francis Ames-Lewis, "Fra Filippo Lippi and Flanders," *Zeitschrift für Kunstgeschichte* 42, no. 4 (1979): 255–73.

25. Much has been written on this subject, but I refer to Meiss's seminal "Light as

Form and Symbol in Some Fifteenth-Century Paintings," 175–81, and more recently Brian Madigan, "Van Eyck's Illuminated Carafe," *Journal of the Warburg and Courtauld Institutes* 49 (1986): 227–30.

26. Alberti originally prepared two versions of *De pictura*, in Latin and in Italian. I cite Grayson's critical edition of the Latin text; see Leon Battista Alberti, *On Painting and on Sculpture: The Latin Texts of "De pictura" and "De statua,"* ed. and trans. Cecil Grayson (London: Phaidon, 1972).

27. Alberti, *On Painting and on Sculpture*, ed. Grayson, 63.

28. It was because he wished to surpass the ancients that Vasari introduced the invention of oil paint to the *Vite*. See chapter 1 above.

29. The discovery of a new technique was also important to Vasari because he could not directly compare the paintings of his own day with those of the ancients. If sixteenth-century painters used a superior technique not known to Apelles and Zeuxis, this could be an argument for having surpassed them.

30. See chapter 1 above, esp. pp. 44–46 and n. 58.

31. The "invention of perspective" has been studied extensively in art history. For a discussion of the history of the use of linear perspective in painting, see Martin Kemp, *The Science of Art* (New Haven, CT: Yale University Press, 1990), and for an overview and discussion of sources see Kirsti Andersen, *The Geometry of an Art: The History of the Mathematical Theory of Perspective from Alberti to Monge* (New York: Springer Science and Business Media, 2007).

32. Alberti, *"On Painting" and "On Sculpture,"* ed. Grayson, 62–63.

33. Alberti, *"On Painting" and "On Sculpture,"* ed. Grayson, 81–87.

34. Alberti, *"On Painting" and "On Sculpture,"* ed. Grayson, 89.

35. Alberti, *"On Painting" and "On Sculpture,"* ed. Grayson, 90–91.

36. Alberti, *"On Painting" and "On Sculpture,"* ed. Grayson, 93.

37. Alberti, *"On Painting" and "On Sculpture,"* ed. Grayson, 62–63.

38. Alberti, *"On Painting" and "On Sculpture,"* ed. Grayson, 62–63.

39. Alberti, *"On Painting" and "On Sculpture,"* ed. Grayson, 3–5.

40. Alberti, *"On Painting" and "On Sculpture,"* ed. Grayson, 81.

41. Nuttall, *From Flanders to Florence*, 32–35. This was recorded by Cardinal Jean Jouffroy in 1468 and thus postdates Alberti's treatise; see also chapter 1 above.

42. For Georges Hulin de Loo's discussion of Hand G, see *Heures de Milan: Troisième partie des "Très-belles heures de Notre-Dame" enluminées par les peintres de Jean de France, duc de Berry et par ceux du duc Guillaume de Bavière* (Brussels: G. van Soest, 1911), 30–36; for a recent critical overview of the state of the discussion, see Dominique Vanwijnsberghe, "The Eyckian Miniatures in the Turin-Milan Hours," in *Van Eyck*, ed. Martens et al., 296–315.

43. Beurs, *Groote waereld*, 124.

Bibliography

Adams, Noël. "Rethinking the Sutton Hoo Shoulder Clasps and Armour." In *Intelligible Beauty: Recent Research on Byzantine Jewellery*, 83–112. London: British Museum Press, 2010.

———. "The Garnet Millennium: The Role of Seal Stones in Garnet Studies." In *"Gems of Heaven": Recent Research on Engraved Gemstones in Late Antiquity, AD 200–600*, edited by Christopher Entwistle and Noël Adams, 10–24. London: British Museum, 2011.

Aëtius of Amida. *Aetii medici Graeci contractae ex veteribus medicinae sermones XVI.* Edited and translated by Janus Cornarius. Basel: Froben, 1542.

Aikema, Bernard, and Beverly Louise Brown. "Painting in Fifteenth-Century Venice and the *Ars Nova* of the Netherlands." In *Renaissance Venice and the North: Crosscurrents in the Time of Bellini, Dürer, and Titian*, edited by Bernard Aikema, Beverly Louise Brown, and Giovanna Nepi Sciré, 176–85. Venice: Bompiani, 1999.

Aikema, Bernard, Beverly Louise Brown, and Giovanna Nepi Sciré, eds. *Renaissance Venice and the North: Crosscurrents in the Time of Bellini, Dürer, and Titian.* New York: Rizzoli, 2000.

Akdogan, Cemil. "Optics in Albert the Great's *De sensu et sensato*: An Edition, English Translation, and Analysis." PhD diss., University of Wisconsin–Madison, 1978.

Alberti, Leon Battista. *"On Painting" and "On Sculpture": The Latin Texts of "De pictura" and "De statua."* Edited and translated by Cecil Grayson. London: Phaidon, 1972.

Albertus Magnus. *De mineralibus (Mineralium libri quinque). Opera omnia Alberti Magni*, vol. 5. Edited by Auguste Borgnet. Paris: Apud Ludovicum Vives, 1890.

———. *De meteoris, Lib. IV. Opera omnia Alberti Magni*, vol. 4. Edited by Auguste Borgnet. Paris: Apud Ludovicum Vives, 1890.

———. *Book of Minerals.* Edited and translated by Dorothy Wyckoff. Oxford: Oxford University Press, 1967.

Alexander-Skipnes, Ingrid, ed. *Cultural Exchange between the Low Countries and Italy (1400–1600).* Turnhout: Brepols, 2007.

Allerhand Farben/und mancherlay weyse/Dünten zůbereyten. Augsburg: Heynrich Steyner, 1533.

Ames-Lewis, Francis. "Fra Filippo Lippi and Flanders." *Zeitschrift für Kunstgeschichte* 42, no. 4 (1979): 255–73.

Andersen, Kirsti. *The Geometry of an Art: The History of the Mathematical Theory of Perspective from Alberti to Monge*. New York: Springer Science and Business Media, 2007.

Armenini, Giovanni Battista. *De' veri precetti della pittura: Opera non solo utile, e necessaria à tutti gli artefici per cagion del disegno; lume, e fondamento di tutte l'altre arti minori, ma anco à ciascun' altra persona intendente di cosi nobile professione*. Ravenna: Francesco Tebaldini, 1586.

———. *On the True Precepts of the Art of Painting*. Edited and translated by Edward J. Olszewski. New York: Franklin, 1977.

Artioli, Gilberto. *Scientific Methods and Cultural Heritage: An Introduction to the Application of Materials Science to Archaeometry and Conservation Science*. Oxford: Oxford University Press, 2010.

Artliche kunste—mancherley weyse Dinten und allerhand farben zubereyten. Mainz: Peter Jordan, 1531.

Atkinson, Catherine. *Inventing Inventors in Renaissance Europe: Polydore Vergil's "De inventoribus rerum."* Tübingen: Mohr Siebeck, 2007.

Augustine. *De Genesi ad litteram*. Edited by Joseph Zycha. Corpus scriptorum ecclesiasticorum latinorum, vol. 28. Vienna: Verlag der Osterreichischen Akademie der Wissenschaften, 1894.

———. *De Genesi contra Manichaeos*. Vol. 9. Edited by Dorothea Weber. Vienna: Austrian Academy of Sciences, 1998.

———. *On Genesis: A Refutation of the Manichees; Unfinished Literal Commentary on Genesis; The Literal Meaning of Genesis*. Vol. 13. Edited and translated by Edmund Hill. New York: New City Press, 2002.

Augustyniak, Anne-Sophie, and Laure Mortiaux. "Conservation and Restoration Treatment. The Frames: In Search of Lost Unity." In *The Ghent Altarpiece: Research and Conservation of the Exterior*, edited by Bart Fransen and Cyriel Stroo, 169–87. Turnhout: Brepols, 2020.

Averlino, Antonio di Pietro (Filarete). *Antonio Averlino detto il Filarete: Trattato di architettura*. Edited by Anna Maria Finoli and Liliana Grassi. 2 vols. Classici italiani di scienze tecniche e arti. Milan: Il Polifilo, 1972.

Azémard, Clara, Matthieu Ménager, and Cathy Vieillescazes. "On the Tracks of Sandarac: Review and Chemical Analysis." *Environmental Science and Pollution Research* 24, no. 36 (2017): 27746–54.

Baroni, Sandro Giuseppe Pizzigoni, and Paola Travaglio, eds. and trans. *"Mappae clavicula": Alle origini dell'alchinia in Occidente*. Saonara: Il Prato, 2013.

Bartholomeus Anglicus, *De proprietatibus rerum*. Lyon: Petrus Ungarus, 1482.

Bartl, Anna, Christoph Krekel, Manfred Lautenschlager, and Doris Oltrogge, eds. and trans. *Der "Liber illuministarum" aus Kloster Tegernsee*. Edition, Übersetzung und Kommentar der kunsttechnologischen Rezepte. Stuttgart: Franz SteinerVerlag, 2005.

Baxandall, Michael. "Bartholomaeus Facius on Painting: A Fifteenth-Century Manuscript of the *De viris illustribus*." *Journal of the Warburg and Courtauld Institutes* 27 (1964): 90–107.

Becker, Jochen. "Zur niederländischen Kunstliteratur des 16. Jahrhunderts: Lucas De Heere." *Simiolus-Netherlands Quarterly for the History of Art* 6 (1972–73): 113–27.

Bernsmeier, Uta. "Die *Nova Reperta* des Jan Van Der Straet-Ein Beitrag zur Problemgeschichte der Entdeckungen und Erfindungen im 16. Jh." Diss., University of Hamburg, 1984.

Berthelot, Marcellin. *Introduction à l'étude de la chimie des anciens et du Moyen Âge*. Paris: George Steinheil, 1889.

Berthelot, Marcellin, and Charles Émile Ruelle, eds. and trans. *Collection des anciens alchimistes grecs*. 3 vols. Paris: Georges Steinheil, 1888.

Beruni, Al-. *Al-Beruni's Book on Mineralogy: The Book Most Comprehensive in Knowledge on Precious Stones*. Edited and translated by Hakim Mohammad Said. Islamabad: Pakistan Hijra Council, 1989.

Betts, Pamela, and Glenn Gates, "Dressed in Tin: Analysis of the Textiles in the *Abduction of Helen* Series." *Journal of the Walters Art Museum* 74 (2019): https://journal.thewalters.org/volume/74/essay/pressbrokat/.

Beurs, Willem. *De groote waereld in't kleen geschildert, of Schilderagtig tafereel van 's weerelds schilderyen. Kortelijk vervat in ses boeken*. Amsterdam: Johannes en Gillis Jansonius van Waesberge, 1692.

———. *Willem Beurs' "The Big World Painted Small."* Edited by Jeroen Stumpel and Ann-Sophie Lehmann. Translated by Myra Scholz. Los Angeles: Getty Publications, forthcoming.

Bingen, Hildegard von. *Physica: Liber subtilitatum diversarum naturarum creaturarum*. Edited and translated by Reiner Hildebrandt and Thomas Gloning. 2 vols. Berlin: De Gruyter, 2010.

———. *Hildegard von Bingen's "Physica": The Complete English Translation of Her Classic Work on Health and Healing*. Edited and translated by Priscilla Throop. Rochester, VT: Healing Arts Press, 1998.

Biringuccio, Vannoccio. *De la "Pirotechnia" Libri X . . . Composti per il Vannoccio Biringuccio Sennese*. Venice: Venturino Roffinello, 1540.

———. *The "Pirotechnia" of Vannoccio Biringuccio: The Classic Sixteenth-Century Treatise on Metals and Metallurgy*. Edited and translated by Cyril Stanley Smith and Martha Teach Gnudi. 1942. Reprint, Cambridge, MA: MIT Press, 1959.

Bischoff, Bernard. "Die Überlieferung der technischen Literatur." In *Mittelalterliche Studien: Ausgewählte Aufsätze zur Schriftkunde und Literaturgeschichte*, 3:277–97. Stuttgart: A. Hiersemann, 1981.

Bol, Marjolijn. "Seeing through the Paint. The Dissemination of Technical Terminology between Three Métiers: *Pictura Translucida*, Enameling and Glass Painting." In *Zwischen Kunsthandwerk und Kunst: Die "Schedula Diversarum artium,"* edited by Andreas Speer, Maxime Mauriège, and Hiltrud Westermann-Angerhausen, 145–62. Berlin: De Gruyter, 2013.

———. "Coloring Topaz, Crystal and Moonstone: Gems and the Imitation of Art and Nature, 300–1500." In *Fakes!?: Hoaxes, Counterfeits and Deception in Early Modern Science*, edited by Marco Beretta and Maria Conforti, 108–29. Sagamore Beach, MA: Science History Publications, 2014.

———. "Gems in the Water of Paradise: The Iconography and Reception of Heavenly Stones in the Ghent Altarpiece." In *Van Eyck Studies: Papers Presented at the*

Eighteenth Symposium for the Study of Underdrawing and Technology in Painting, Brussels, 19–21 September 2012, edited by Christina Currie and Bart Fransen, 34–48. Leuven: Peeters, 2017.

———. "Technique and the Art of Immortality, 1800–1900." *History of Humanities* 2, no. 1 (2017): 179–99.

———. "The Emerald and the Eye: On Sight and Light in the Artisan's Workshop and the Scholar's Study." In *Perspective as Practice: Renaissance Cultures of Optics*, edited by Sven Dupré, 71–101. Turnhout: Brepols, 2019.

———. "*Polito et claro*: The Art and Knowledge of Polishing, 1200–1500." In *Gems in the Early Modern World: Materials, Knowledge and Global Trade, 1450–1800*, edited by Michael Bycroft and Sven Dupré, 223–57. Cham: Palgrave Macmillan, 2019.

Bol, Marjolijn, and Emma C. Spary, eds. *The Matter of Mimesis: Studies on Mimesis and Materials in Nature, Art and Science*. Leiden: Brill, 2023.

Boltz, Valentin. *Illuminier Buoch, Wie man allerley Farben bereitten, mischen, schattieren unnd ufftragen soll: Allen jungen angehnden Molern und Illuministen, nutzlich und fürderlich: Mit flysz und arbeit ersuocht, geuebt und zuosammen bracht*. Basel: Jacob Kündig, 1549.

———. *Illuminierbuch: Wie man allerlei Farben bereiten, mischen und auftragen soll . . .* Edited by C. J. Benziger. Munich: Georg D. W. Callwey, 1913.

Bomford, David. "Picture Cleaning: Positivism and Metaphysics." In *The Conservation of Easel Paintings*, edited by Joyce Hill Stoner and Rebecca Anne Rushfield. London: Routledge, 2012.

Bomford, David, and Mark Leonard, eds. *Issues in the Conservation of Paintings*. Los Angeles: Getty Conservation Institute, 2004.

Bonnardot, François, and René de Lespinasse, eds. *Les métiers et corporations de la ville de Paris, XIIIe siècle: Le livre des métiers*. Paris: Imprimerie national, 1879.

Borst, Arno. *Das Buch der Naturgeschichte: Plinius und seine Leser im Zeitalter des Pergaments*. 1994. Reprint, Heidelberg: Universitätsverlag C. Winter, 1995.

Braekman, Willy L., ed. *Dat Batement van recepten: Een secreetboek uit de zestiende eeuw*. Brussels: Omirel UFSAL, 1990.

Brandi, Cesare. "The Cleaning of Pictures in relation to Patina, Varnish, and Glazes." *Burlington Magazine* 91, no. 556 (1949): 183–88.

Brill, Thomas B. *Light: Its Interaction with Art and Antiquities*. New York: Plenum Press, 1980.

Brinkman, Pieter Willem Frederik. *Het geheim van Van Eyck: Aantekeningen bij de uitvinding van het olieverven*. Zwolle: Waanders, 1993.

Brown, G. Baldwin, and Louisa S. Maclehose, eds. *Vasari on Technique: Being the Introduction to the Three Arts of Design, Architecture, Sculpture and Painting, Prefixed to the Lives of the Most Excellent Painters, Sculptors and Architects*. London: Dent, 1907.

Brown, Katharine Reynolds, Dafydd Kidd, and Charles T. Little, eds. *From Attila to Charlemagne: Arts of the Early Medieval Period in the Metropolitan Museum of Art*. New York: Metropolitan Museum of Art, 2000.

Brown, Michelle P. *The British Library Guide to Writing and Scripts*. Toronto: University of Toronto Press, 1998.

Bryce, Judith. *Cosimo Bartoli (1503–1572): The Career of a Florentine Polymath*. Geneva: Librairie Droz, 1983.

Bucklow, David. "The Description of Craquelure Patterns." *Studies in Conservation* 43, no. 3 (1997): 129–40.

Buckton, David. "Enamelling on Gold: A Historical Perspective." *Gold Bulletin* 15, no. 3 (1982): 101–9.

———. "The Oppenheim or Fieschi-Morgan Reliquary in New York, and the Antecedents of Middle Byzantine Enamel." *Byzantine Studies Conference Abstracts* 8 (1982): 35–36.

Bulatkin, Eleanor Webster. "The Spanish Word 'Matiz': Its Origin and Semantic Evolution in the Technical Vocabulary of Medieval Painters." *Traditio* 10 (1954): 459–527.

Burioni, Matteo. "Vasari's *Rinascita*: History, Anthropology or Art Criticism?" In *Renaissance?: Perceptions of Continuity and Discontinuity in Europe, c. 1300–c. 1550*, edited by Alexander Lee, Pit Péporté, and Harry Schnitker, 115–27. Leiden: Brill, 2010.

Burke, Peter. *What Is the History of Knowledge?* Cambridge: Polity Press, 2015.

Caley, Earle Radcliffe. "The Leyden Papyrus X: An English Translation with Brief Notes." *Journal of Chemical Education* 3, no. 10 (1926): 1149–66.

———. "The Stockholm Papyrus: An English Translation with Brief Notes." *Journal of Chemical Education* 4, no. 8 (1927): 979–1002.

Campbell, Lorne. "Robert Campin, the Master of Flémalle and the Master of Mérode." *Burlington Magazine* 116, no. 860 (1974): 634–46.

Campbell, Lorne, Susan Foister, Ashok Roy, and Rachel Billinge, eds. "Methods and Materials of Northern European Painting in the National Gallery, 1400–1550." *National Gallery Technical Bulletin* 18 (1997): 6–55.

Cardon, Dominique. *Natural Dyes: Sources, Tradition, Technology and Science*. London: Archetype, 2007.

Carlyle, Leslie. *The Artist's Assistant: Oil Painting Instruction Manuals and Handbooks in Britain, 1800–1900, with Reference to Selected Eighteenth-Century Sources*. London: Archetype, 2001.

Carrara, Eliana. "Reconsidering the Authorship of the 'Lives': Some Observations and Methodological Questions on Vasari as a Writer." *Studi di memofonte* 15 (2015): 53–90.

Cats, Jacob. *Alle de wercken, so ouden als nieuwen*. Amsterdam: Jan Jacobsz. Schipper, 1658.

Causey, Faya. *Amber and the Ancient World*. Los Angeles: Getty Publications, 2011.

Cennini, Cennino. *Cennino Cennini's "Il libro dell'arte": A New English Translation and Commentary with Italian Transcription*. Edited and translated by Lara Broecke. London: Archetype, 2015.

———. *A Treatise on Painting by Cennino Cennini Written in the Year 1437* . . . Edited and translated by Mary Philadelphia Merrifield. London: Edward Lumley, 1844.

Charleston, R. J. "Lead in Glass." *Archeometry* 3, no. 1 (1960): 1–4.

Chawla, Krishan K. *Fibrous Materials*. Cambridge: Cambridge University Press, 2016.

Chibnall, Marjorie. "Pliny's *Natural History* and the Middle Ages." In *Empire and After-*

math: Silver Latin II, edited by Thomas A. Dorey, 57–78. London: Routledge and Kegan Paul, 1975.

Clarke, Mark, ed. and trans. *Mediaeval Painters' Materials and Techniques: The Montpellier "Liber diversarum arcium."* London: Archetype, 2011.

———. "The Earliest Technical Recipes: Assyrian Recipes, Greek Chemical Treatises and the *Mappae clavicula* Text Family." In *Craft Treatises and Handbooks: The Dissemination of Technical Knowledge in the Middle Ages*, edited by Ricardo Córdoba, 9–31. Turnhout: Brepols, 2013.

Clarke, Mark, Joyce H. Townsend, and Ad Stijnman, eds. *Art of the Past: Sources and Reconstructions. Proceedings of the First Symposium of the Icom-CC Working Group for Art Technological Source Research.* London: Archetype, 2005.

Conti, Alessandro. *History of the Restoration and Conservation of Works of Art.* 1973. Reprint, Oxford: Butterworth-Heinemann, 2007.

Coremans, Paul. *L'agneau mystique au laboratoire: Examen et traitement.* Les Primitifs flamands: III, Contributions à l'étude des Primitifs flamands, no. 2. Anvers: De Sikkel, 1953.

Coremans, Paul, R. J. Gettens, and J. Thissen. "La technique des Primitifs flamands: Étude scientifique des matériaux, de la structure et de la technique picturale." *Studies in Conservation* 1, no. 1 (1952): 1–29.

Cruz, António João, and Luís Urbano Afonso. "On the Date and Contents of a Portuguese Medieval Technical Book on Illumination: *O Livro de como se fazem as cores.*" *Medieval History Journal* 11, no. 1 (2008): 1–28.

Currie, Cristina. "Genesis of a Pre-Eyckian Masterpiece: Melchior Broederlam's Painted Wings for the Crucifixion Altarpiece." In *Pre-Eyckian Panel Painting in the Low Countries*, edited by Cyriel Stroo, 1:23–86. Turnhout: Brepols, 2009.

Dalby, Andrew. *Food in the Ancient World from A to Z.* London: Routledge, 2003.

Darrah, Josephine. "White and Golden Tin Foil in Applied Relief Decoration: 1240–1530." In *Looking through Paintings: The Study of Painting Techniques and Materials in Support of Art Historical Research*, edited by Erma Hermens, Annemiek Ouwekerk, and Annemiek Costaras, 49–79. Baarn: Archetype, 1995.

Daston, Lorraine. "The History of Science and the History of Knowledge." *KNOW: A Journal on the Formation of Knowledge* 1 (2017): 131–54.

Davis, Lucy. "Renaissance Inventions: Van Eyck's Workshop as a Site of Discovery and Transformation in Jan Van der Straet's *Nova Reperta.*" *Nederlands Kunsthistorisch Jaarboek* 59 (2010): 223–48.

De Hollanda, Francisco. *On Antique Painting.* Edited by Hellmut Wohl. Translated by Alice Sedgwick Wohl. University Park: Pennsylvania State University Press, 2015.

Delumeau, Jean. *History of Paradise: The Garden of Eden in Myth and Tradition.* Translated by Matthew O'Connell. 1992. Reprint, Urbana: University of Illinois Press, 2000.

Devolder, Bart, Hélène Dubois, Jochen Ketels, and Simon Laevers, eds. *Behandeling van het Lam Godsveelluik. Verslag van fase 1: Het gesloten altaarstuk (2012–2016).* Brussels: Koninklijk Instituut voor het Kunstpatrimonium, 2017.

D'Heere, Lucas. *Den hof en boomgaerd der poësien door Lucas d'Heere.* Edited by Werner Waterschoot. 1565. Reprint, Zwolle: Tjeenk Willink, 1969.

Dioscorides. *De Materia Medica: Being an Herbal with Many Other Medicinal Materials Written in Greek in the First Century of the Common Era*. Edited and translated by Tess Anne Osbaldeston and Robert P. A. Wood. Johannesburg: Ibidis Press, 2000.

Distelberger, Rudolf, and Manfred Leithe-Jasper. *Kunsthistorisches Museum, Vienna: The Imperial and Ecclesiastical Treasury*. Munich: C. H. Beck and Scala, 2009.

Duijn, Esther van. "All That Glitters Is Not Gold: The Depiction of Gold-Brocaded Velvets in Fifteenth- and Early Sixteenth-Century Netherlandish Paintings." Diss., University of Amsterdam, 2013.

Duits, Rembrandt. *Gold Brocade and Renaissance Painting: A Study in Material Culture*. London: Pindar Press, 2008.

Dunkerton, Jill, Susan Foister, Dillian Gordon, and Nicholas Penny. *Giotto to Dürer: Early Renaissance Painting in the National Gallery*. New Haven, CT: Yale University Press, 1991.

Dunkerton, Jill, and Raymond White. "The Discovery and Identification of an Original Varnish on a Panel by Carlo Crivelli." *National Gallery Technical Bulletin* 21 (2000): 70–76.

Dürer, Albrecht. *Dürer: Schriftlicher Nachlaß*. Edited by Hans Rupprich. 3 vols. Berlin: Deutscher Verlag für Kunstwissenschaft, 1956–69.

Eamon, William. "Arcana Disclosed: The Advent of Printing, the Books of Secrets Tradition and the Development of Experimental Science in the Sixteenth Century." *History of Science* 22, no. 2 (1984): 111–50.

———. *Science and the Secrets of Nature: Books of Secrets in Medieval and Early Modern Culture*. Princeton, NJ: Princeton University Press, 1996.

Eamon, William, and Françoise Paheau. "The Accademia Segreta of Girolamo Ruscelli: A Sixteenth-Century Italian Scientific Society." *Isis* 75, no. 2 (1984): 327–42.

Eastaugh, Nicholas, Valentine Walsh, Tracey Chaplin, and Ruth Siddall. *Pigment Compendium: A Dictionary and Optical Microscopy of Historical Pigments*. Abingdon, UK: Routledge, 2013.

Eastlake, Charles Lock. *Methods and Materials of Painting of the Great Schools and Masters* (original title *Materials for a History of Oil Painting*). 2 vols. 1847. Reprint, New York: Dover, 1960.

Effmann, Elise. "Theories about the Eyckian Painting Medium from the Late-Eighteenth to the Mid-Twentieth Centuries." *Reviews in Conservation* 7 (2006): 17–26.

Emmens, Jan. "De uitvinding van de olieverf: Een kunsthistorisch probleem in de zestiende eeuw." In *Kunsthistorische Opstellen*, edited by Jan Emmens, 1:191–203. Amsterdam: Van Oorschot, 1981.

Engelsman, Bartholomeus. *Van den Proprieteyten der Dinghen*. Haarlem: Jacob Bellaert, 1485.

Feller, Robert L., Nathan Stolow, and Elizabeth H. Jones. *On Picture Varnishes and Their Solvents*. Revised and enlarged edition. Cleveland: Press of Case Western Reserve University, 1971.

Fleming, Ronald W., Henrik Wann Jensen, and Heinrich H. Bülthoff. "Perceiving Translucent Materials." *APGV '04 Proceedings of the 1st Symposium on Applied Perception in Graphics and Visualization* (2004), 127–34.

Forbes, Robert James. *Studies in Early Petroleum History*. Leiden: Brill, 1958.

Frangenberg, David. "Bartoli, Giambullari and the Prefaces to Vasari's *Lives* (1550)." *Journal of the Warburg and Courtauld Institutes* 65, no. 15 (2002): 244–58.

Fransen, Bart, and Cyriel Stroo, eds. *The Ghent Altarpiece: Research and Conservation of the Exterior*. Turnhout: Brepols, 2020.

Freeman, M. H., C. C. Hull, and W. N. Charman. *Optics*. 11th ed. New York: Butterworth-Heinemann, 2003.

Frezzato, Fabio, and Claudio Seccaroni, eds. *Segreti d'arti diverse nel regno di Napoli. Il manoscritto It. III.10 della Biblioteca Marciana di Venezia*. Saonara: Prato, 2010.

Frinta, Mojmír Svatopluk. *Punched Decoration: On Late Medieval Panel and Miniature Painting*. 2 vols. Prague: Maxdorf, 1998.

Frison, Guido, and Giulia Brun. "*Compositiones lucenses* and *Mappae clavicula*: Two Traditions or One? New Evidence from Empirical Analysis and Assessment of the Literature." *Heritage Science* 6, no. 1 (2018): 1–24, https://doi.org/10.1186/s40494-018-0189-y.

Fuchs, Leonhart. *New Kreüterbuch*. Basel: Michael Isingrin, 1543.

Gabrieli, Bernardo Oderzo. "L'inventario della spezieria di Pietro Fasolis e il commercio dei materiali per la pittura nei documenti piemontesi (1332–1453): Parte prima." *Bollettino della Società storica pinerolese* 29 (2012): 7–43.

———. "L'inventario della spezieria di Pietro Fasolis e il commercio dei materiali per la pittura nei documenti piemontesi (1332–1453): Parte seconda." *Bollettino della Società storica pinerolese* 30 (2013): 7–53.

Gage, John. *Colour and Culture: Practice and Meaning from Antiquity to Abstraction*. 1993. Reprint, London: Thames and Hudson, 2009.

———. *Goethe on Art*. Berkeley: University of California Press, 1980.

Gardette, Jean-Luc, Jacques Lemaire, and Jacky Mallégol. "Long-Term Behavior of Oil-Based Varnishes and Paints. Photo- and Thermooxidation of Cured Linseed Oil." *Journal of the American Oil Chemists' Society* 77, no. 3 (2000): 257–63.

———. "Yellowing of Oil-Based Paints." *Studies in Conservation* 46, no. 2 (2001): 121–31.

Gaye, Johann Wilhelm, ed. *Carteggio inedito d'artisti dei secoli XIV. XV. XVI*. Florence: Giuseppe Molini, 1840.

Geelen, Ingrid, and Delphine Steyaert. *Imitation and Illusion: Applied Brocade in the Art of the Low Countries in the Fifteenth and Sixteenth Centuries*. Brussels: Koninklijk Instituut voor het Kunstpatrimonium, 2011.

Gettens, Rutherford J., and Elisabeth West Fitzhugh. "Azurite and Blue Verditer." *Studies in Conservation* 11, no. 2 (1966): 54–61.

———. "Malachite and Green Verditer." *Studies in Conservation* 19, no. 1 (1974): 2–23.

Gifford, E. Melanie. "Van Eyck's Washington Annunciation: Technical Evidence for Iconographic Development." *Art Bulletin* 81, no. 1 (1999): 108–16.

Goethe, Johann Wolfgang von. *Über Kunst und Alterthum in den Rhein und Mayn Gegenden*. 3 vols. Stuttgart: Cottaischen Buchhandlung, 1816.

Gombrich, Ernst H. *Art and Illusion: A Study in the Psychology of Pictorial Representation*. 1960. Reprint, Oxford: Phaidon Press, 1983.

———. "Vasari's *Lives* and Cicero's *Brutus*." *Journal of the Warburg and Courtauld Institutes* 23, no. 3–4 (1960): 309–11.

———. "Dark Varnishes: Variations on a Theme by Pliny." *Burlington Magazine* 104, no. 707 (1962): 51–55.

———. "Light, Form and Texture in XVth Century Painting." *Royal Society for the Encouragement of Arts* 112, no. 5099 (1964): 826–49.

———. "Light, Form and Texture in Fifteenth-Century Painting North and South of the Alps." In *The Heritage of Apelles: Studies in the Art of the Renaissance*, 19–35. Ithaca, NY: Cornell University Press, 1976.

Goovaerts, Alphonse. "Les ordonnances données en 1480, à Tournai, aux métiers des peintres et des verriers . . ." *Compte-rendu des séances de la Commission royale d'histoire* 65, no. 6 (1896).

Gordon, Dillian. "A Sienese *Verre Eglomisé* and Its Setting." *Burlington Magazine* 123, no. 936 (1981): 148–53.

Gordon, Dillian, Caroline M. Barron, Ashok Roy, and Martin Wyld. *The Wilton Dyptich: Making and Meaning*. London: National Gallery, 1993.

Graaf, Johannes Alexander van de. *Het De Mayerne Manuscript als bron voor schildertechniek van de barok, British Museum, Sloane 2052*. Mijdrecht: Drukkerij Verweij, 1958.

———. "The Interpretation of Old Painting Recipes." *Burlington Magazine* 104, no. 716 (1962): 471–75.

Graham, Jenny. *Inventing Van Eyck: The Remaking of an Artist for the Modern Age*. Oxford: Berg, 2007.

Grosseteste, Robert. *Hexaëmeron*. Edited by Richard C. Dales and Servus Gieben. London: Oxford University Press, 1982.

Halleux, Robert, ed. and trans. *Les alchimistes grecs: Papyrus de Leyde–Papyrus de Stockholm–Recettes*. Vol. 1. Paris: Belles Lettres, 1981.

Halleux, Robert, and Paul Meyvaert. "Les origines de la *Mappae clavicula*." *Archives d'histoire doctrinale et littéraire du Moyen-Âge* 54 (1987): 7–58.

Harley, Rosamond D. *Artists' Pigments c. 1600–1835*. London: Butterworths, 1970.

Healy, John F. *Pliny the Elder on Science and Technology*. Oxford: Oxford University Press, 1999.

Heard, Clare, and Spike Bucklow. "Reconstruction of the Westminster Retable." In *The Westminster Retable: History, Technique, Conservation*, edited by Paul Binski, Ann Massing, and Marie Louise Sauerberg, 392–400. Cambridge: Brepols, 2009.

Hedemann, Anne D. "Making the Past Present: Visual Translation in Jean Lebègue's 'Twin' Manuscripts of Sallust." In *Patrons, Authors and Workshops: Books and Book Production in Paris around 1400*, edited by Godfried Croenen and Peter F. Ainsworth, 173–96. Louvain: Peeters, 2006.

Hedfors, Hjalmar, ed. and trans. *Compositiones ad tingenda musiva [Codex lucensis 490]*. Uppsala: Almqvist och Wiksells, 1932.

Hills, Paul. *Venetian Colour: Marble, Mosaic, Painting and Glass, 1250–1550*. New Haven, CT: Yale University Press, 1999.

Hoogstraten, Samuel van. *Inleyding tot de hooge schoole der schilderkonst: Anders de zichtbaere werelt*. Rotterdam: Francois van Hoogstraten, 1678.

Hope, Charles. "Vasari's *Vite* as a Collaborative Project." In *The Ashgate Research Companion to Giorgio Vasari*, edited by David. J. Cast, 11–23. London: Routledge, 2016.

Howes, F. N. "Age-Old Resins of the Mediterranean Region and Their Uses." *Economic Botany* 4, no. 4 (1950): 307–16.

Hulin de Loo, Georges. *Heures de Milan: Troisième partie des Très-belles heures de Notre-Dame enluminées par les peintres de Jean de France, duc de Berry et par ceux du Duc Guillaume de Bavière*. Brussels: G. van Soest, 1911.

Hunt, R. W. G., and M. R. Pointer. *Measuring Colour*. Chichester, UK: John Wiley, 2011.

Husband, Timothy. *The Art of Illumination: The Limbourg Brothers and the "Belles Heures" of Jean de France, Duc de Berry*. New York: Metropolitan Museum of Art, 2008.

Imperato, Ferrante. *Dell'historia naturale di Ferrante Imperato napolitano. Libri XXVIII. Nella quale ordinatamente si tratta della diversa condition di miniere, e pietre. Con alcune historie di piante et animali; sin' hora non date in luce*. Naples: Costantino Vitale, 1599.

Isidore of Seville. *Isidori Hispalensis episcopi etymologiarum sive originum libri XX*. Edited by Wallace Martin Lindsay. 2 vols. Oxford: Oxford University Press, 1911.

———. *The Etymologies of Isidore of Seville*. Edited and translated by Stephen A. Barney, W. J. Lewis, J. A. Beach, and O. Berghof. Cambridge: Cambridge University Press, 2006.

Johnson, Rozelle P. "Notes on Some Manuscripts of the *Mappae clavicula*." *Speculum* 10 (1935): 72–81.

Jones, Frank N., Mark E. Nichols, and Socrates Peter Pappas, eds. *Organic Coatings: Science and Technology*. Hoboken, NJ: John Wiley, 2017.

Jones, Richard. *The Medieval Natural World*. London: Routledge, 2013.

Jones, Susan Frances, Anne-Sophie Augustyniak, and Hélène Dubois. "The Authenticity of the Frames of the Quatrain and the Other Frame Inscriptions." In *The Ghent Altarpiece: Research and Conservation of the Exterior*, edited by Bart Fransen and Cyriel Stroo, 173–308. Turnhout: Brepols, 2020.

Jongen, Ludo, Cees Schotel, and Josephine Franken, eds. and trans. *Het leven van Liedewij, de maagd van Schiedam: De Middelnederlandse tekst naar de bewaarde bronnen uitgegeven . . .* Schiedam: Fonds historische publikaties Schiedam, 1989.

Kammerling, Robert C., John I. Koivula, Robert E. Kane, Patricia Maddison, James E. Shigley, and Emmanuel Fritsch. "Fracture Filling of Emeralds: Opticon and Traditional 'Oils.'" *Gems and Gemology* 27, no. 2 (1991): 70–85.

Kargère, Lucretia, and Adriana Rizzo. "Twelfth-Century French Polychrome Sculpture in the Metropolitan Museum of Art: Materials and Techniques." *Metropolitan Museum Studies in Art, Science, and Technology* 1 (2010): 39–72.

Keen, Elizabeth. *Journey of a Book: Bartholomew the Englishman and the Properties of Things*. Canberra: ANU E Press, 2007.

Kemp, Martin. *The Science of Art*. New Haven, CT: Yale University Press, 1990.

Kemperdick, Stefan. "Robert Campin, Jacques Daret, and Rogier van der Weyden: The Written Sources." In *The Master of Flémalle and Rogier van der Weyden*, edited by Stefan Kemperdick and Jochen Sander, 53–61. Ostfildern: Hatje Catz Verlag, 2008.

Kirby, Jo, Maarten van Bommel, and André Verhecken. *Natural Colorants for Dyeing and Lake Pigments: Practical Recipes and Their Historical Sources*. London: Archetype, 2014.

Kirby, Jo, Marika Spring, and Catherine Higgit. "The Technology of Red Lake Pigment Manufacture: Study of the Dyestuff Substrate." *National Gallery Technical Bulletin* 26 (2005): 71–87.

Kirby, Jo, and Raymond White. "The Identification of Red Lake Pigment Dyestuffs and a Discussion of Their Use." *National Gallery Technical Bulletin* 17 (1996): 56–80.

Kirsh, Andrea, and Rustin S. Levenson. *Seeing through Paintings: Physical Examination in Art Historical Studies*. New Haven, CT: Yale University Press, 2000.

Klein, Ursula, and Emma C. Spary, eds. *Materials and Expertise in Early Modern Europe: Between Market and Laboratory*. Chicago: University of Chicago Press, 2010.

Kleiner, Fred. *Gardner's Art through the Ages: Backpack Edition, Book D: Renaissance and Baroque*. Boston: Wadsworth, 2015.

Koch, Anke. *Darstellung von Seidenstoffen in der Kölner Malerei der ersten Hälfte des 15. Jahrhunderts*. Weimar: Verlag und Datenbank für Geisteswissenschaften, 2010.

Kollandsrund, Katja, and Jilleen Nadolny, eds. *Medieval Painting in Northern Europe: Techniques, Analysis, Art History*. Studies in Commemoration of the 70th Birthday of Unn Plahter. London: Archetype, 2006.

Kroustallis, Stefanos. "Theophilus Matters: The Thorny Question of the 'Schedula diversarium artium' Authorship." In *Zwischen Kunsthandwerk und Kunst: Die "Schedula diversarium artium,"* edited by Andreas Speer and Günther Binding, 52–71. Berlin: De Gruyter, 2013.

Kühn, Hermann. "Verdigris and Copper Resinate." In *Artists' Pigments: A Handbook of Their History and Characteristics*, edited by Ashok Roy, 131–58. Washington, DC: National Gallery of Art, 1997.

Kunicki-Goldfinger, Jerzy J., Ian C. Freestone, Iain McDonald, Jan A. Hobot, Heather Gilderdale-Scott, and Tim Ayers. "Technology, Production and Chronology of Red Window Glass in the Medieval Period—Rediscovery of a Lost Technology." *Journal of Archaeological Science* 41 (2014): 89–105.

Lagercrantz, Otto, ed. and trans. *Papyrus graecus holmiensis (P. holm.): Recepte für Silber, Steine und Purpur*. Vol. 13. Uppsala, 1913.

Lairesse, Gerard de. *Groot Schilderboek*. 2 vols. Amsterdam: Erfgenamen van Willem de Coup, 1707.

Lampsonius, Dominicus. *Les effigies des peintres célèbres des Pays-Bas*. Edited and translated by Jean Puraye and Marie Delcourt. Paris: Brouwer, 1956.

Langenheim, Jean H. *Plant Resins: Chemistry, Evolution, Ecology, and Ethnobotany*. Portland, OR: Timber Press, 2003.

Lautenschlager, Manfred. "Kodikologische Angaben zur Handschrift." In *Der "Liber illuministarum" aus Kloster Tegernsee: Edition, Übersetzung und Kommentar der kunsttechnologischen Rezepte*. Edited and translated by Anna Bartl, Christoph Krekel, Manfred Lautenschlager, and Doris Oltrogge, 49–54. Stuttgart: Franz Steiner Verlag, 2005.

Leber, Jean Michel Constant, ed. "Statuts des tailleurs d'images, sculpteurs, peintres et elumineurs de la ville de Paris." In *Collection des meilleurs dissertations, notices et traités relatifs à l'histoire de France*, vol. 19. Paris: G. A. Dentu, 1838.

Lessing, Gotthold Ephraim. *Sämmtliche Schriften*. Edited by Karl Gotthelf Lessing. 30 vols. Berlin: Nicolaischen Buchhandlung, 1793.

———. "Vom Alter der Oelmalerei, aus dem Theophilus Presbyter." In *G. E. Lessing's Gesammelte Werke: In zwei Bänden*. Leipzig: G. J. Göschen, 1855.

Lomazzo, Giovan Paolo. *Idea del tempio della pittura*. Milan: Paolo Gottardo Pontio, 1590.

———. *Trattato dell'arte de la pittura, scoltura, et architettura*. Milan: Paolo Gottardo Pontio, 1584.

———. *Idea of the Temple of Painting*. Edited and translated by Jean Julia Chai. University Park: Pennsylvania State University Press, 2013.

Long, Pamela O. *Openness, Secrecy, Authorship: Technical Arts and the Culture of Knowledge from Antiquity to the Renaissance*. Baltimore: Johns Hopkins University Press, 2003.

———. *Artisan/practitioners and the Rise of the New Sciences, 1400–1600*. Corvallis: Oregon State University Press, 2011.

Lindberg, David C. *Theories of Vision from Al-Kindi to Kepler*. Chicago: University of Chicago Press, 1976.

Maclaren, Neil, and Anthony Werner. "Some Factual Observations about Varnishes and Glazes." *Burlington Magazine* 92, no. 568 (1950): 189–92.

Madigan, Brian. "Van Eyck's Illuminated Carafe." *Journal of the Warburg and Courtauld Institutes* 49 (1986): 227–30.

Mander, Karel van. *The Lives of the Illustrious Netherlandish and German Painters, from the First Edition of the Schilder-boeck (1603–1604): Preceded by the Lineage, Circumstances and Place of Birth, Life and Works of Karel van Mander, Painter and Poet and Likewise His Death and Burial, from the Second Edition of the Schilder-boeck (1616–1618)*. Edited and translated by Hessel Miedema. 6 vols. Doornspijk: Davaco, 1994–99.

———, ed. and trans. *Mediaeval Painter's Materials and Techniques Het Schilder-Boeck*. Haarlem: Jacob de Meester for Passchier van Westbusch, 1604.

Maroger, Jacques. "The Secret of Van Eyck Regained: The Long-Lost Emulsions of the First Great Oil Painter." *Magazine of Art* 35 (1943): 219–22.

Martelli, Matteo. *The Four Books of Pseudo-Democritus*. Leeds: Maney, 2013.

———. "Translating Ancient Alchemy: Fragments of Graeco-Egyptian Alchemy in Arabic Compendia." *Ambix* 64, no. 4 (2017): 326–42.

Martens, Maximiliaan. "Jan van Eyck's Optical Revolution." In *Van Eyck*, edited by Maximiliaan Martens, Till-Holger Borchert, Jan Dumolyn, Johan De Smet, and Frederica Van Dam, 141–79. London: Thames and Hudson, 2020.

Martens, Maximiliaan, Till-Holger Borchert, Jan Dumolyn, Johan De Smet, and Frederica Van Dam, eds. *Van Eyck*. London: Thames and Hudson, 2020.

Maryon, Herbert. *Metalwork and Enamelling*. 1912. Reprint, New York: Dover, 1971.

Masschelein-Kleiner, Liliane. *Ancient Binding Media, Varnishes and Adhesives*. Rome: ICCROM, 1995.

Massing, Ann, ed. *The Thornham Parva Retable: Technique, Conservation and Context of an English Medieval Painting*. Cambridge: Brepols, 2003.

Mayerne, Theodore Turquet de. *Beiträge zur Entwicklungs-Geschichte der Maltechnik*. Part 4, *Quellen für Maltechnik während der Renaissance und deren Folgezeit (16.–18. Jahrhundert) in Italien, Spanien, den Niederlanden, Deutschland, Frankreich und*

England nebst dem de Mayerne Manuskript. Edited and translated by Ernst Berger. Munich: G. D. W. Callwey, 1901.

———. *Lost Secrets of Flemish Painting, Including the First Complete English Translation of the De Mayerne Manuscript, BM Sloane 2052*. Edited and translated by Donald C. Fels, Joseph H. Sulkowski, Richard Bedell, and Rebecca A. McClung. Floyd, VA: Alchemist, 2004.

McHam, Sarah Blake. *Pliny and the Artistic Culture of the Italian Renaissance: The Legacy of the "Natural History."* New Haven, CT: Yale University Press, 2013.

Mededeelingen van de vereeniging ter beoefening der geschiedenis van 's-Gravenhage. Vol. 1. Den Haag, 1863.

Meier, Christel. *Gemma spiritalis: Methode und Gebrauch der Edelsteinallegorese vom frühen Christentum bis ins 18. Jahrhundert*. Munich: Wilhelm Fink Verlag, 1977.

Meiss, Millard. "Light as Form and Symbol in Some Fifteenth-Century Paintings." *Art Bulletin* 27, no. 3 (1945): 176–81.

Mérimée, Jean François Léonor. *De la peinture à l'huile, ou Des procédés matériels employés dans ce genre de peinture depuis Hubert et Jan van Eyck jusqu'à nos jours*. Paris: Huzard, 1830.

———. *The Art of Painting in Oil, and in Fresco: Being a History of the Various Processes and Materials Employed, from Its Discovery, by Hubert and John van Eyck, to the Present Time: Translated from the Original Treatise of M. J. F. L. Mérimée*. Edited and translated by Benjamin Sarsfield. London: Whittaker, 1839.

Merrifield, Mary Philadelphia, ed. and trans. *Original Treatises Dating from the XIIth to XVIIIth Centuries on the Arts of Painting*. 2 vols. London: John Murray, 1849.

Metzger, Cathy, and Griet Steyaert. "Painting, a Distinct Profession." In *Rogier van der Weyden, 1400–1464: Master of Passions*. Edited by Lorne Campbell and Jan Van der Stock. Leuven: Davidsfonds, 2009.

Miglio, Massimo, ed. *Storiografia pontificia del quattrocento*. Bologna: Patron, 1975.

Migne, Jacques-Paul, ed. *Patrologiae cursus completus: Seu bibliotheca universalis . . .* Vol. 106. Paris: J. P. Migne, 1863.

Mills, John S., and Raymond White. *The Organic Chemistry of Museum Objects*. 2nd ed. Oxford: Butterworth-Heinemann, 1994.

Monnas, Lisa. *Merchants, Princes and Painters: Silk Fabrics in Italian and Northern Paintings, 1300–1550*. New Haven, CT: Yale University Press, 2008.

Mowat, J. L. G., ed. and trans. *Sinonoma Bartholomei: A Glossary from a Fourteenth-Century Manuscript in the Library of Pembroke College, Oxford*. Oxford: Clarendon Press, 1882.

———. *Alphita, a medico-botanical glossary from the Bodleian manuscript, Selden B.35*. Oxford: Clarendon Press, 1887.

Nadolny, Jilleen. "The Techniques and Use of Gilded Relief Decoration by Northern European Painters, c. 1200–1500." PhD diss., University of London, Courtauld Institute of Art, 2000.

———. "The First Century of Published Scientific Analyses of the Materials of Historical Painting and Polychromy, c. 1780–1880." *Reviews in Conservation* 4 (2003): 39–51.

———. "A Problem of Methodology: Merrifield, Eastlake and the Use of Oil-Based

Media by Medieval English Painters." *14th Triennial Meeting The Hague Reprints* 2 (2005): 1028–33.

Nash, Susie. "'Pour couleurs et autres choses prise de lui . . .': The Supply, Acquisition, Cost and Employment of Painters' Materials at the Burgundian Court, c. 1375–1419." In *Trade in Artists' Materials: Markets and Commerce in Europe to 1700*, edited by Jo Kirby, Susie Nash, and Joanna Cannon, 97–182. London: Archetype, 2010.

Neven, Sylvie, ed. and trans. *The Strasbourg Manuscript: A Medieval Tradition of Artists' Recipe Collections*. London: Archetype, 2016.

Newman, William R. *Promethean Ambitions: Alchemy and the Quest to Perfect Nature*. Chicago: University of Chicago Press, 2004.

Nuttall, Paula. *From Flanders to Florence: The Impact of Netherlandish Painting, 1400–1500*. New Haven, CT: Yale University Press, 2004.

Oltrogge, Doris. "Writing on Pigments in Natural History and Art Technology." In *Early Modern Color Worlds*, edited by Tawrin Baker, Sven Dupré, Sachiko Kusukawa, and Karin Leonhard, 47–69. Leiden: Brill, 2015.

Oppenheim, A. L., R. H. Brill, and A. von Saldern. *Glass and Glassmaking in Ancient Mesopotamia: An Edition of the Cuneiform Texts Which Contain Instructions for Glassmakers with a Catalogue of Surviving Objects*. Corning, NY: Corning Museum of Glass Press, 1970.

Pächt, Otto. *Van Eyck: Die Begründer der altniederländischen Malerei*. Munich: Prestel, 1989.

———. *Van Eyck and the Founders of Early Netherlandish Painting*. London: Harvey-Miller, 1994.

Paillot de Montabert, Jacques-Nicolas. *Traité complet de la peinture*. 10 vols. Paris: Delion, 1829.

Panofsky, Erwin. *Early Netherlandish Painting: Its Origins and Character*. 2 vols. 1953. Reprint, New York: Harper and Row, 1971.

Pauchard, Ludovic, and Frédérique Giorgiutti-Dauphiné. "Craquelures and Pictorial Matter." *Journal of Cultural Heritage* 46 (2020): 361–73.

Pauling, Linus. *General Chemistry*. 1947. Reprint, from 1970 edition, Newburyport, MA: Dover, 2014.

Phillipps, Thomas. "Letter from Sir Thomas Phillipps, . . . Communicating a Transcript of a MS. Treatise . . . , Written in the Twelfth Century, and Entitled *Mappae clavicula*." In *Archaeologia, or Miscellaneous Tracts relating to Antiquity*. London: Society of Antiquaries of London, 1770.

Piemontese, Alessio. *De' secreti del reverendo donno Alessio Piemontese, prima parte, divisa in sei libri*. Milan: Valerio, e Hieronymo fratelli da Meda, 1557.

———. *The Secrets of the Reverend Maister Alexis of Piemont, . . . Translated out of French into English by William Ward*. London: Peter Short for Thomas Wight, 1595.

Plahter, Unn. "*Líkneskjusmíð*: 14th-Century Instructions for Painting from Iceland, Norwegian Medieval Altar Frontals and Related Material." In *Norwegian Medieval Altar Frontals and Related Material: Papers from the Conference in Oslo 16th to 19th December 1989*, edited by Magne Malmanger, Laszlo Berczelly, and Signe Horn Fuglesang, 151–71. Rome: Giorgio Bretschneider, 1995.

Plahter, Unn, Erla B. Hohler, Nigel J. Morgan, and Anne Wichstrøm, eds. *Painted Altar Frontals of Norway, 1250–1350*. 3 vols. London: Archetype, 2004.

Platt, Hugh. *The Jewell House of Art and Nature . . . , by Hugh Platte*. London: Peter Short, 1594.

Plesters, Joyce. "Dark Varnishes-Some Further Comments." *Burlington Magazine* 104, no. 716 (1962): 452–60.

———. "Ultramarine Blue, Natural and Artificial." *Studies in Conservation* 11, no. 2 (1966): 62–91.

Pliny the Elder. *Natural History, Volume 4: Books 12–16*. Edited and translated by H. Rackham. Loeb Classical Library 370. Cambridge, MA: Harvard University Press, 1968.

———. *Natural History, Volume 7: Books 24–27*. Edited and translated by W. H. S. Jones and A. C. Andrews. Loeb Classical Library 393. Cambridge, MA: Harvard University Press, 1956.

———. *Natural History, Volume 9: Books 33–35*. Edited and translated by H. Rackham. Loeb Classical Library 394. Cambridge, MA: Harvard University Press, 1961.

———. *Natural History, Volume 10: Books 36–37*. Edited and translated by D. E. Eichholz. Loeb Classical Library 419. Cambridge, MA: Harvard University Press, 1962.

Ploss, Ernst. "Studien zu den deutschen Maler und Färberbüchern des Mittelalters: Ein Beitrag zur Deutschen Altertumskunst und Wortforschung." Diss., Ludwig-Maximilian Universität, 1952.

———. "Das Amberger Malerbüchlein: Zur Verwandtschaft der spätmhd. Farbrezepte." In *Festschrift für Hermann Hempel zum 70. Geburtstag*, edited by den Mitarbeitern des Max-Planck-Instituts für Geschichte, 3:693–703. Göttingen: Vandenhoeck und Ruprecht, 1972.

Raspe, Rudolph Erich. *A Critical Essay on Oil-Painting: Proving That the Art of Painting in Oil Was Known before the Pretended Discovery of John and Hubert van Eyck; to Which Are Added Theophilus, "De arte pigendi," Eraclius, "De artibus Romanorum," and a Review of Farinator's "Lumen animae."* London: Printed for the author by H. Goldney, 1781.

Rees-Jones, Stephen G. "Early Experiments in Pigment Analysis." *Studies in Conservation* 35, no. 2 (1990): 93–101.

Rie, E. René de la. "The Influence of Varnishes on the Appearance of Paintings." *Studies in Conservation* 32 (1987): 1–13.

Rizzo, Adriana, Nobuko Shibayama, and Daniel P. Kirby. "A Multi-analytical Approach for the Identification of Aloe as a Colorant in Oil-Resin Varnishes." *Analytical and Bioanalytical Chemistry* 399, no. 9 (2011): 3093–107.

Roosen-Runge, Hans. "Einführung: Entwicklung der Forschung über die Technik der Buchmalerei." In *Farbmittel, Buchmalerei, Tafel- und Leinwandmalerei des Mittelalters*, edited by Hermann Kühn, Heinz Roosen-Runge, Rolf E. Straub, and Manfred Koller, 1:58–123. 1984. Reprint, Stuttgart: Reclam, 2002.

Rott, Hans, ed. *Quellen und Forschungen zur südwestdeutschen und schweizerischen Kunstgeschichte im XV. und XVI. Jahrhundert (Band 3, 1): Quellen I (Baden, Pfalz, Elsass)*. 3 vols. Stuttgart: Strecker und Schröder Verlag, 1936.

Roy, Ashok. "Van Eyck's Technique: The Myth and the Reality, I." In *Investigating Jan van Eyck*, edited by Susan Foister and Sue Jones, 97–100. Turnhout: Brepols, 2000.

Ruhemann, Helmut. "Leonardo's Use of Sfumato." *British Journal of Aesthetics* 1, no. 4 (1961): 231–37.

Ruscelli, Girolamo. *Secreti nuovi di maravigliosa virtù*. Venice: Marchiò Sessa, 1567.

Sander, Jochen. "*Ars Nova* and European Painting in the Fifteenth Century." In *The Master of Flémalle and Rogier van der Weyden*, edited by Stefan Kemperdick and Jochen Sander, 31–37. Ostfildern: Hatje Cantz Verlag, 2008.

———. "Reconstructing Artists and Their Oeuvres." In *The Master of Flémalle and Rogier van der Weyden*, edited by Stefan Kemperdick and Jochen Sander, 75–93. Ostfildern: Hatje Catz Verlag, 2008.

Santi, Giovanni. *La vita e le gesta di Federico di Montefeltro, duca d'Urbino: Poema in terza rima (codice Vat. Ottob. lat. 1305)*. Edited by Luigi Michelini Tocci. Vatican City: Bibliotheca Apostolica Vaticana, 1985.

Sauerberg, Marie Louise, Ashok Roy, Marika Spring, Spike Bucklow, and Mary Kempski. "Materials and Techniques." In *The Westminster Retable: History, Technique, Conservation*, edited by Paul Binski, Ann Massing, and Marie Louise Sauerberg, 233–51. Cambridge: Brepols, 2009.

Savoia, Andrea Ubrizsy, and Luís Ramón-Laca. "The *Libri picturati* Watercolours and Early Botanical Illustrations." In *Drawn After Nature: The Complete Botanical Watercolours of the 16th-Century "Libri picturati,"* edited by J. de Koning, Gerda van Uffelen, Alicja Zemanek, and Bogdan Zemanek, 60–67. Zeist: KNV, 2008.

Scafi, Alessandro. *Mapping Paradise: A History of Heaven on Earth*. London: British Library, 2006.

Schneider, Laurie. *Giotto in Perspective*. Englewood Cliffs, NJ: Prentice-Hall, 1974.

Schreyberey: Mancherley weyse Dinten und allerhand farben zubereyten. Mainz: Peter Jordan, 1532.

Scott, David A. *Copper and Bronze in Art: Corrosion, Colorants, Conservation*. Los Angeles: Getty Publications, 2002.

Seneca the Younger. *Seneca: Epistles 1–65*. Edited and translated by Richard M. Gummere. Loeb Classical Library 75. 3 vols. Cambridge, MA: Harvard University Press, 1917.

Smith, Cyril Stanley, and John G. Hawthorne, eds. and trans. "*Mappae Clavicula*: A Little Key to the World of Medieval Techniques." *Transactions of the American Philosophical Society* 64, no. 4 (1974): 1–128.

Smith, Jeffrey Chipps. *The Northern Renaissance*. Berlin: Phaidon Press, 2004.

Smith, John Thomas. *Antiquities of Westminster . . .* London: T. Bensley, 1807.

Smith, Pamela H. *The Body of the Artisan: Art and Experience in the Scientific Revolution*. Chicago: University of Chicago Press, 2004.

Speer, Andreas, Maxime Mauriège, and Hiltrud Westermann-Angerhausen, eds. *Zwischen Kunsthandwerk und Kunst: Die "Schedula diversarium artium."* Berlin: De Gruyter, 2013.

Spring, Marika. "Colourless Powdered Glass as an Additive in Fifteenth- and Sixteenth-Century European Paintings." *National Gallery Technical Bulletin* 33 (2012): 4–26.

Stockum, Wilhelmus P. van, ed. *'S-Gravenhage in den loop der tijden*. The Hague: W. P. van Stockum, 1889.

Straelen, Jan Baptist van der. *Jaerboek der vermaerde en kunstryke gilde van Sint Lucas binnen de stad Antwerpen*. Antwerp: Pieter-Theodor Moons-Van der Straelen, 1855.

Stratford, Neil. *Catalogue of Medieval Enamels in the British Museum*. Vol. 2, *Northern Romanesque Enamel*. London: British Museum, 1993.

Striebel, Ernst. "Das Augsburger Kunstbuechlin von 1535: Eine kunsttechnologische Quellenschrift der deutschen Renaissance." Diss., Technische Universität, Munich, 2007.

Strolovitch, Devon L. "Old Portuguese in Hebrew Script." PhD diss., Cornell University, 2005.

Stroo, Cyriel, ed. *Pre-Eyckian Panel Painting in the Low Countries*. 2 vols. Turnhout: Brepols, 2009.

Suger, Abbot, von Saint-Denis. *Abbot Suger on the Abbey Church of St.-Denis and Its Art Treasures*. Edited and translated by Erwin Panofsky. Princeton, NJ: Princeton University Press, 1979.

———. *Ausgewählte Schriften: Ordinatio, De consecratione, De administratione*. Edited and translated by Andreas Speer and Günther Binding. Darmstad: Wissenschaftliche Buchgesellschaft, 2000.

Taubes, Frederic. *The Mastery of Oil Painting*. London: Thames and Hudson, 1953.

Taylor, Paul. *Condition: The Ageing of Art*. London: Paul Holberton, 2015.

Terzi, Ernesto. *I prodotti delle conifere memoria di Ernesto Terzi*. Milan: Galleria Vittorio Emanuele, 1871.

The Making and Knowing Project, Pamela H. Smith, Naomi Rosenkranz, Tianna Helena Uchacz, Tillmann Taape, Clément Godbarge, Sophie Pitman, Jenny Boulboullé, Joel Klein, Donna Bilak, Marc Smith, and Terry Catapano, eds. and trans. *Secrets of Craft and Nature in Renaissance France: A Digital Critical Edition and English Translation of BnF MS Fr.* 640. New York: Making and Knowing Project, 2020: https://doi.org/10.7916/78yt-2v41.

Theophilus. *Theophili qui est Rugerus, presbyteri et monachi, libri III. De diversis artibus: seu Diversarium artium schedula* . . . Edited and translated by Robert Hendrie. London: Johannes Murray, 1847.

———. *"On Divers Arts": The Foremost Medieval Treatise on Painting, Glassmaking and Metalwork*. Edited and translated by John G. Hawthorne and Cyril Stanley Smith. 1963. Reprint, New York: Dover, 1979.

———. *The Various Arts: De diversis artibus*. Edited and translated by Charles Reginald Dodwell. 1961. Reprint, Oxford: Clarendon Press, 1986.

———. *Theophilus Presbyter und das mittelalterliche Kunsthandwerk: Gesamtausgabe der Schrift "De diversis artibus" in zwei Bänden*. Edited and translated by Erhard Brepohl. 2 vols. Cologne: Böhlau, 1999.

———. *The "Schedula diversarum artium" . . . a Digital Critical Edition*. Edited by Andreas Speer and Hiltrud Westermann-Angerhausen. Cologne: Thomas-Institute, University of Cologne, 2010: http://schedula.uni-koeln.de/.

Thomas of Cantimpré. *Thomas Cantimpratentsis Liber de natura rerum: Editio princeps secundum codices manuscriptos*. Edited by Helmut Boese. Berlin: Walter de Gruyter, 1973.

Thompson, Daniel Varney. "Trial Index to Some Unpublished Sources for the History of Mediaeval Craftsmanship." *Speculum* 10, no. 4 (1935): 410–31.

Thompson, David L. *Mummy Portraits in the J. Paul Getty Museum*. Malibu, CA: J. Paul Getty Museum, 1982.

Thompson, R. Campbell. "A Survey of the Chemistry of Assyria in the Seventh Century BC." *Ambix* 2, no. 1 (1938): 3–16.

Thornton, Jonathan. "Use of Dyes and Colored Varnishes in Wood Polychromy." In *Painted Wood: History and Conservation*, edited by Valerie Dorge and F. Carey Howlett, 226–41. Los Angeles: Getty Conservation Institute, 1998.

Tramelli, Barbara. *Giovanni Paolo Lomazzo's "Trattato dell'arte della pittura."* Leiden: Brill, 2017.

Tumosa, Charles S., and Marion F. Mecklenburg. "The Influence of Lead Ions on the Drying of Oils." *Reviews in Conservation* 6 (2005): 39–47.

Vaan, Michiel de. *Etymological Dictionary of Latin and the Other Italic Languages.* Leiden: Brill, 2008.

Vaernewyck, Marcus van. *Den spieghel der Nederlandscher audtheyt.* Ghent: Gheeraert van Salenson, 1568.

Vanwijnsberghe, Dominique. "The Eyckian Miniatures in the Turin-Milan Hours." In *Van Eyck*, edited by Maximiliaan Martens, Till-Holger Borchert, Jan Dumolyn, Johan De Smet, and Frederica Van Dam, 296–315. London: Thames and Hudson, 2020.

Vasari, Giorgio. *Le vite de' più eccellenti architetti, pittori, et scultori italiani, da Cimabue insino a' tempi nostri, descritte in lingua Toscana.* Florence: Appresso Lorenzo Torrentino, 1550.

———. *Le Vite de' più eccellenti pittori, scultori, e architettori . . .* Florence: Appressa i Giunti, 1568.

———. *Le vite de' più eccellenti pittori, scultori, e architettori scritte da Giorgio Vasari.* Edited by Gaetano Milanesi. 9 vols. Florence: Sansoni, 1878–85.

———. *Lives of the Painters, Sculptors, and Architects: Giorgio Vasari.* Edited and translated by Gaston du C. De Vere, with an introduction by David Ekserdjian. 2 vols. 1912–15. Reprint, New York: Alfred A. Knopf, 1996.

———. *Il carteggio di Giorgio Vasari.* Edited by Karl Frey and Herman Walther Frey. Munich: Georg Müller, 1923.

———. *Le vite de' più eccellenti pittori, scultori, e architettori nelle redazioni del 1550 e 1568.* Edited by Paola Barocchi and Rosanna Bettarini. 6 vols. Florence: Sansoni, 1966–87.

———. *Le vite de' più eccellenti architetti, pittori, et scultori italiani, da Cimabue, insino a' tempi nostri: Nell'edizione per i tipi di Lorenzo Torrentino, Firenze 1550.* Edited by Luciano Bellosi and Aldo Rossi. 2 vols. Turin: Einaudi, 1991.

Villani, Philippi [Filippo]. *De origine civitatis Florentie et de eiusdem famosis civibus.* Edited by Giuliano Tanturli. Padua: Antenore, 1997.

Villela-Petit, Inès, "La peinture médiévale vers 1400 autour d'un manuscrit de Jean Lebègue: Édition des *Libri colorum.*" Diss., École Nationale des Chartes, 1995.

Voelkle, Wilhelm, and Charles Ryskamp. *The Stavelot Triptych: Mosan Art and the Legend of the True Cross.* London: Pierpont Morgan Library, 1980.

Walpole, Horace. *Anecdotes of Painting in England: With Some Account of the Principal Artists.* London: J. Dodsley, 1762.

Walton, K. L. *Mimesis as Make-Believe: On the Foundations of the Representational Arts.* Cambridge, MA: Harvard University Press, 1990.

Ward, Gerald W. R., ed. *The Grove Encyclopedia of Materials and Techniques in Art*. New York: Oxford University Press, 2008.

Wardwell, Anne E. "The Stylistic Development of 14th- and 15th-Century Italian Silk Design." *Aachener Kunstblätter* 47 (1976): 177–226.

Waźbiński, Zygmunt. "L'idée de l'histoire dans la première et la seconde édition des 'Vies' de Vasari." In *Il Vasari storiografo e artista: Atti del congresso internazionale nel IV centenario della morte . . .*, 1–25. Florence: Istituto nazionale di studi sul Rinascimento, 1976.

Weigert, Laura. *Weaving Sacred Stories: French Choir Tapestries and the Performance of Clerical Identity*. Ithaca, NY: Cornell University Press, 2004.

Wetering, Ernst van de. *Rembrandt: The Painter at Work*. Amsterdam: Amsterdam University Press, 1997.

White, Raymond. "Van Eyck's Technique: The Myth and the Reality, II." In *Investigating Jan van Eyck*, edited by Susan Foister and Sue Jones, 101–5. Turnhout: Brepols, 2000.

White, Raymond, and Jo Kirby. "Rembrandt and His Circle: Seventeenth-Century Dutch Paint Media Re-examined." *National Gallery Technical Bulletin* 15 (1994): 64–78.

———. "Medium Analysis." In *The Westminster Retable: History, Technique, Conservation*, edited by Paul Binski, Ann Massing, and Marie Louise Sauerberg, 252–59. Cambridge: Brepols, 2009.

Woudhuysen-Keller, Renate. "Leinöl als Malmittel. Rekonstruktionsversuche nach Rezepten aus dem 13. bis 19. Jahrhundert." *Maltechnik Restauro* 2 (1973): 74–105.

Woudhuysen-Keller, Renate, and Paul Woudhuysen. "A Short History of Eggwhite Varnishes." In *Firnis: Material-Ästhetik-Geschichte . . .*, edited by A. Harmssen, 80–86. Stegen: AdR-Arbeitsgemeinschaft der Restauratoren, 1999.

Index

The letter *f* following a page number denotes a figure.